ELECTRONIC DISPLAY MEASUREMENT

ELECTRONIC DISPLAY MEASUREMENT
CONCEPTS, TECHNIQUES, AND INSTRUMENTATION

Peter A. Keller
Tektronix, Inc.
Society for Information Display

Published in Association with the *Society for Information Display*
A WILEY-INTERSCIENCE PUBLICATION
JOHN WILEY & SONS, INC.
New York · Chichester · Weinheim · Brisbane · Singapore · Toronto

This text is printed on acid-free paper. ∞

Copyright © 1997 by Peter A. Keller. All rights reserved.

Published by John Wiley & Sons, Inc.

Published simultaneously in Canada.

No part of this publication may be reproduced, stored in a retrieval system or transmitted in any form or by any means, electronic, mechanical, photocopying, recording, scanning or otherwise, except as permitted under Sections 107 or 108 of the 1976 United States Copyright Act, without either the prior written permission of the Publisher, or authorization through payment of the appropriate per-copy fee to the Copyright Clearance Center, 222 Rosewood Drive, Danvers, MA 01923, (508) 750-8400, fax (508) 750-4744. Requests to the Publisher for permission should be addressed to the Permissions Department, John Wiley & Sons, Inc., 605 Third Avenue, New York, NY 10158-0012, (212) 850-6011, fax (212) 850-6008, E-Mail: PERMREQ@WILEY.COM.

Library of Congress Cataloging In Publication Data:

Keller, Peter A., 1937–
    Electronic display measurement : concepts, techniques and
  instrumentation / Peter A. Keller.
        p.   cm.
    Includes index.
    ISBN 0-471-14857-1 (cloth : alk. paper)
    1. Information display systems—Evaluation. 2. Information
  display systems—Measurement. I. Title.
TK7882.I6K45   1997
621.3815′422′0287—dc21   97-1774

Printed in the United States of America

10 9 8 7 6 5 4 3 2 1

To Irene for her encouragement, assistance, and providing the incentive of spending more time together upon its completion!

CONTENTS

FOREWORD xi
PREFACE xiii
ACKNOWLEDGMENTS xv

1. Light and Color 1

1.1	Electromagnetic Spectrum	2
1.2	The Eye	3
1.3	Radiometry	12
1.4	Photometry	14
1.5	Colorimetry	17
1.6	Color Temperature	21
1.7	Lambert's Law (Cosine Law)	25
1.8	Inverse Square Law	25
1.9	Units	26
	References	26

2. Light Sources, Filters, and Detectors 29

2.1	Light Sources	29
2.2	Filters	52
2.3	Light Detectors	69
	References	84

3. Displays 87

3.1	Display Formats and Characteristics	87
3.2	Cathode-Ray Tube Displays	90
3.3	Vacuum Fluorescent Displays	105
3.4	Liquid Crystal Displays	107
3.5	Plasma Displays	115
3.6	Plasma-Addressed Liquid Crystal Displays	118
3.7	Electroluminescent Displays	119
3.8	Light-Emitting Diode Displays	120
3.9	Other Displays	122
	References	125

4. Measurement Instrumentation — 129

- 4.1 Photometers — 129
- 4.2 Radiometers — 146
- 4.3 Filter Colorimeters — 147
- 4.4 Spectroradiometers — 149
- 4.5 Spectrophotometers — 151
- 4.6 Goniophotometers — 151
- 4.7 Integrating Spheres — 152
- 4.8 Optical Comparitors — 153
- 4.9 Convergence Measurement Instruments — 153
- References — 155

5. Luminance and Contrast Measurement — 156

- 5.1 General Background — 156
- 5.2 Luminance — 158
- 5.3 Luminance Uniformity — 162
- 5.4 Luminance Warmup and Aging — 162
- 5.5 Luminance Stability versus Fill Factor — 163
- 5.6 Luminance Linearity — 164
- 5.7 Gamma — 166
- 5.8 Contrast — 167
- 5.9 Diffuse Reflectance — 169
- 5.10 Specular Gloss — 173
- 5.11 Projection System Measurements — 173
- 5.12 Visible LED Output — 177
- References — 180

6. Color Measurement — 182

- 6.1 General Background — 182
- 6.2 Purity — 183
- 6.3 Chromaticity — 184
- 6.4 Color Tracking — 185
- 6.5 Color Uniformity — 186
- 6.6 Color Gamut — 187
- 6.7 Spectral Output — 187
- 6.8 Color Anisotropy — 189
- References — 189

7. Resolution Measurement — 192

- 7.1 General Background — 192
- 7.2 Spot Characteristics — 196
- 7.3 Ways of Expressing Resolution — 197
- 7.4 Visual Methods — 197
- 7.5 Photometric Methods — 202
- 7.6 Modulation Transfer Function — 213
- 7.7 Standards — 220
- References — 220

8. Geometry Measurements — 224

    8.1 General Background — 224
    8.2 Size and Aspect Ratio — 225
    8.3 Centering or Positioning — 226
    8.4 Nonlinearity and S Distortion — 226
    8.5 Tilt — 228
    8.6 Orthogonality — 229
    8.7 Trapezoidal Distortion — 230
    8.8 Pincushion and Barrel Distortion — 231
    8.9 Misconvergence — 232
    8.10 Moire Patterns — 236
    8.11 Jaggies — 236
    8.12 Jitter — 237
    8.13 Magnetic Aberrations — 237
    References — 239

9. Time-Related Measurements — 241

    9.1 General Background — 241
    9.2 With Photoamperic Silicon Photodiodes — 241
    9.3 With Photoconductive Silicon Photodiodes — 244
    9.4 With Photomultiplier Tubes — 245
    9.5 Phosphor Persistence Measurement — 251
    9.6 Flicker (Modulation) — 254
    9.7 Sensor Response Time — 254
    References — 257

10. Calibration — 258

    10.1 General Background — 258
    10.2 Metrology — 258
    10.3 Illuminance — 260
    10.4 Luminance — 264
    10.5 Chromaticity — 266
    10.6 Light-Emitting Diodes — 267
    References — 268

11. Display and Related Standards — 269

    11.1 ANSI — 269
    11.2 ASTM — 269
    11.3 CIE — 270
    11.4 CORM — 270
    11.5 DIN — 271
    11.6 EBU — 271
    11.7 EIA — 271
    11.8 EIAJ — 272
    11.9 EN — 272
    11.10 HFES — 272

11.11	IEC	272
11.12	IEEE	273
11.13	IESNA	273
11.14	ISCC	273
11.15	ISO	274
11.16	IS&T	274
11.17	ITSB	275
11.18	MIL	275
11.19	NAPM	275
11.20	NIDL	275
11.21	NIST	276
11.22	SAE	276
11.23	SID	277
11.24	SMPTE	277
11.25	SPIE	277
11.26	VESA	278
11.27	Other	278
References		278

APPENDICES 286

A	Unit Abbreviations	286
B	SI Prefixes	287
C	Laws, Formulas, and Constants	288
D	Radiometric and Photometric Conversions	293
E	1 nm CIE Tristimulus Values	294
F	Chromaticity Conversions	302
G	Standard Illuminants (Normalized)	303
H	Standard Fluorescent Lamp References (Normalized)	305
I	WTDS Phosphor Designations	309
J	JEDEC-to-WTDS Phosphor Equivalents	310
K	Instrument Manufacturers	311
L	Calibration Services	316
M	Standards Organizations	317

INDEX 319

FOREWORD

Electronic Display Measurement by Peter Keller is the second volume in a new series, Display Technology, published by John Wiley in Collaboration with the Society for Information Display.

The intended readership of this volume is not primarily the expert in display measurement, though, such experts will find much of use here. Rather, it is aimed at scientists, engineers, and technicians working in the field of display technology, display product development, and display performance specification who need to evaluate and measure displays reliably, repeatably, and with confidence that their measurements are valid and free from artifacts. Written in a conversational and easy to understand style, *Electronic Display Measurement* contains a wealth of information and advice that its readership will find invaluable.

Peter Keller presents each major topic in a self contained chapter so that all the relevant information is easily accessible without cross-referencing. In addition to presenting each topic from a basic to a working level, this book is full of the type of practical advice often required by readers, but seldom offered by standard texts. Extensive bibliographies are provided for the reader who wishes to delve more deeply into any topic. In keeping with the practical style of the book, equations in the body of the text are kept to a minimum, but a full set is provided in one of several appendices, which also contain the CIE Standard Observer tables and provide information on international and national standards organizations, national standards laboratories, equipment suppliers, and calibration services.

In short, this volume is intended to be a hands-on practical work book, to be used, not confined to the bookshelf, and contains all the information and data that will be required in daily use. I hope, with some confidence, that it will fulfill both this singular aim and the aim of the series, which is to provide those working in display technology and its associated fields with volumes which are not only technically rigorous, but full of useful practical information.

Anthony C. Lowe
President, Society for Information Display

Greenock, UK
May 1997

PREFACE

Electronic displays in their many forms are a major contributor to our employment, entertainment, education, convenience, comfort, and safety. From television to computers to coffee makers to aircraft cockpits and more, displays are a significant part of many aspects of our everyday life. Display devices are almost as varied as their uses. The liquid crystal, cathode-ray, light-emitting diode, field emission, vacuum fluorescent, digital micromirror, electroluminescent, and plasma devices are just part of the alphabet soup of displays. Whole industries have been developed based on various display technologies.

Despite such widespread usage, display measurements have long been a source of difficulty for display device designers, manufacturers, users, and service technicians. Conflicts over performance specifications are a daily occurrence, and much time, effort, and money are spent resolving such differences. To add to the difficulty, the human eye, with all of its nonlinearities, and the human brain, with its preferences, are the ultimate judge of whether a display can be correctly and rapidly interpreted. Measurements do not always accurately reflect how the user actually sees the display. Finally, since light measurements are involved in most display characterizations, the combined effects of the displays' spectral (color), spatial (geometric), and temporal (time) characteristics result in measurement accuracy that is appalling to the electrical engineer long accustomed to voltage, current, time, and frequency measurements in the small-fraction-of-a-percent realm. Light measurements of 5% accuracy are considered excellent, but errors of 50% or more are all too common. And this does not include the differences between the assumed characteristics of the human eye and the instrumentation that attempts to simulate it or the differences from individual to individual and even between two eyes on the same individual. Outside of textbooks on light and color, little material has been published on the art of display measurements. It is hoped that this book will fill this void by providing practical information on such measurements along with sufficient references to aid in the location of more detailed information and pertinent industry standards.

This book will describe practical techniques and instrumentation used for display measurement and the common pitfalls that result in errors. Awareness of the sources of measurement errors are very useful in assessing and questioning the validity of one's measurements as well as published specifications. Ed Kelley of the National Institute of Standards and Technology (NIST) summed it up well in a recent paper[*] he presented to the Society for Information Display in Orlando. He stated, "Metrology is more an attitude than procedures—an attitude of skepticism, even cynicism!" This brings us to the subject of display standards. Many display standards formulating organizations exist today as well as a bewildering array of standards. Some of these may be conflicting and often do

---

[*]E. F. Kelley et al., "A Survey of the Components of Display Measurement Standards," SID International Symposium, Orlando FL, May 23–25, 1995.

not reflect current technology. Despite the fact that display technology has been with us for 100 years, since the first CRT by Karl Ferdinand Braun in 1897, advances in display technology are being made at an ever-increasing rate. Lack of timely standards forces manufacturers and users to develop their own methods for specifying performance. Unfortunately, this often results in a difficult-to-resolve "apples-and-oranges" situation when comparing specifications between different manufacturers, especially when they are in different countries.

The foregoing should not be interpreted as too-negative sounding. It is merely intended to sensitize the reader to the problems underlying measurements of light where interpretation by the somewhat subjective and non-linear human visual system is the final objective. Indeed, intensive work is under way within the industry by organizations such as NIST, the Society for Information Display (SID), the National Information Display Laboratory (NIDL), the Video Electronics Standards Association (VESA), and so on, to improve the situation. It is anticipated that this book will aid in resolving some of the differences that occur. The awareness of what problems others have encountered and their solutions should go a long way toward that end. If nothing else, a little of Ed Kelley's recommended skepticism regarding measurements is a most useful tool!

PETER A. KELLER

Aloha, Oregon
January 1997

ACKNOWLEDGMENTS

The author would like to thank the following individuals and companies for their assistance and support in the preparation of this book. Many others have contributed in an indirect manner over the years through discussions and meetings. Another, less obvious, contribution are the many questions asked by electronic engineers and technicians making display measurements, a subject largely passed over in their education. These questions have aided in two ways: They caused further research to find the answer and they emphasized the need for more practical information to be made available by the number of times the same questions were heard. The latter is the real reason for this book.

Dennis Bechis, David Sarnoff Research Center

Stan Buckstad, Microvision

Tom Buzak, Technical Visions

Ken Futornick, Tektronix Metrology Laboratory

Ken Hillen*, Tektronix

Jim Hopkins, Northwest Technical Company

Luhr Jensen, Klein Optical Instruments

Edward Kelley, NIST

Tony Lowe, IBM UK, Society for Information Display

Jamie Moritz, Tektronix Technical Standards

Kathleen Muray, InPhoRa, Inc.

Frank Rochow, LMT

Bob Ruff, Opto-Cal, Inc.

Terry Scheffer, In Focus Systems

Aris Silzars, SI Diamond Technology

Joann Taylor, Color Technology Solutions

Ken Werner, Editor, *Information Display*

*Who resuscitated Chapter 2 following a particularly nasty disk crash.

ELECTRONIC DISPLAY MEASUREMENT

CHAPTER 1

Light and Color

This book deals with characterizing the appearance of displays primarily by light measurements. This involves concepts often unfamiliar to the average electronics engineer or technician. Radio engineers may find it somewhat easier to understand light measurement because of the similarity of light to radio waves. It is just a higher frequency (terahertz) portion of the electromagnetic spectrum that behaves more like microwaves. Actually, many of the devices and instruments discussed in this book are *electro-optical* in nature, that is, they combine electronics and optics in devices that interface between the two worlds.

Three fundamental concepts should be borne in mind at all times when working with light measurements. Learn these well and you will go a long way toward solving most of the light measurement problems that are encountered. These concepts are the *spectral, spatial,* and *temporal* characteristics of the light source, an intermediate element such as a filter or lens, a nearby object that may emit or reflect light, and a light sensor that is part of the test configuration. A solid understanding of the three concepts cannot be emphasized too heavily.

The spectral considerations of electro-optical elements and instruments are closely related to the frequency bandpass characteristics of devices and systems in electronics. It is just the different effects and interactions of these that will need to be understood. Spectral characteristics also must be considered in the initial decision of whether to measure light as intended for the human eye (photometry), as power in the familiar watt (radiometry), or for film, video camera, or other such device that does not fit in either of the first two categories.

Spatial characteristics are the geometric characteristics affecting the emission, reflection, absorption, transmission, and sensing of light. Many of the different units in use in photometry and radiometry are derived from the base units of the candela and watt, respectively, combined with geometric and angular units such as the meter and steradian.

Temporal considerations are merely the time-related characteristics of active electro-optical components. As in electronic devices, light sources and sensors usually have risetimes, frequency bandwidths, and so on, which must be considered. It is not just a direct-current (DC) world in electro-optics. All of the complications of alternating current (AC) may be encountered, although these should not cause excessive grief to the electronically oriented individual.

You will find some repetition of the important concepts throughout the book. This is for three reasons. The first is for emphasis; the second is because most users probably will not read this book "cover to cover" but will instead use it as a reference resource as the need arises. It is suggested that the first two chapters should be read as thoroughly as possible to develop a practical understanding of the three basic concepts.

2 LIGHT AND COLOR

The third reason for some repetition is to avoid the need to refer to other chapters for related information.

1.1 ELECTROMAGNETIC SPECTRUM

The electromagnetic spectrum[1-4] (Figure 1.1) depicts the range of electromagnetic radiation from DC to cosmic rays. Most readers with a background in electronics will be familiar with measurements in the portion of the spectrum extending from DC to

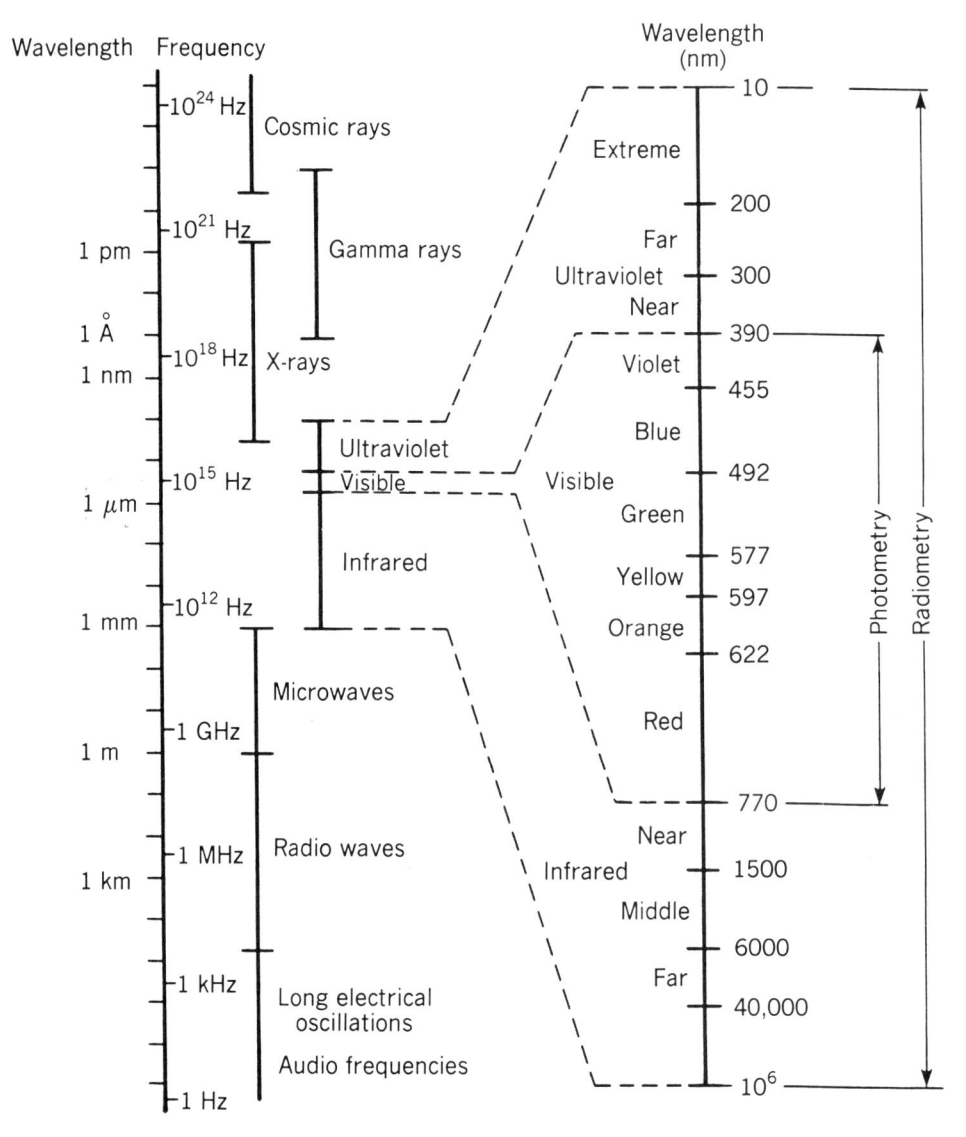

Figure 1.1. Electromagnetic spectrum from DC to cosmic rays.

about 1 GHz. Above that, the microwave region gradually transitions to infrared, visible light, ultraviolet, X-rays, gamma rays, and finally cosmic rays. Since displays are designed for viewing by the human eye, we will be most concerned with the visible light spectrum having wavelengths from 380 to 780 nm. From the electronics engineer's perspective, this corresponds to frequencies of 790–385 THz (10^{15} Hz). Wavelength in nanometers is the common unit used to express the location within the electromagnetic spectrum for visible, ultraviolet, and near-infrared radiation. Note that in some older publications, wavelength may be expressed in angstroms or angstrom units (Å). Just divide angstroms by 10 to convert to nanometers. For infrared wavelengths, the *micron,* or *micrometer,* is often used. Multiply by 1000 to convert micrometers to nanometers.

1.2 THE EYE

The human eye is an extremely versatile detector. It can deliver high-resolution, full-color, stereoscopic images to the brain. The eye functions over a useful range of 11–14 decades, from about 0.0000003 to 30,000 candelas per square meter (nits) or more. Focusing can be accomplished for distances from about 75–100 mm (3–4 in.) to infinity. It can detect subtle color differences between thousands of individual hues. An image can be captured from a single light flash of less than 1 μs duration.

Physiology

The human eye is the ultimate receptor for most displays; thus most measurements of displays attempt to measure the ability of the display to convey information to the viewer in an efficient, accurate, and comfortable manner. The eye is a highly complex optical, mechanical, electrical, and chemical system for transfer of visible information to the brain. Otto H. Schade, Sr., of RCA presented a paper[5] in 1955 describing an electromechanical analog model of the eye (Figure 1.2). His model is probably more clearly understandable to the engineer than the more conventional physiologically based descriptions.

Nevertheless, a more conventional description of the eye and its characteristics is still in order here.[6-13] Figure 1.3 shows a cross section of the human eye. The human eye is nearly spherical in shape with a diameter of about 24 mm. The principal elements are the optical system and the sensory system that converts photons of light to electrical signals that are transferred to the brain while retaining the spatial or positional information of the image formed by the optical elements.

The cornea is a protective transparent layer covering the key optical element, the lens. Between the cornea and the lens is a waterlike fluid known as the aqueous humor that helps maintain proper pressure on the eye. The lens is not rigid, as in a camera, but is compressible via the ciliary muscles around its perimeter that serve to alter its focal length and thus focus. With the muscles at rest, the eye will focus images near infinity on the retina. As the ciliary muscles compress the lens, the focal distance shortens to as little as 75–100 mm. The iris, located at the surface of the lens, acts to reduce the effective diameter of the lens or *pupil* at higher light levels, which reduces the total amount of light

4 LIGHT AND COLOR

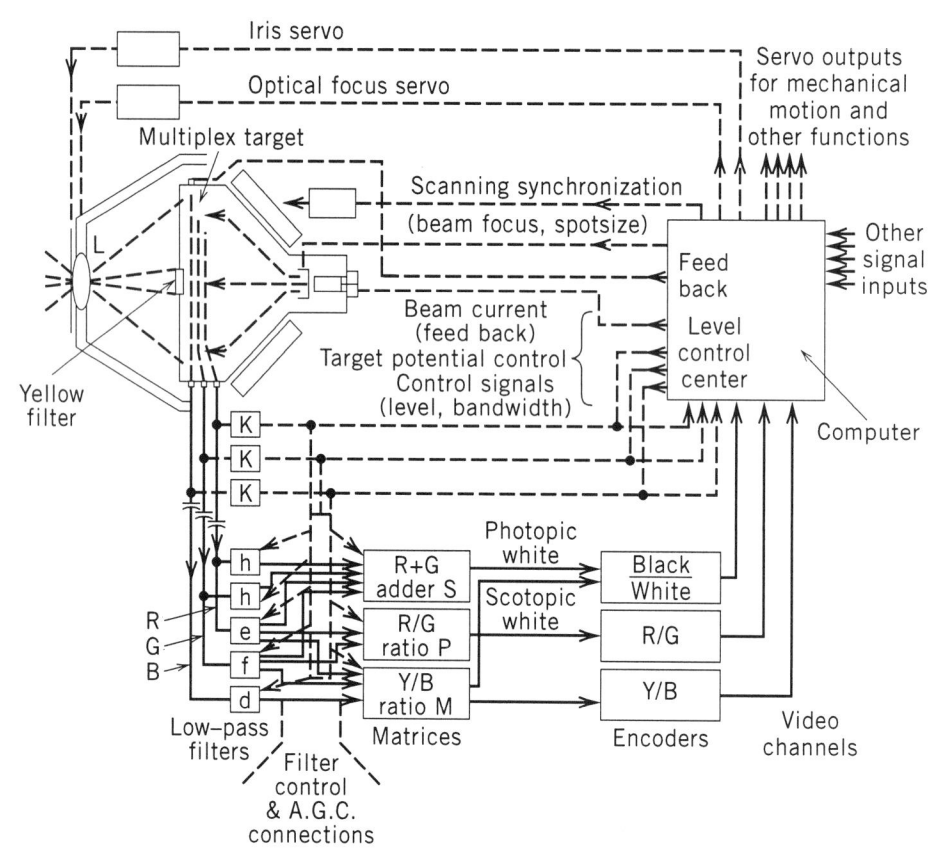

Figure 1.2. Electromechanical analog of the human eye. (From O. H. Schade, "Optical and Photoelectric Analog of the Eye," *J. Opt. Soc. Am.*, Vol. 46, No. 9, pp. 721, Sept. 1956. Reprinted by permission.)

reaching the retina, thus providing a form of automatic gain control. Reducing the pupil size also increases the depth of field similar to using higher f/number settings with a camera lens. This accounts for the improved legibility that results at higher illumination levels. The pupil size is normally about 8 mm diameter at low light levels for maximum sensitivity. The focal length of the eye is approximately 17 mm so the focal ratio is somewhere around $f/2.1$ in low light ambient. At high illumination levels it may stop down to about 2 mm for a focal ratio of $f/8.5$. This represents a change in light-gathering ability of up to 16 times. The remainder of the sensitivity variation is due to chemical changes within the retina.

The interior of the eye is filled with a jellylike substance known as the vitreous humor that provides support for the delicate retina. The retina is the surface upon which the image formed by the lens is focused. Two types of light-sensitive elements populate the retina, cones and rods, so named because of their physical appearance. Cones are sensitive to color and provide the highest resolution but require moderately high

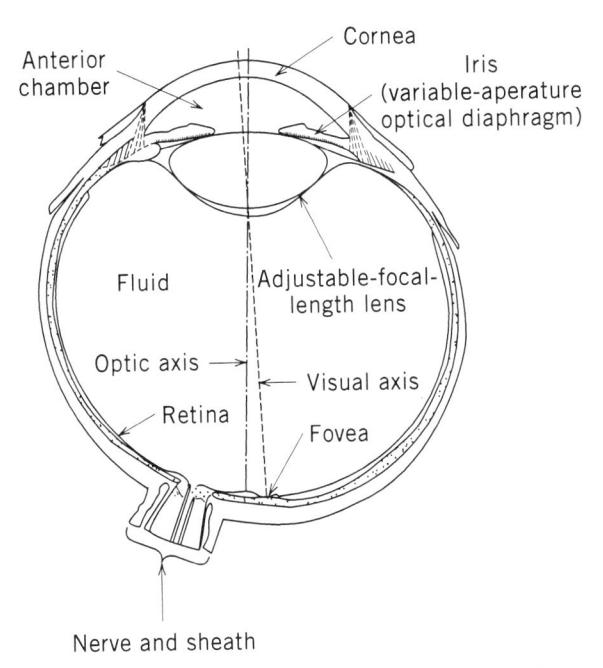

Figure 1.3. Cross section of the human eye. (From Navships Display Illumination Guide, p. I-4, 1973.)

light levels to function. They provide most of our normal color vision and, according to the trichromatic theory, are subdivided into three types sensitive primarily to red, green, and blue with peak sensitivities at about 575, 535, and 445 nm, respectively. Actually, 575 nm is more correctly yellow than red, but it contributes the eye's red sensitivity and is usually referred to as red sensitive.[14] Conversely, the rods have no color discrimination capability but provide the high sensitivity necessary to see in near darkness. They are most sensitive in the blue-green region of about 560 nm. The cones are located mostly in the central portion of the retina, known as the fovea, while the far more numerous rods are mostly located in the area surrounding the fovea. A slightly off-center area where the optic nerve exits the eye on its way to the brain causes an insensitive area on the retina popularly referred to as the "blind spot." Because of the lack of rods in the central region of the retina, the best sensitivity to faint objects is obtained using "averted vision." This is a technique whereby the observer looks slightly to the side of the object, thus using the higher sensitivity region surrounding the fovea.

At very low light levels, the photochemical *rhodopsin,* or "visual purple," is produced in the rods, which greatly enhances their sensitivity. The process is known as dark adaption, and the result is what we call "night vision." It takes about 30 min in darkness to obtain full dark adaptation for night vision. Figure 1.4 demonstrates the typical sensitivity characteristic changes that occur with the shift from cone to rod receptors with time. At its most extreme sensitivity the human eye can detect light sources emitting just a few photons with a quantum efficiency of about 10%.

6 LIGHT AND COLOR

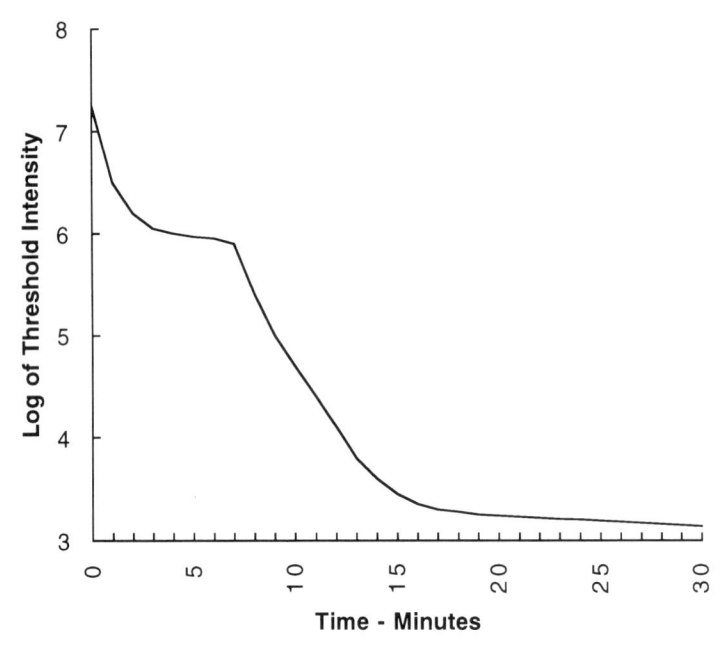

Figure 1.4. Dark-adaption curves of the human eye versus time. (After S. Hecht and S. Shiker, "An Adaptometer for Measuring Human Dark Adaption," *J. Opt. Soc. Am.*, Vol. 28, pp. 269–275, July 1938.)

Spectral Sensitivity

The spectral, or color, response of the average human eye extends from about 380 to 780 nm. In 1924, the Commission Internationale de l'Eclairage (CIE), sometimes referred to in early publications by its English translation, the International Commission on Illumination (ICI), adopted an internationally agreed-upon spectral response curve representing the average human eye as measured for a number of individuals.[15] The curve, which was derived from work by the Bureau of Standards published in 1923,[16] is known variously as the CIE 1924 Standard Observer, the $V_{(\lambda)}$ curve, or the *photopic* response curve (Figure 1.5) and represents the central 2° field of vision. It was also refined and incorporated into the Y function of the 1931 CIE system for color specification to be discussed later.[17] Maximum sensitivity of the eye occurs at about 555 nm in the yellowish-green region. Sensitivity falls off rapidly toward the red and blue ends of the spectrum. Much debate has ensued over the years as to the validity of the photopic curve, and it is known to underrepresent the blue sensitivity of the eye below 460 nm, especially for heterochromatic light sources such as displays where the light is composed of mixtures of colors.[18–23] While the debate continued, D. B. Judd proposed a modification of the curve in 1951 that has been accepted primarily among vision researchers. It was not until 1988 that the CIE officially adopted a slightly refined version (Figure 1.6) of the Judd modification identified as the CIE 2° Spectral Luminous Efficiency Function for Photopic Vision, with the notation $V_{M(\lambda)}$.[24] Although the CIE designates the 1988 version as preferred over the original curve, instrumentation using the newer curve has been slow in coming to the marketplace, and the 1924 version continues to be the industry standard. It will cause serious descrepancies with some existing long-established data if and when a change is finally made.

1924 CIE Photopic Curve

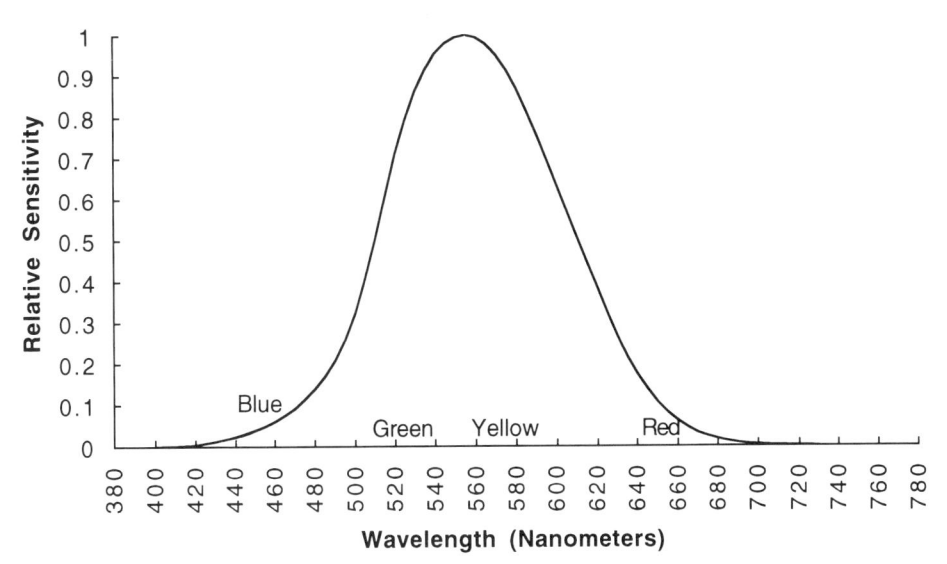

Figure 1.5. 1924 CIE Standard Observer, also known as the photopic curve. (From data in ref. 15.)

1988 CIE Modified Photopic Curve

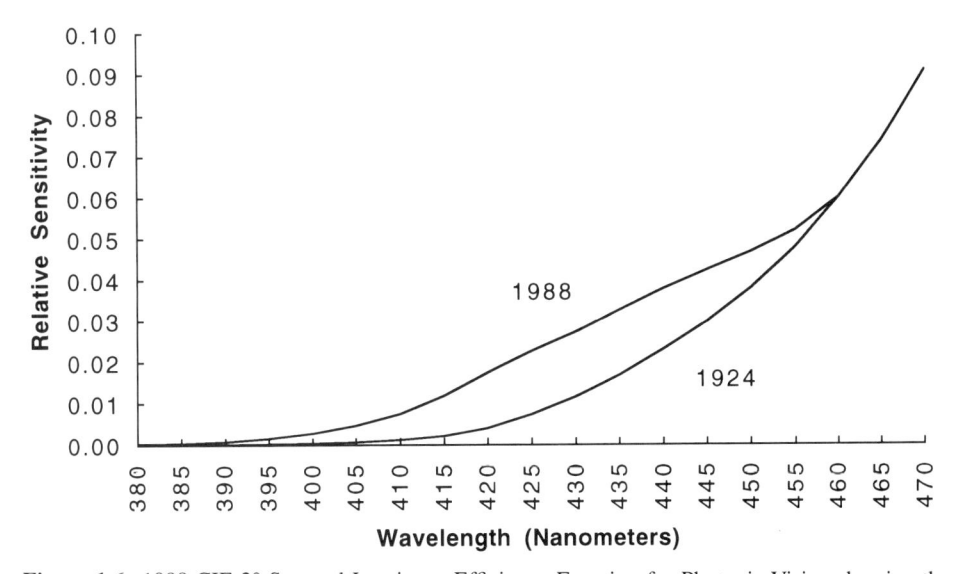

Figure 1.6. 1988 CIE 2° Spectral Luminous Efficiency Function for Photopic Vision showing the deviation from the 1924 CIE Standard Observer in the blue region. (From data in ref. 24.)

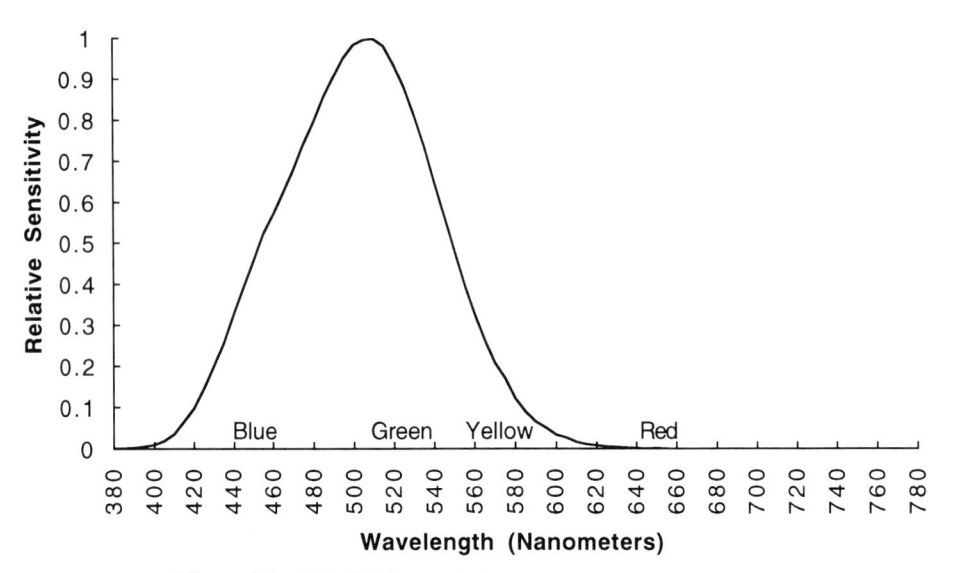

Figure 1.7. 1951 CIE Scotopic Observer. (From data in ref. 22.)

The remaining spectral characteristic of the eye is its response at low illumination levels. Below 3.4 cd/m² a gradual shift occurs from the eye's cones, which are the color-sensing elements, to its rods, which have greater sensitivity. Vision at light levels below luminances of about 0.034 cd/m² is referred to as *scotopic* (night, or dark-adapted) vision and exhibits higher sensitivity to blue light with a peak at 507 nm (Figure 1.7). The scotopic response curve was officially adopted by the CIE in 1951 as the CIE 1951 Scotopic Function and is designated $V'_{(\lambda)}$. The range between 3.4 and 0.034 cd/m² is known as mesopic vision and has a resultant spectral response curve somewhere between the photopic and scotopic curves.[25]

Limitations

The eye, with all of its remarkable capabilities, is not without some shortcomings. These often cause problems for users of displays. Eyestrain and headaches are routinely reported by computer display users. Prolonged viewing hours, improper corrective lenses for the distance of the screen, glare from incorrectly positioned displays, excessive room lighting, and screens that are too small, poorly focused, or too dim or exhibit low contrast all contribute to the problem. Up to 25% of the human body's energy budget has been reported to be used for the visual system, so it is easy to understand why visual problems can have a substantial impact on comfort and well-being.

Correction for focusing defects such as near sightedness (myopia and hyperopia) and far sightedness (presbyopia) as well as astigmatism is required for many individuals. Contrary to popular belief, 20/20 vision is not perfect eyesight, just average. Vision of 20/15 or even 20/13 sometimes occurs or is often attainable with correction. Combat pilots having such vision in the preradar gunsight days had a distinct advantage in either target acquisition or enemy aircraft avoidance over those with average 20/20 vision.

The eye normally focuses near infinity, and the ciliary muscles distort the shape of the lens for greater refraction to focus the image on the retina at closer distances. This is known as *accommodation*. The range of accommodation rapidly decreases with age (Figure 1.8), as the eye muscles lose strength, and by about age 40 usually requires correction.[26] This leads to the familiar complaint of "the arms not being long enough to read or do fine work comfortably any more." Prescription eyeglasses specifically corrected for normal display viewing distances are often used by computer users who must spend considerable time in front of the screen. They can greatly improve viewing comfort by allowing the eye muscles to be in a relaxed condition for that particular distance.

Color-defective vision, more commonly called "color blindness," occurs in about 8% of the male population and 0.5% of the female population. Usually red and green are confused while blue and yellow are not.[27] Only about 0.002–0.003% of the population have no color perception at all (monochromatism).[28] Defective color vision may cause difficulty where important information is color coded on the display. It goes without saying that normal color vision is required for individuals making visual color adjustments of displays. Use of a colorimeter will partially alleviate the problem, although it would

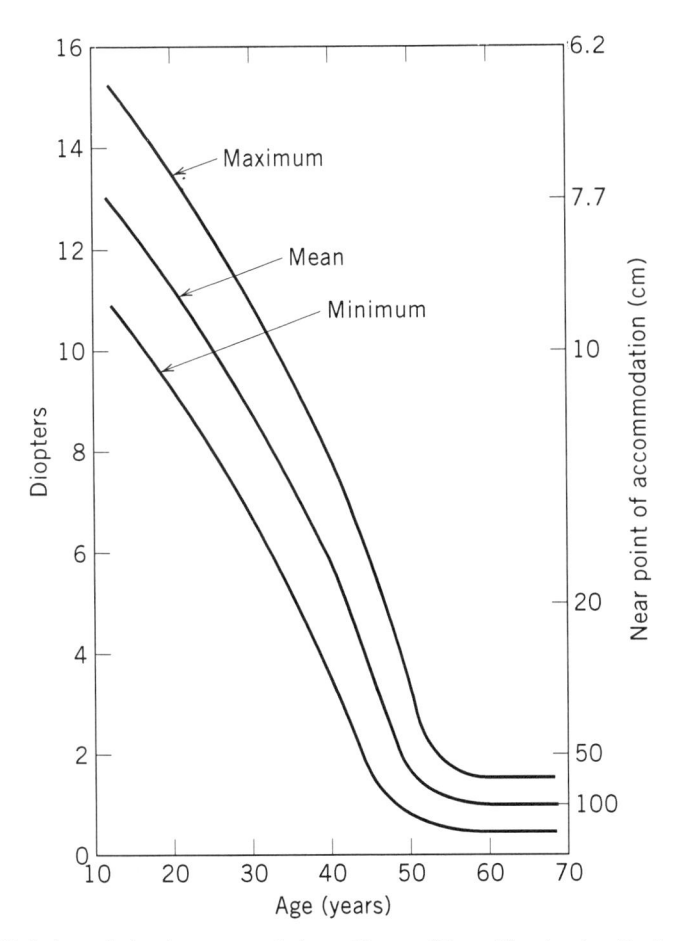

Figure 1.8. Variation of visual accommodation with age. [From Illuminating Engineering Society of North America (IESNA), *Lighting Handbook,* 8th ed., 1993, p. 75. Reprinted by permission.]

probably still be more difficult for someone with defective color vision to adjust the display or recognize problems. Adjustment by eye is often used to roughly set the display to a visual estimate of the correct color balance as a starting point so that fewer measurements and adjustments are required.

Even "normal" individuals will interpret colors slightly differently due to differing spectral sensitivities. Furthermore, each eye of one individual is likely to differ slightly from the other. It is not until an attempt is made to match adjacent colors with one eye at a time that it becomes obvious.

Discrimination

Comparison of light intensities or colors of adjacent areas demonstrates the excellent discrimination abilities of the human eye. Yet, trying to match or compare colors or light levels in different locations or at different times is very difficult at best. The eye and brain are influenced by surroundings and colors that have been viewed most recently. Stare at a colored paper for a few moments and the opposite, or *complimentary*, color will be seen when shifting the eye to a white surface. The eye/brain tends to shift its color balance reference point to make the colored paper appear more nearly white. The same applies when viewing objects in daylight, incandescent light, and fluorescent light, all of which are very different in shades of white.

Flicker

Certain repeating patterns and flash rates of pulsed light cause viewing discomfort and, in the extreme, epileptic seizures.[29] Flicker is a related annoyance. The eye is sensitive to repetitive flashing such as the cathode-ray tube (CRT) raster that produces a display. Flicker is usually most pronounced at high brightnesses and/or when located near the periphery of the visual field. Increasing the frame refresh rate to 70–75 Hz or higher eliminates the effect for most observers by operating above the eye's *flicker fusion frequency,* the frequency at which the eye no longer perceives flicker.[30-34] Figure 1.9 illustrates the number of subjects in one test that were bothered by flicker versus refresh rate using both monochrome and color displays operated at 100 cd/m^2. The differences between monochrome and color displays are probably due to the differing phosphor decay characteristics.

Interlacing has long been used for television and some information displays to reduce flicker while maintaining the horizontal frequency and video bandwidth within practical limits.[35-39] In interlaced scanning, the even-numbered lines are scanned in the first field while the odd-numbered lines are scanned in the second field (Figure 1.10). For NTSC (National Television System Committee) television, the two fields, each consisting of 262.5 lines, combine to produce a complete frame of 525 lines due to the persistence of vision. Each field is refreshed at a rate of 60 times per second to display one complete picture 30 times per second without the severe flicker normally associated with a 30 Hz refresh rate.[40] This is known as 2:1 interlace, as opposed to noninterlaced scanning, also referred to as progressive or sequential scanning, where a conventional single field displays the entire image.

For photometric measurement instruments designed to measure as the eye responds, a time constant of 100 ms or more is used. This avoids flickering readings due to the pulsed nature of raster displays and other AC-operated light sources.

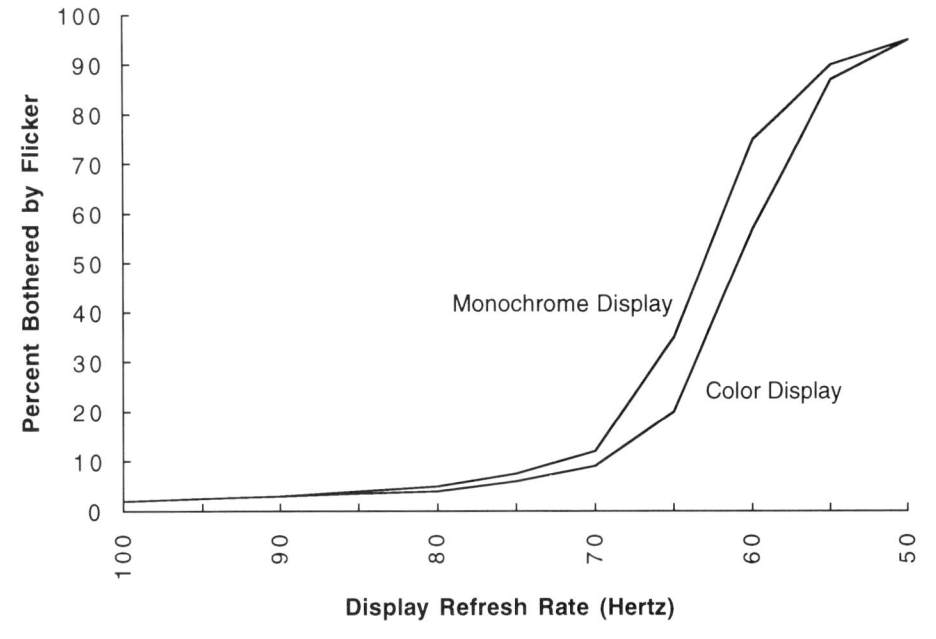

Figure 1.9. Percentage of sample viewers bothered by flicker versus raster refresh rate. (From C. Sigel, "CRT Refresh Rate and Perceived Flicker," *SID Dig.,* p. 302, 1989. Reprinted by permission.)

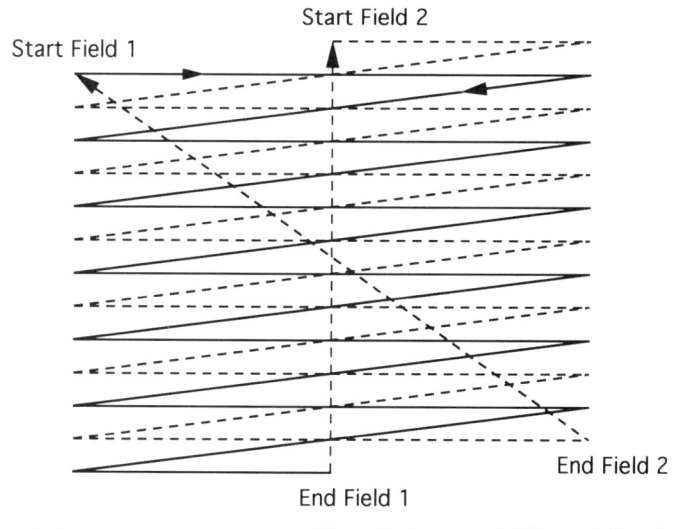

Figure 1.10. A 2:1 interlaced scanning (simplified). Field 1 is solid line and field 2 is dashed line.

12 LIGHT AND COLOR

1.3 RADIOMETRY

Radiometry is the basis for all light measurement. It is defined by the Institute of Electrical and Electronics Engineers (IEEE) as "the measurement of quantities associated with radiant energy"[41] and has sometimes included the measurement of radio frequency radiation as well as optical radiation. For our purposes, a better definition is supplied by the Illuminating Engineering Society of North America (IESNA). They define radiometry as "the measurement of optical radiation."[42] In radiometry as well as photometry, most confusion is caused by a lack of understanding of the units of measure. Unfortunately, most books on the subject are written at a level that only adds to the confusion. We will try to explain the basic units as concisely as possible with emphasis on the commonly encountered units and where they should be used. Those readers concerned only with the visual performance of displays or lighting may skip this section and proceed to the next section, Photometry, instead. All of the concepts of radiometry have parallels in photometry that will be covered in a similar manner.

Following are descriptions of the commonly used radiometric quantities. Note the use of the prefix *rad-* for all quantities related to radiometry.

Radiant Flux

The *watt,* the unit of *radiant flux,* is the fundamental unit of radiometry. All other radiometric units are derived by combining the watt with units of area, distance, solid angle, and/or time.

The watt is essentially the unit of power so familiar to every electrical engineer. A standard 100-W light bulb, or more properly a "lamp," primarily produces light radiation as well as heat in the form of infrared radiation. The total combined *radiant power* produced by the lamp is about 100 W since almost all of the conduction losses, for example, are eventually radiated as infrared radiation. Only about 10% of the total radiant power radiated by the lamp is within the visible spectrum, and even less (2%) is useful to the human eye because of the latter's insensitivity to red and blue wavelengths. This will be discussed further in the following section on photometry. Lasers, infrared light-emitting diodes (IR LEDs), and other highly directional light sources also have total radiant power output commonly specified in watts, since it is relatively simple to capture their total light output.

Radiant Intensity

For practical purposes we will assume that 100% of the 100 W is emitted as radiant flux. This 100 W of radiant flux is emitted almost equally in all directions, and the lamp is very nearly a *point source* when viewed from a distance. A true point source is an *isotropic radiator;* that is, it radiates equally in all directions (ignoring the lamp base) and has a finite size. If the distance from a light source exceeds 5–10 times the maximum dimension of the emitting area, it may be assumed to approximate a point source. Again, for practical purposes, we will assume that our 100-W lamp is an isotropic radiator emitting light uniformly into an imaginary sphere. If we form a cone of 57.296°, or one steradian (the unit of solid angle which encloses a surface area on the sphere equal to the square of the radius), with its origin at the lamp and extending to the surface of the sphere, the total radiation flowing through the cone will be the *radiant intensity*. A full sphere contains a total of 4π (12.566) sr; thus any 1 sr in our example will contain 100 W/12.566 sr, or about

8 W. The diameter of the sphere is immaterial; as the sphere diameter increases and the circle formed on the surface of the sphere increases, the total radiation within the circle remains the same. Radiant intensity is sometimes used to characterize IR LEDs, which emit light in a cone rather than isotropically.

Irradiance

The most commonly encountered radiometric unit is *irradiance*. Irradiance is simply the amount of optical radiation *incident* upon (falling upon) a specified surface area. The preferred unit is the watt per square meter although the watt per square centimeter is still sometimes encountered. Watts per square centimeter are easily converted to watts per square meter merely by multiplying by 10,000, which is the number of square centimeters in a square meter. Of course, a wide variety of prefixes are attached to the watt to accommodate the range of power levels encountered. See Appendix B for a listing of SI prefixes from 10^{-24} to 10^{+24}. In our previous example of the 100-W lamp, a sphere radius of 1 m would result in an irradiance of 8 W/m^2 (100 W/12.566 sr). For any other radius, the *inverse square law* determines the irradiance. The irradiance will change inversely with the square of the distance, as shown in Figure 1.11. If the radiation source is moved to twice the distance, the same amount of light will be spread over four times the area and the irradiance will be reduced to one-fourth of the original amount. At three times the distance the light will be reduced by a factor of 9. Whether the surface is reflective or absorbing, the amount of radiation incident upon it will remain the same.

Examples of typical irradiance levels are sunlight on the earth's surface at approximately 1000 and 1353 W/m^2 just above the earth's atmosphere.[43,44] The latter is referred to as the *solar constant*.

Radiant Exitance

Radiant exitance, measured in watts per square meter, in common with irradiance, is used to indicate the total radiation per unit of area emitted, reflected, or transmitted by a 1-m^2 surface regardless of direction.

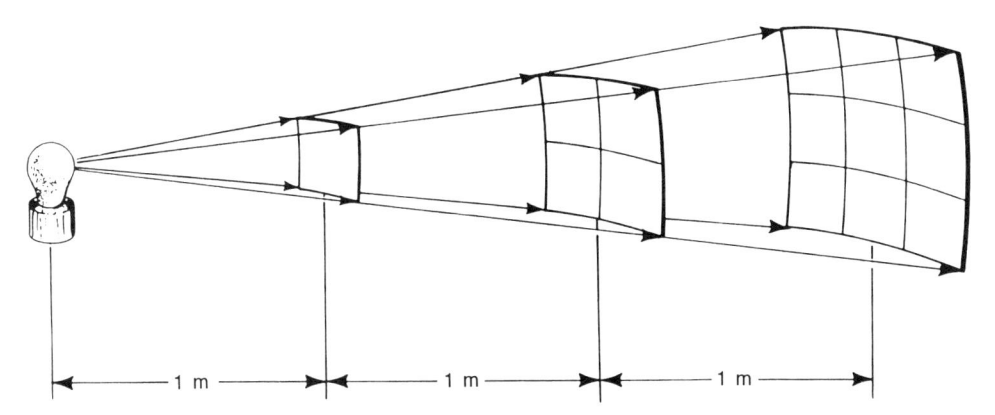

Figure 1.11. Inverse square law. At twice the distance the light is spread over four times the area, which results in one-quarter of the illuminance, and at three times the distance the light is spread over nine times the area, which reduces the illuminance to one-ninth.

Radiance

Radiance, in watts per steradian per square meter, is also used to indicate the light reflected, transmitted, or emitted by a diffusing surface. For a uniformly diffusing surface it will be equal to the radiant excitance in watts per square meter divided by π.

Radiant Energy

Radiant energy is encountered in pulsed laser specifications. It is merely the power in watts multiplied by the pulse duration in seconds or decimals of a second and is expressed in joules (watt seconds). A one watt (peak power) pulsed laser producing one microsecond duration pulses at a one-kilohertz repetition rate would have an average power of one milliwatt and a radiant energy of one millijoule (one milliwatt per second). The total dose in 10 s would be 10 mJ.

Radiometric Detectors

An ideal detector for radiometric measurements would have equal sensitivity to all wavelengths (flat response) from the far infrared through visible light to the vacuum ultraviolet region. Often, a narrower wavelength range of interest may be specified and a detector suitable to that range used for the measurement. Filters may be used to reject wavelengths outside the range of interest. Most practical radiometric detectors have somewhat limited useful wavelength ranges and must be used with this limitation kept in mind. Also, they will not usually have equal response to all wavelengths within their useful spectral range. See the section on light detectors in Chapter 2 for further information.

1.4 PHOTOMETRY

Photometry is a subset of radiometry. Each radiometric unit has a photometric equivalent. Whereas radiometric measurements are made with a detector having "flat" spectral response, photometric measurements are intended to represent the amount of light useful to the human visual system. To accomplish that, the detector should be closely matched to the spectral response curve of the average human eye (photopic response) as defined by the CIE 1931 Standard Observer previously described in Section 1.2.

Following are descriptions of commonly used photometric quantities. Note the use of the prefix *lum-* for all photometric quantities.

Luminous Flux

The *lumen* is essentially a unit of power useful to the human eye. It is related closely to the watt and is defined as the spectral luminous efficacy (K_m) for monochromatic light at the peak visual response wavelength of 555 nm. It has been standardized at 683 lm/W.[45] At all other wavelengths it is lower in proportion to the factors in the *Y* column of the tristimulus table in Appendix E. The maximum visual efficiency that could be produced by a lamp would occur if it produced only narrow-band yellowish-green light at 555 nm. Unfortunately, other factors such as the need to differentiate colors and a desire for "white" illumination similar to sunlight enter into the picture. A standard 100-W light bulb, or more properly lamp, primarily produces broadband light radiation as well as heat in the form of infrared radiation. The total combined *radiant flux* produced by the lamp is nearly 100 W since almost all of the conduction losses, for example, are eventually radiated as infrared radiation. If all of its radiation were concentrated at 555 nm, it would have an output of about 68,300 lm (683 lm/W · 100 W). However, only about 10% of the total radiant power radiated by the lamp is within the visible spectrum and even less (2%) is useful to the human eye because of the eye's insensitivity to red and blue wavelengths. The resultant visual output for the common 100-W lamp is specified at 1750 lm. From these two numbers, we may compute the *luminous efficacy* of the lamp in lumens per watt. In this case it is 1750 lm/100 W, or 17.5 lm/W. This is about 2.5% of the theoretical maximum possible efficacy of 68,300 lm. Most of the other 97.5% is lost as heat. Luminous efficacies of other light sources may be much higher. Examples include some LEDs that are approaching 100 lm/W, flourescent lamps with up to 90 lm/W, and low-pressure sodium lamps that are about 150 lm/W.

Luminous Intensity

Ignoring the lamp base, the 1750 lm of luminous flux is emitted almost equally in all directions and the lamp is very nearly a *point source* when viewed from a distance. A true point source is an *isotropic radiator;* that is, it radiates equally in all directions and has finite size. If the distance from a light source exceeds 5–10 times the maximum dimension of the emitting area, it may be assumed to approximate a point source. For practical purposes, we will assume that our 100-W lamp is an isotropic radiator emitting light uniformly into an imaginary sphere. If we form a cone of 57.296°, or 1 sr (the unit of solid angle that encloses a surface area on the sphere equal to the square of the radius), with its origin at the lamp and extending to the surface of the sphere, the total visible light flowing through the cone will be the *luminous intensity*. Luminous intensity is expressed in *candelas* (lumens per steradian). A full sphere contains a total of 4π (12.566) sr; thus any 1 sr of our example will contain 1750 lm/12.566 sr, or about 139 cd (or lumens per steradian). The diameter of the sphere is immaterial; as the sphere diameter increases and the circle formed on the surface of the sphere increases, the total radiation within the circle remains the same. Luminous intensity is often used to characterize visible LEDs, which emit light in a cone rather than isotropically.

The luminous intensity in candelas is the fundamental unit of photometry. All other photometric units are derived by combining the candelas with units of area, distance, solid angle, and/or time.

Illuminance

The most commonly encountered photometric unit is *illuminance*. Illuminance is simply the amount of visible radiation *incident* upon (falling upon) a specified surface area. The preferred unit is the lux (lumen per square meter). The deprecated but still widely used footcandle (lumen per square foot) is easily converted to lux merely by multiplying by 10.764, which is the number of square feet in a square meter. In our previous example of the 100-W lamp with 1750 lm, a sphere radius of 1 m would result in an illuminance of 139 lx (1750 lm/12.566 m). For any other radius, the *inverse square law* determines the illuminance. The illuminance will change inversely with the square of the distance, as shown in Figure 1.11. If the light source is moved to twice the distance, the same amount of light will be spread over four times the area and the illuminance will be

16 LIGHT AND COLOR

TABLE 1.1 Natural Illuminance Levels (Approximate)

Direct sunlight	100,000 lx
Full daylight (excluding direct sunlight)	10,000 lx
Overcast sky	1,000 lx
Heavy overcast	100 lx
Twilight[a]	10 lx
Late twilight[a]	1 lx
Full moon	0.1 lx
Quarter moon	0.01 lx
Clear night sky (no moon or light pollution)	0.001 lx
Overcast night sky (no moon or light pollution)	0.0001 lx

[a]At 45° latitude, the illuminance level decreases by approximately half every 5 min during the evening twilight period and approximately doubles every 5 min during the morning twilight period.

reduced to one-fourth of the original amount. At three times the distance the light will be reduced by a factor of 9. Whether the surface is reflective or absorbing, the amount of radiation incident upon it will remain the same. Examples of commonly encountered illuminance levels are direct sunlight on the earth's surface at approximately 100,000 lx and typical office illumination of 500 lx. Table 1.1 lists typical illuminance levels for natural illumination.

Luminance

Luminance, in candelas per square meter, also called the *nit* and equivalent to lumens per steradian per square meter, is preferred to indicate the light reflected, transmitted, or emitted by a diffusing surface. The deprecated unit, the footlambert ($1/\pi$ cd/ft^2), is still in common usage in the United States and is easily converted to candelas per square meter by multiplying by 3.426, which is the number of square feet in a square meter divided by π. The term *brightness* is often confused with luminance. Luminance is the measurement of light from a surface while brightness is the subjective appearance of the surface. Table 1.2 lists typical luminance levels that may be encountered.

TABLE 1.2 Luminance Levels (Approximate)

Sun's disc	1,500,000,000 cd/m^2
100-W soft white lamp	30,000 cd/m^2
White paper, scattered clouds, and snow in sunlight	25,000 cd/m^2
Fluorescent lamp surface	10,000 cd/m^2
Overcast sky	3,000 cd/m^2
Moon's disc	2,500 cd/m^2
Blue sky	1,000 cd/m^2
White paper on desk (office)	100 cd/m^2
CRT display	60–150 cd/m^2
White paper on desk (home)	30 cd/m^2

Luminous Exitance

Luminous exitance, also measured in lumens per square meter, in common with illuminance, is sometimes used to indicate the total light per unit of area emitted, reflected, or transmitted by a surface regardless of direction. For a uniformly diffusing surface it will be equal to the luminance in candelas per square meter multiplied by π.

1.5 COLORIMETRY

Color (chromaticity) specification and measurement are formidable subjects for many display designers and users. These individuals, knowledgeable in their own fields, often feel intimidated by the bewildering terminology and concepts used for color. To add to the difficulties, the terminology and concepts of photometry, radiometry, and spectroradiometry are thoroughly embedded in the subject. And do not forget the elements of human visual perception that distort measurement linearities and relationships of different colors under differing viewing conditions.

We will discuss the terminology, concepts, pitfalls, and standards commonly encountered in flat panel and CRT color specifications.

First, the issue of color primaries. Ask an artist what the color primaries are and they will quickly respond with red, yellow, and blue. This is correct for *subtractive* color such as paints and dyes where each primary *absorbs* a portion of the white light illuminating the object or painting. With no pigmentation present, nearly all light is reflected from the object with a resultant white appearance. If all of the three primary colors are applied to the same surface, most of the light will be absorbed and it will appear black. Mixtures of the three primary pigment colors may be used to produce many other colors and shades such as green, orange, and so on. Most displays, however, are based on *additive* color. In this case, the three primaries are red, green, and blue. Instead of absorbing light, the primaries are *emitters* of light. With no emission of the primaries, as when the device is turned off (and no ambient light is present), black is seen. When all three are emitting in the proper ratios, white light is produced. This is just the opposite of subtractive color. Intermediate combinations of the three additive primaries result in the ability to produce many other colors and shades.

Two systems are usually used to specify display chromaticity: the 1931 CIE system and the 1976 CIE Uniform Chromaticity Scale (UCS) system.[46–48] Each has its own advantages (and faults). Fortunately, measurements and specifications in either system may be converted to the other, often at the press of a button on the colorimeter or spectroradiometer used to take the measurements. At worst, simple linear equations may be used for the conversion (see Appendix F). Many other color systems have been devised to aid color specification and matching. Other industries, particularly those based on subtractive color such as paints, printing, and textiles, use color specification systems suited to their particular needs, such as Munsell, Pantone, and the CIELAB systems.

1931 CIE Chromaticity System

The 1931 CIE Chromaticity System[49–51] was devised by the CIE, an international organization devoted to the specification of light and color. The 1931 CIE Chromaticity Diagram (Plate 1) allows any color that may be seen by the average human eye to be

18 LIGHT AND COLOR

specified by a pair of coordinates, x and y. The area bounded by the horseshoe-shaped perimeter defines the limits of the colors viewable by the human eye and is termed the *color space*. The interior color boundaries and names were added in 1943 by K. L. Kelly[52] of the National Bureau of Standards (NBS), now the National Institute of Standards and Technology [NIST]. The boundaries are not actually sharply defined but gradually blend from one color to the next and are approximations for the purpose of visualizing the meaning of a pair of numerical coordinates. Color saturation or purity increases as it approaches the outer periphery of the chart and shifts to pastels and finally to neutral white as the center is approached. Color mixtures, particularly applicable to color displays, are easily derived from the 1931 diagram. A 50–50 mixture of a particular green and red will lie exactly halfway on a straight line between the two points. The principal drawback of the 1931 system is that due to nonlinearities in the human eye, equal distances on the chart do not represent equal perceived color differences. Thus a small change in color coordinates in the red region would require a much greater change in color coordinates in the green region to produce the same degree of perceived color difference.

1976 CIE UCS System

The 1976 CIE UCS[48] system was developed to minimize the limitations of the 1931 system previously mentioned. The chart was squeezed and stretched via linear transformations (see Appendix F for the equations) to show approximately equal perceived color differences by equal distances on the chart while still allowing display on a flat two-dimensional surface. The trade-off is the lack of ability to show the resultant color of a mixture of two colors by a proportional distance along a straight line between them. The 1976 CIE UCS diagram (Plate 2) uses u' and v' coordinates. The symbols u' and v' were chosen to differentiate from the short-lived 1960 CIE UCS system using u and v which was very similar except that the length of the v axis was only two-thirds that of the v' axis.

The three color primaries (red, green, and blue) of a display, when excited individually, may be measured and plotted on the chromaticity diagram (either 1931 or 1976). The three resultant points form a triangle (Figure 1.12). The display can only reproduce

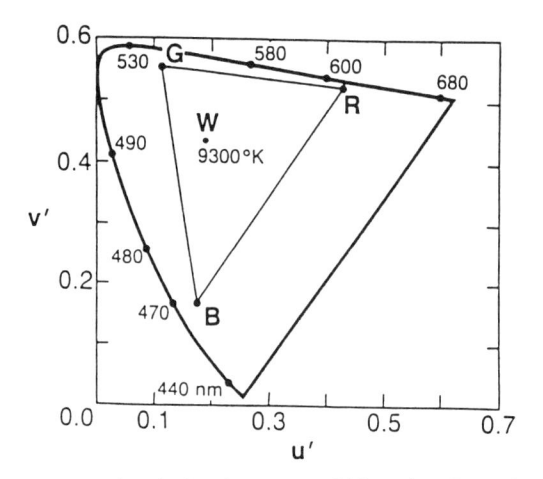

Figure 1.12. The color gamut of typical red, green, and blue phosphor primaries plotted on 1976 CIE UCS chart. Any color within the triangle may be reproduced by controlling the proportions of excitation of each of the phosphors.

colors within these bounds. Ambient light added to that of the display will dilute the color purity, giving it a "washed-out" appearance. The displayed color will be somewhere between that of the display and the ambient illumination depending on the ratio of intensities of each. This will reduce the overall size of the triangle formed by the color primaries and hence the color gamut or range of colors that may be reproduced.

CIE Tristimulus Functions

Color measurement is based on three sets of data, the X, Y, and Z tristimulus values. (Note that these are uppercase X and Y, as opposed to the previously described lowercase x and y color coordinates.) These are three carefully prescribed spectral sensitivity curves[48] (Figure 1.13) that simulate the sensitivity of the average human eye receptors to red, green, and blue, respectively. They may be measured using three or four separate detectors, each closely matched to the tristimulus functions in the case of the filter colorimeter or computed by multiplying the measured intensities at each wavelength by the tristimulus functions for that wavelength and summing them to determine the area under the three curves. The X, Y, and Z values are then summed and divided into the X value to determine x and divided into the Y value to determine y (see Appendix F).

Color Difference

The just-noticeable difference (JND) is used to denote the minimum perceivable difference between two closely spaced colors.[53] This difference may vary considerably between different regions of the 1931 CIE chromaticity diagram and to a lesser degree on the 1976 CIE UCS diagram. There will also be significant differences in distance with angle. Thus 1 JND in the x axis, in the y axis, or for any intermediate vector may differ by 2 or more

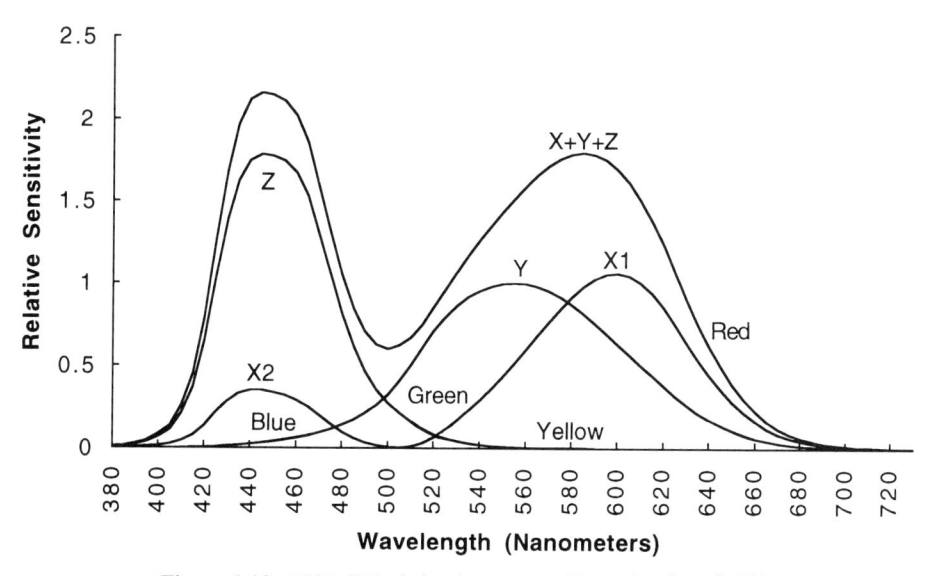

Figure 1.13. 1931 CIE tristimulus curves. (From data in ref. 48.)

times the distance of 1 JND in any other axis. Plotting distances of 10 JNDs at various points on a chromaticity chart will result in a series of ellipses known as MacAdam ellipses.[54–56] Figures 1.14 and 1.15 demonstrate MacAdam and Stiles ellipses in the 1931[57] CIE color space and the similar Stiles ellipses in the 1960[58] CIE color space. Unfortunately, no similar diagram appears to have been computed for the 1976 CIE UCS system, but it would be similar to the 1960 diagram with the v axis multiplied by 1.5.

The JND is a good indication of how closely matched two displays must be if they are to be used in close proximity or for color matching critical applications. For applications where displays are used in different locations, they need not be as closely matched as the white reference point of the eye will adjust of its own accord to the combination of display white and that of surrounding lighting. Two displays adjusted to have exactly the same white reference will appear quite different when one is viewed under incandescent light and the other viewed under fluorescent light or daylight.

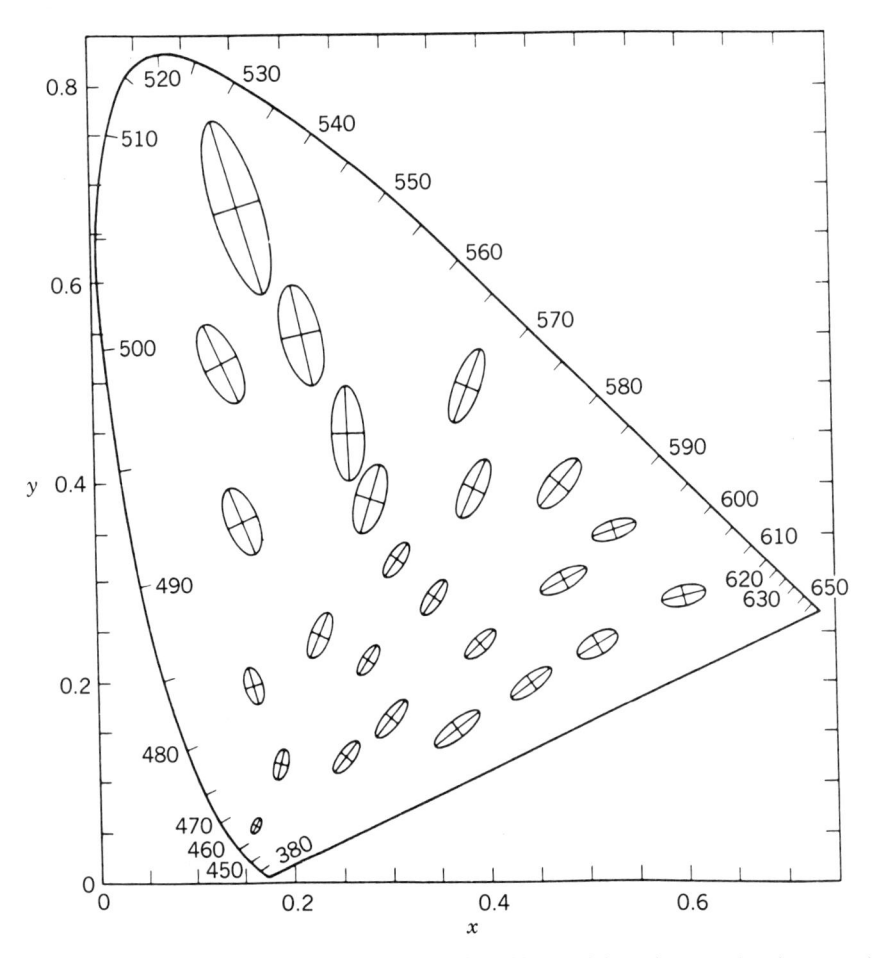

Figure 1.14. MacAdam ellipses plotted on 1931 CIE Chromaticity Diagram showing perceived color differences for selected areas. The ellipses each represent 10 JNDs for clarity. (From G. Wyszecki and W. S. Stiles, *Color Science,* Wiley, New York, 1982, p. 308. Reprinted by permission.)

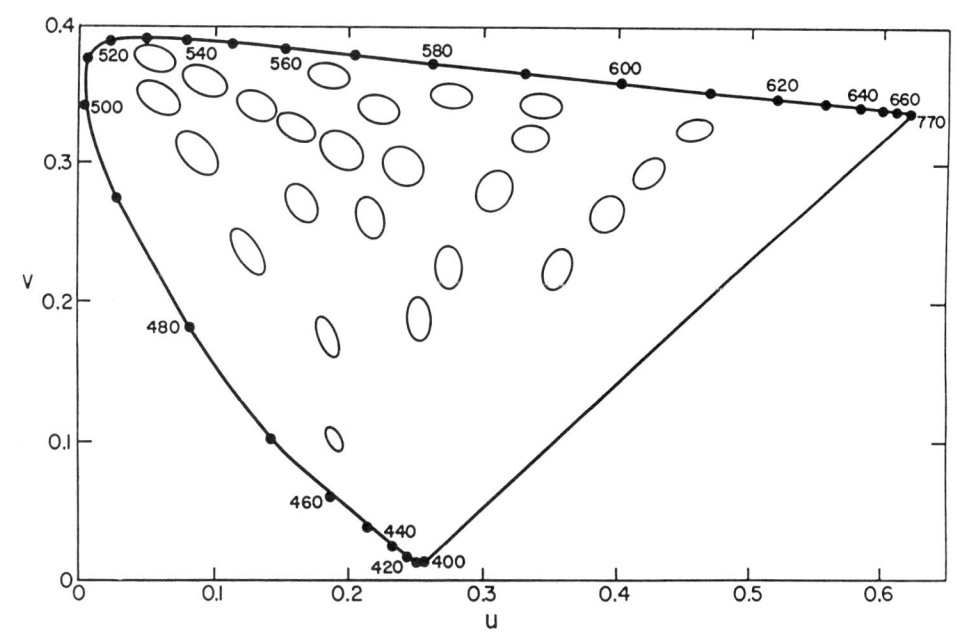

Figure 1.15. Stiles ellipses plotted on 1960 CIE Chromaticity Diagram showing perceived color differences for selected areas. The ellipses each represent 10 JNDs for clarity. (From F. Grum and C. J. Bartleson, *Optical Radiation Measurements,* Vol. 2: *Color Measurement,* Academic, New York, 1980, p. 126. Reprinted by permission.)

Typically a difference of the vector distance between any pair of color coordinates of 0.002–0.006 in either system will be one JND. The closely-related just-perceptible difference (JPD) and, earlier, the minimum-perceptible color difference (MPCD) have also been used for the same purpose.

The 1976 CIE UCS system is best used to measure color difference for displays. To determine the total difference in appearance between two displays including the effect of luminance, use the equation

$$\Delta L\, u'v' = [(L_1 - L_2)^2 + (u'_1 - u'_2)^2 + (v'_1 - v'_2)^2]^{1/2}$$

where

L = luminance

Since different monitors are often adjusted to different luminances because of different viewing conditions, their inherent drive capabilities, or luminance fall-off between the screen center and corners, it is often desirable to omit the luminance portion of the equation in order to isolate the color difference.

1.6 COLOR TEMPERATURE

An object heated to any temperature above approximately 650–800 K will produce a broad-spectrum emission of light with its color directly related to its temperature. This is

22 LIGHT AND COLOR

termed *blackbody* radiation. The color progresses from a very deep red through orange, yellow, white, and finally bluish-white as the temperature is increased. This path may be plotted on the chromaticity chart (Figure 1.16) and is referred to as the Planckian locus. Most natural incandescent light sources such as stars, the sun, and fire fall very close to the Planckian locus. Due to its familiar nature, the same appearance is used as a goal for man-made light sources as well as displays. The commonly accepted incandescent lamp near 2854 K, warm-white fluorescent lamp at 3000 K, cool-white fluorescent lamp at 4100 K, daylight fluorescent lamps at 6500 K, and trichromatic rare-earth phosphor fluorescent lamps of 3000, 4000, and 5000 K are all typical examples. Other light sources not falling on the Planckian locus, such as sodium and mercury lamps, are used for their high visual efficiency, but not without considerable debate and complaints about their unnatural appearance. Cathode-ray tube displays have become loosely standardized at 9300 K (approximate coordinates $x = 0.294$ and $y = 0.294$) for home television and some computer displays and 6500 K (referred to as D_{65} and having coordinates $x = 0.313$, $y = 0.329$) for television studio monitors and critical display monitor applications. A tem-

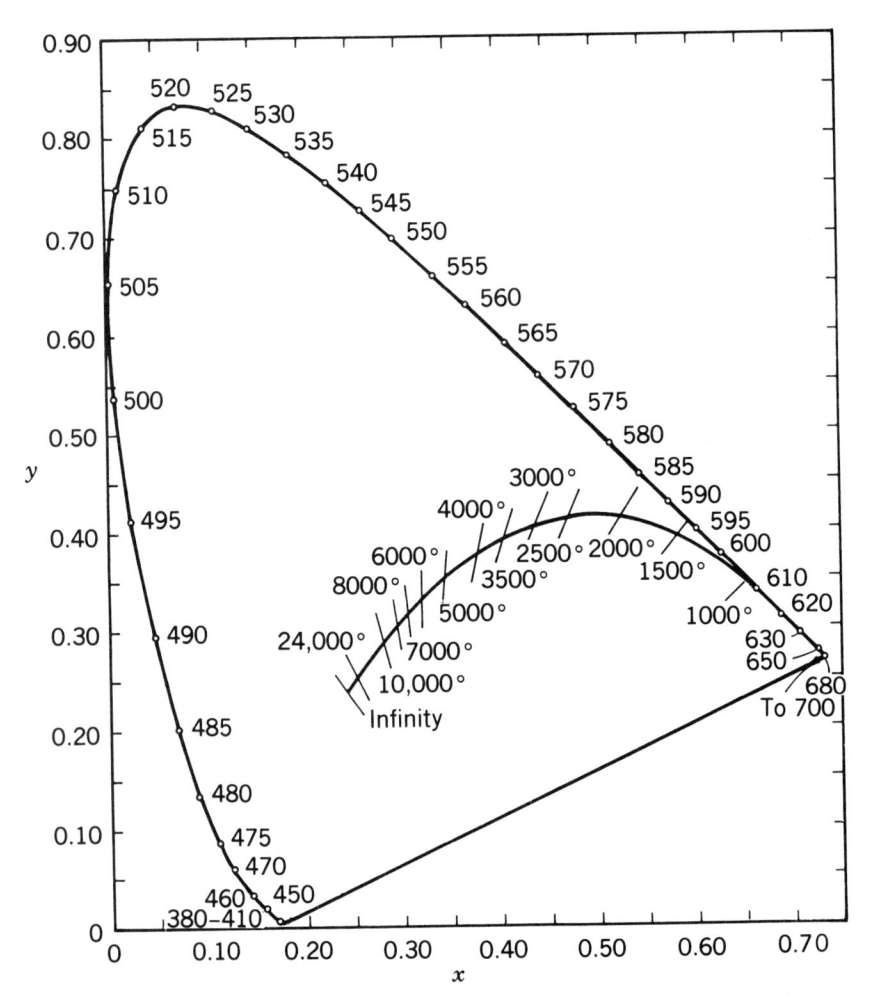

Figure 1.16. Planckian locus plotted on 1931 CIE Chromaticity Diagram. Temperatures are shown in Kelvin. (From Navships Display Illumination Guide, p. I-13, 1973.)

perature of 6500 K also corresponds to the appearance of natural daylight. Standardization of backlit liquid crystal display (LCD) color temperatures has not yet occurred, and miniature fluorescent lamps for backlighting LCDs and having color temperatures of at least 3000–6500 K are advertised. Most readily available are those using standard warm-white (3000 K), cool-white (4100 K), and daylight (6500 K) fluorescent lamp phosphors as well as narrow-band red, green, and blue phosphors similar to the color primaries for color CRTs.

The bluish-white color preferences (at least in North America) of both 9300 and 6500 K are partially due to a similar human preference in laundry detergents. Detergent manufacturers have long added fluorescent "bluing" to their products to make clothes appear "cleaner" and "brighter." Otherwise, washed items would tend to look yellowish ("not clean") under the comparatively low color temperature of incandescent lamps used in the home environment. Try looking at a freshly washed white shirt under long-wave ultraviolet light to see the effect of the bluing or, better yet, look at some of the detergent itself. During the 1950s, Hoffman "Easy-Vision" television receivers with sepia-toned phosphor were a market failure. "Crisp, clean" pictures were and still are preferred by the public.

Correlated Color Temperature

Since fluorescent lamps and display screens usually depend on mixtures of light emitted by two or three different phosphors (yellow and blue or red, green, and blue), they often emit light of a color near the Planckian locus but either above or below it. To determine what color temperature they approximate, a concept known as *correlated color temperature* (CCT) was devised.[59,60] This consists of a series of straight lines representing various color temperatures drawn diagonal to the corresponding color temperature points on the Planckian locus (Figure 1.17). Light sources falling anywhere along the line are said to have a correlated temperature equivalent to that of its intersection with the Planckian locus. Herein lies a frequently misused aspect of depending on color temperature for specifying a display. A display adjusted to 6500 K that happens to fall above the Planckian locus will have a decidedly greenish cast, especially at greater extremes along the line. Another display also adjusted to 6500 K but below the Planckian locus will have a purplish tint, yet the colorimeter indicated that both were set to the manufacturer's specification of 6500 K. If one of the displays had been misadjusted to 6300 K but both were on the Planckian locus, they could have appeared better matched than having both at 6500 K but off the Planckian locus. To easily avoid this trap, always use the actual color coordinates of the desired color temperature at the point at which the correlated color temperature line intersects the Planckian locus. Never rely on color temperature alone for anything other that an incandescent light source. Even that is subject to some error due to emissivity of the filament and colorimeter accuracy but that is another subject by itself.

Dominant Wavelength

Around the periphery of the chromaticity diagrams, a series of numbers is often shown. These represent *dominant wavelengths* in nanometers or micrometers. The dominant wavelength is the wavelength of a monochromatic source that when mixed with a reference white results in the same appearance as the color being analyzed. A practical example of the use of dominant wavelength for displays would be a display that produces a D_{65} white at $x = 0.313$ and $y = 0.329$ and its green primary at, say, $x = 0.300$ and $y = 0.600$.

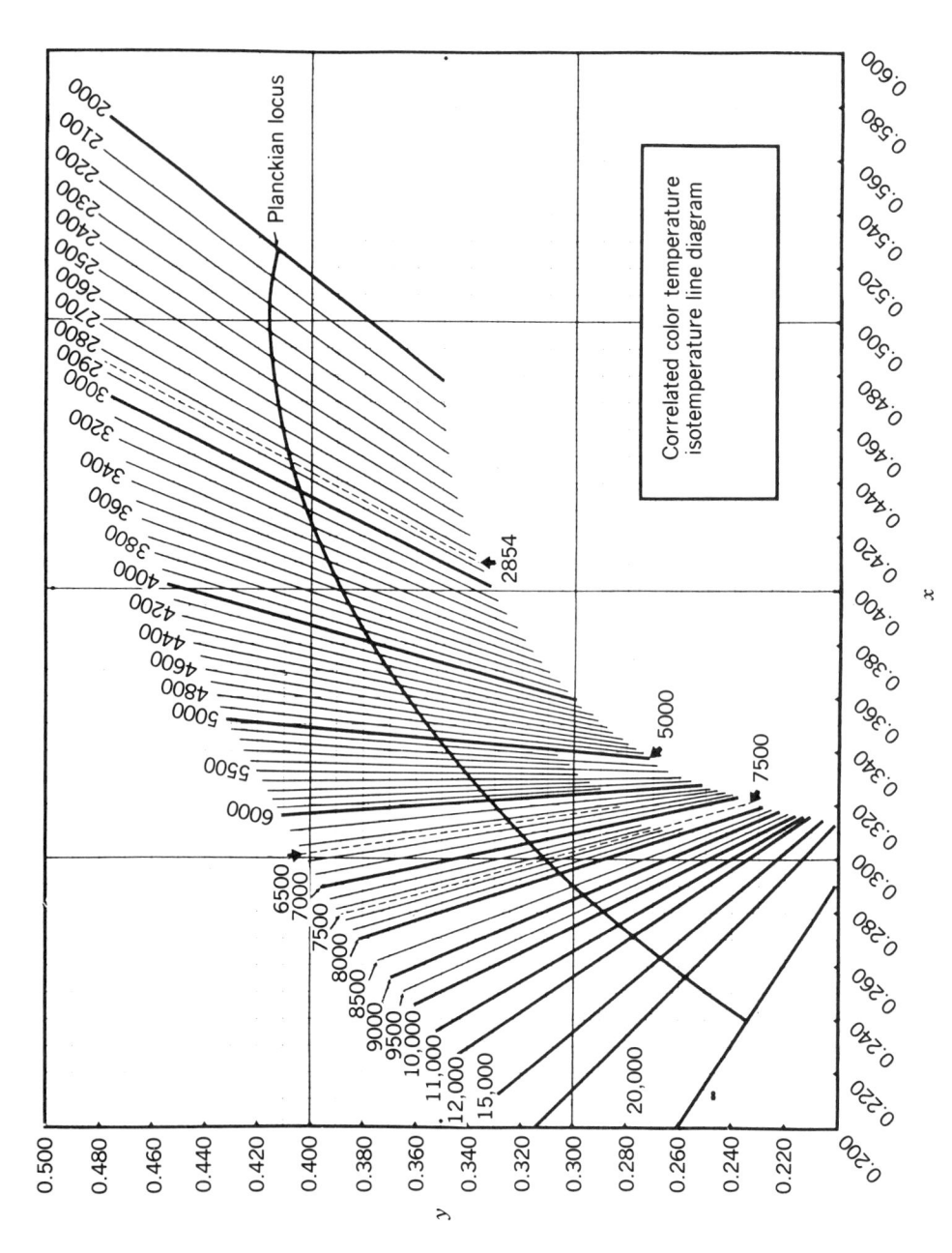

Figure 1.17. Central portion of the 1931 CIE Chromaticity Diagram expanded to show the Planckian locus and correlated temperature (isotemperature) lines. (From Navships Display Illumination Guide, addendum IV, p. 14, 1973.)

The dominant wavelength is found by drawing a straight line from the white point through the green point and extending it to the periphery of the bounded area where the dominant wavelength may be read off. Extrapolation will probably be required in most cases but should be close enough for practical purposes. In the example presented, the dominant wavelength would be about 550 nm, or 0.550 μm. Dominant wavelength is not used very often for display measurements and is described here primarily to answer the frequently asked question, "What do all those little numbers around the edge of the CIE chromaticity diagram mean?"

1.7 LAMBERT'S LAW (COSINE LAW)

Lambert's law states that the illumination or irradiation of any surface varies according to cosine of the angle of incidence. Thus, a surface perpendicular to a point light source will receive maximum illumination, at 45° it will receive 0.707, or 70.7%, as much, and at 60° it will receive 0.5, or 50%, as much (Figure 1.18). The illumination or irradiation will be the same regardless of the nature of the surface. The light may be either absorbed (black surface), reflected (mirrors), diffused (white paper), or a combination of these. A perfectly diffusing surface is known as a Lambertian surface and will scatter any incoming light according to the cosine of the angle at which the surface is viewed. Note that this describes light *scattered from* a surface whereas the previous description was for light *received by* a surface. A piece of matte white paper approximates a Lambertian surface. The surface need not be a passive surface that is diffusing light from an external light source. It may be a self-luminous surface such as a CRT screen, an electroluminescent panel, or a photographic light table.

1.8 INVERSE SQUARE LAW

The inverse square law previously mentioned in the sections on radiometry and photometry states that the illumination or irradiation of a point on a surface varies inversely with the square of the distance from the surface to the light source. Moving a light source

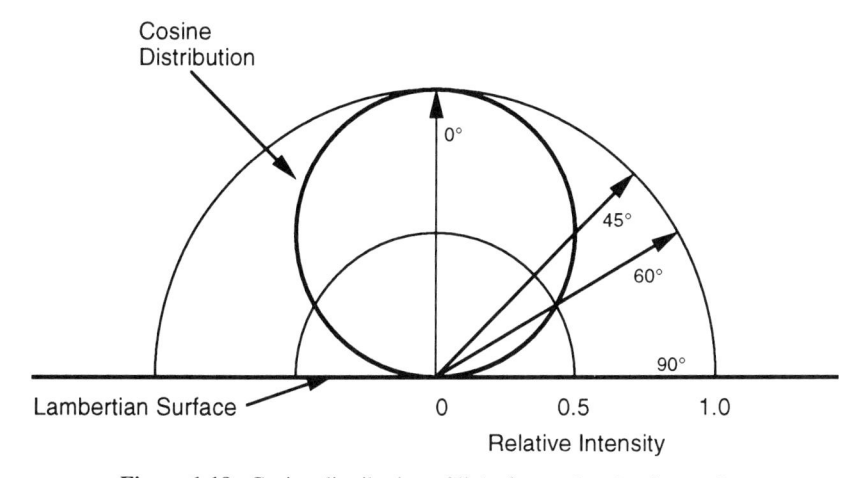

Figure 1.18. Cosine distribution of light from a Lambertian surface.

twice as far away will result in only one-quarter of the light per unit of area. This is primarily a factor for illumination engineering and plays a lesser role in displays where an illuminated surface is being viewed. The inverse square law still operates for an illuminated surface but is exactly compensated for by the fact that the surface area seen for a given narrow viewing angle will increase with the square of the distance. Of course, the resolving capability of the viewer will decrease linearly with distance, and the *total* light output from the display screen will decrease by the square of the distance. In other words, the screen will appear the same brightness but smaller and harder to resolve details on.

1.9 UNITS

Many obsolete or specialized photometric and radiometric units exist that only add to the confusion. Some of the obscure units include the Hefnerkerze, glim, blondel, bril, brill, nox, stilb, apostilb, phot, scot, skot, helios, Troland, lumerg, pharos, and Talbot. Fortunately, most of these have been discarded in recent years, and the units described in the sections on radiometry and photometry have been standardized. Wherever possible, metric units, such as the lux, should be used and foot-based units such as the footcandle and footlambert should be converted to their metric equivalent. Conversion tables are located in Appendix D.

REFERENCES

1. R. McCluney, *Introduction to Radiometry and Photometry,* Artech House, Boston, 1994, pp. 1–2.
2. RCA, *Electro-Optics Handbook,* Harrison, NJ, 1974, pp. 13–14.
3. P. A. Keller, "Optical Sensors and Sources: Measuring Infrared, Visible and Ultraviolet Light," *Test Meas. World,* May 1984.
4. A. Stimson, *Photometry and Radiometry for Engineers,* Wiley, New York, 1974, pp. 4–5.
5. O. H. Schade, "Optical and Photoelectric Analog of the Eye," *Jour. Opt. Soc. Am.,* Vol. 46, No. 9, pp. 721–739, 1956.
6. J. W. T. Walsh, *Photometry,* Constable and Co., London, 1958, pp. 53–89.
7. Illuminating Engineering Society of North America (IESNA), *Lighting Handbook,* 8th ed., IESNA, New York, 1993, pp. 69–105.
8. D. B. Judd and G. Wyszecki, *Color in Business, Science and Industry,* Wiley, New York, 1975, pp. 5–40.
9. G. Wyszecki and W. S. Stiles, *Color Science,* Wiley, New York, 1982, pp. 83–116.
10. W. G. Driscoll, *Handbook of Optics,* McGraw-Hill, New York, 1978, Chap. 12.
11. P. L. Pease, "A General Review of Vision and Visibility," *Optical Industry and Systems Directory,* Laurin, Pittsfield, MA, 1980, pp. B929–B936.
12. S. Fantone, "Visual Imaging Systems Design: The Human Factor," *Photonics Spectra,* pp. 123–127, Oct. 1990.
13. K. P. Bowan, "Vision and the Amateur Astronomer," *Sky and Telescope,* pp. 321–324, Apr. 1984.
14. G. Murch, "The Effective Use of Color: Physiological Principles," *Tekniques,* (Tektronix, Inc.), Vol. 7, No. 4, pp. 13–16, Winter 1983.
15. Commission Internationale de l'Eclairage (CIE), *Compte Rendu,* 6th Session, Geneva, 1924, Cambridge University Press, 1926, p. 67.

16. K. S. Gibson and E. P. T. Tyndall, "Visibility of Radiant Energy," Scientific Papers of the Bureau of Standards, No. 475, Aug. 11, 1923.
17. Commission Internationale de l'Eclairage (CIE), *Compte Rendu,* 8th Session, Cambridge, 1931, Cambridge University Press, 1932, p. 10.
18. D. B. Judd, "Radical Changes in Photometry and Colorimetry Foreshadowed by CIE Actions in Zurich," *J. Opt. Soc. Am.,* Vol. 45, pp. 897–898, 1955.
19. Commission Internationale de l'Eclairage (CIE), "Comité E.1.4.1, Revue des Progrés du Comité Secrétariat," 1965, pp. 152–160.
20. J. A. S. Kinney, "Is Photometry Still Relevant to Illuminating Engineering," *Compte Rendu,* 18th Session, Vol. 36, London, 1975, pp. 70–76.
21. D. H. Alman, "Errors of the Standard Photometric System when Measuring the Brightness of General Illumination Light Sources," *J. IES,* pp. 55–62, Oct. 1977.
22. Commission Internationale de l'Eclairage (CIE), "Light as a True Visual Quantity: Principles of Measurement," CIE Publication No. 41 TC-1.4, 1978.
23. J. A. S. Kinney, "Brightness of Colored Self-Luminous Displays," *Color Res. Appl.,* Vol. 8, No. 2, pp. 82–89, 1983.
24. Commission Internationale de l'Eclairage (CIE), "1988 2° Spectral Luminous Efficiency Function for Photopic Vision," CIE Publication No. 86, 1990.
25. J. A. S. Kinney, "Comparison of Scotopic, Mesopic and Photopic Spectral Sensitivity Curves," *J. Opt. Soc. Am.,* Vol. 48, No. 3, pp. 185–190, 1958.
26. Illuminating Engineering Society of North America (IESNA), *Lighting Handbook,* 8th ed., IESNA, New York, 1993, p. 75.
27. Illuminating Engineering Society of North America (IESNA), *Lighting Handbook,* 8th ed., IESNA, New York, 1993, p. 108.
28. D. B. Judd and G. Wyszecki, *Color in Business, Science and Industry,* Wiley, New York, 1975, pp. 68–81.
29. A. J. Wilkins, "Discomfort and Visual Displays," *Displays,* pp. 101–103, April 1985.
30. R. E. Turnage, "The Perception of Flicker in Cathode-Ray Tube Displays," *Info. Disp.,* pp. 38–52, May/June 1966.
31. D. Bauer, M. Bonacker, and C. R. Cavonius, "Frame Repetition Rate for Flicker-Free Viewing of Bright VDU Screens," *Displays,* pp. 31–33, Jan. 1983.
32. J. E. Farrell, B. L. Benson, and C. R. Haynie, "Predicting Flicker Thresholds for Video Display Terminals," *Proc. SID,* Vol. 28, No. 4, pp. 449–453, 1987.
33. J. E. Farrell, E. J. Casson, C. R. Haynie, and B. L. Benson, "Designing Flicker-Free Video Display Terminals," *Displays,* pp. 115–122, July 1988.
34. C. Sigel, "CRT Refresh Rate and Perceived Flicker," *SID Dig.,* pp. 300–302, 1989.
35. V. K. Zworykin and G. A. Morton, *Television,* 2nd ed., Wiley, 1954, pp. 187–191, 541–542.
36. D. G. Fink, *Television Engineering,* pp. 34–36, McGraw-Hill, New York, 1952.
37. P. A. Keller, "Cathode-Ray Tube Displays for Medical Imaging," *J. Dig. Imag.,* Vol. 3, No. 1, pp. 15–25, 1990.
38. C. W. Tyler, "Interlacing Eliminates CRT Perceptible Flicker," *Info. Disp.,* pp. 14–18, 1986.
39. K. B. Benson, *Television Engineering Handbook,* McGraw-Hill, New York, 1992, pp. 13.14–13.15.
40. P. A. Keller, *The Cathode-Ray Tube: Technology, History and Applications,* Palisades, New York, 1991, p. 40.
41. Institute of Electrical and Electronics Engineers (IEEE), IEEE Std. 100-1972, *IEEE Standard Dictionary of Electrical and Electronics Terms,* IEEE, New York.
42. Illuminating Engineering Society of North America (IESNA), *Lighting Handbook,* 8th ed., IESNA, New York, 1993, p. 27.

43. M. P. Thekaekara, "Solar Energy Outside the Earth's Atmosphere," *J. Solar Energy Sc. Technol.*, Vol. 14, No. 2, pp. 109–127, 1973.
44. Illuminating Engineering Society of North America (IESNA), *Lighting Handbook,* 8th ed., IESNA, New York, 1993, p. 935.
45. "International Body Recommends Value for K_m," *Optical Radiation News,* No. 22, p. 1, 1977.
46. Illuminating Engineering Society of North America (IESNA), *Lighting Handbook,* 8th ed., IESNA, New York, 1993, Chap. 4.
47. G. Wyszecki and W. S. Stiles, *Color Science,* Wiley, New York, 1982.
48. Commission Internationale de l'Eclairage (CIE), Publication No. 15.2, *Colorimetry,* 2nd ed., CIE, Vienna, 1986.
49. T. Smith, "The C.I.E. Colorimetric Standards and Their Use," *Trans. Opt. Soc.,* Vol. 33, No. 3, pp. 5–134, 1931–32.
50. D. B. Judd, "The 1931 I.C.I. Standard Observer and Coordinate System for Colorimetry," *J. Opt. Soc. Am.,* Vol. 23, pp. 359–374, 1933.
51. W. D. Wright, "Experimental Origins of the 1931 CIE System of Colorimetry," *J. Coating Technol.,* Vol. 54, No. 685, pp. 65–71, 1982.
52. K. L. Kelly, "Color Designation for Lights," *J. Res.,* Vol. 31, pp. 271–278 (NBS Res. Paper RP1565), 1943.
53. G. Wyszecki and W. S. Stiles, *Color Science,* Wiley, New York, 1982, p. 489.
54. D. L. MacAdam, "Visual Sensitivities to Color Differences in Daylight," *J. Opt. Soc. Am.,* Vol. 32, No. 5, p. 247, 1942.
55. W. R. J. Brown and D. L. MacAdam, *J. Opt. Soc. Am.,* Vol. 39, p. 808, 1949.
56. G. Wyszecki and G. H. Fielder, "New Color-Matching Ellipses," *J. Opt. Soc. Am.,* Vol. 61, No. 9, pp. 1135–1152, 1971.
57. G. Wyszecki and W. S. Stiles, *Color Science,* Wiley, New York, 1982, p. 308.
58. F. Grum and C. J. Bartleson, *Optical Radiation Measurements, Vol. 2, Color Measurement,* Academic, New York, 1980, p. 126.
59. D. B. Judd, "Estimation of Chromaticity Differences and Nearest Color Temperature on the Standard 1931 ICI Colorimetric Coordinate System," *J. Opt. Soc. Am.,* Vol. 26, pp. 421–426, 1936.
60. K. L. Kelly, "Lines of Constant Correlated Color Temperature Based on MacAdam's (u,v) Uniform Chromaticity Transformation of the CIE Diagram," *J. Opt. Soc. Am.,* Vol. 53, No. 8, pp. 999–1002, 1963.

CHAPTER 2

Light Sources, Filters, and Detectors

2.1 LIGHT SOURCES

Many types of light sources exist, both natural and man-made. These differ widely in their spectral distribution, spatial distribution, and temporal characteristics. The spectral distribution is measured as light intensity versus wavelength and determines the source's color. Transmission through filters as well as reflection from surfaces can alter the spectral distribution of the source through selective absorption or reflection.

Spatial distribution refers to the light output versus angle within a sphere centered on the light source as well as the physical size of the source's light-emitting area. Nearby objects such as lenses and walls can control or alter the spatial distribution of a light source.

The temporal characteristics of the source refer to the light intensity as a function of time. For instance, a candle emits a near DC light output, a conventional fluorescent lamp emits light in 120-Hz pulses (100 Hz where 50-Hz power is employed), and a strobe lamp or laser may emit a single large-amplitude pulse of light for just a microsecond or less.

The following section will briefly discuss these characteristics for light sources of interest and concern to the display market. Some are characteristic of the display itself while others are forms of ambient illumination that may degrade the viewability of a display or may be employed to view the display in the case of nonemitting displays such as LCDs.

Thermal Sources

Any incandescent or heated object will radiate energy according to the temperature to which it is heated. Any object that can radiate most of its energy equally well at all wavelengths is termed a *blackbody* or *Planckian radiator*. The amount of radiation it produces will be different at each wavelength, but a perfect blackbody has the ability to radiate all of the radiation that it produces at each wavelength.

At any temperature above absolute zero (0 K = –273.16°C) an object will emit blackbody radiation. This is in the form of infrared radiation at any temperature up to 700–800 K. As the temperature is raised above that, the object first begins to emit a visible dull red color.[1] Further increases in temperature will cause it to become bright red, orange, yellow, white, and finally bluish-white in color with the quantity of visible light increasing with temperature. Both of these characteristics, color and light output, are defined by *Wien's displacement law,* which states that as the temperature increases, more energy is emitted and the increase in emission is proportionally greater at shorter wavelengths (Figure 2.1).[2,3] Further, the shape of the spectral output curve and emission at each wavelength is defined by *Planck's radiation law*.[4] Another law, the *Stefan–Boltzmann law*, is used to

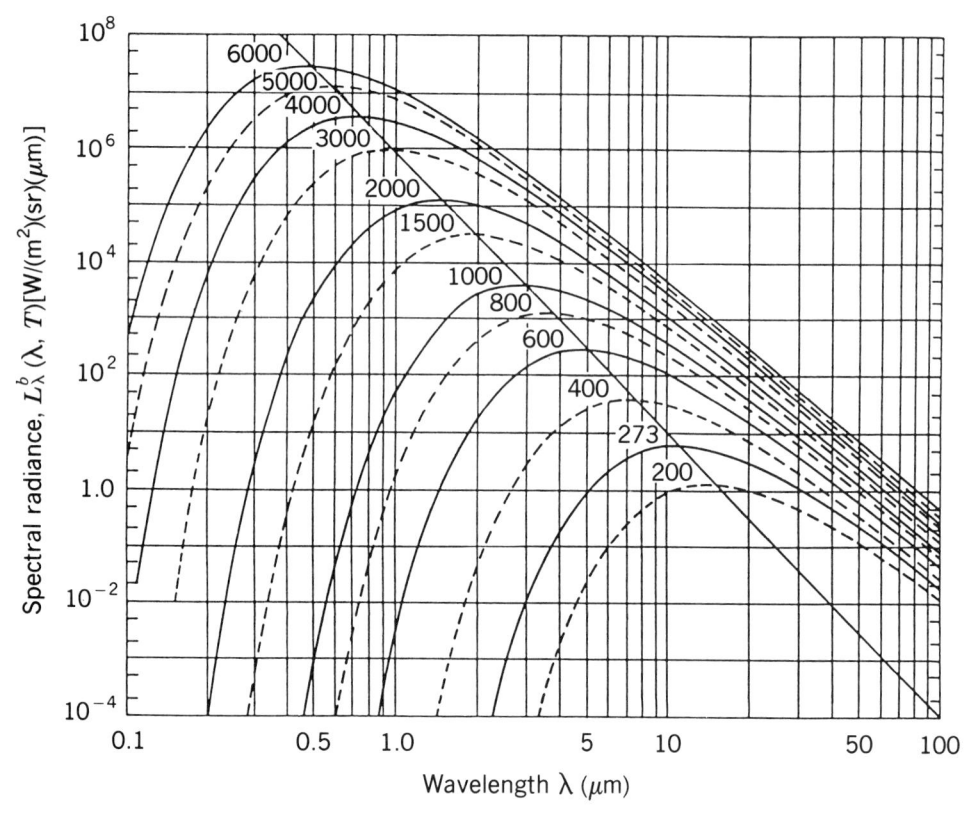

Figure 2.1. Planckian radiator spectral emission versus temperature. (From Optical Society of America, *Handbook of Optics*, McGraw-Hill, New York, 1978, pp. 1–15. Reprinted by permission.)

compute the total power radiated at a given temperature. This increases with the fourth power of the absolute temperature in Kelvin.[5]

All of the three previous laws pertain to a true blackbody emitter, which in reality does not exist in nature. Every material has a characteristic *emissivity* (ϵ) that will vary with its surface and temperature and is defined as the ratio of the emission of that material to that of a true blackbody radiator. A black surface will emit radiation more efficiently than a shiny metallic surface. A perfect blackbody radiator would have an emissivity of 1.00. Carbon black comes close to an ideal blackbody radiator with an emissivity of 0.95 at 77°C and is used to coat the thermal junctions in thermopile radiometric detectors for near-total absorption of incident radiation. The latter property is suggested by Kirchhoff's law, which stipulates that the absorbance of a material is proportional to its emissivity. See Appendix C for the equations related to the blackbody laws. Surfaces with lower emissivities are termed "graybody" rather than blackbody radiators.

Sun and Stars

The sun is an example of a light source that is both a help and a hindrance for display viewing. Liquid crystal displays become very readable in direct sunlight while most CRT and many other displays may become unusable. Sunlight is a major consideration for aircraft cockpit, aviation control tower, and automotive displays where light levels of about

100,000 lx (10,000 fc) and irradiance levels of 1000 W/m² are present. Above the earth's atmosphere, 40% higher levels are encountered and are a consideration for spacecraft displays. The luminance of the surface of the sun is about 1.5 Gcd/m² from the earth's surface and 2.1 Gcd/m² above the atmosphere. (This is a unit you do not see every day, the gigacandela per square meter or *giganit,* which has a nice ring to it!)

Spectrally, sunlight consists of a broadband emission ranging from radio waves to X-rays. The National Aeronautics and Space Administration (NASA) standards of solar irradiance outside the earth's atmosphere and at the surface are shown in Figure 2.2.[6,7] Maximum output occurs at about 500 nm, near the center of the visible spectrum, which probably accounts for the human eye evolving to a peak sensitivity in that vicinity. Because of the good match of the solar spectrum to the human eye, the luminous efficacy of the sun is quite high, around 100 lm/W.

Direct sunlight at noon at the earth's surface was sometimes encountered in the literature as a CIE reference standard called *Illuminant B* (now deprecated) and has a correlated color temperature of 4900 K.[8] An artificial approximation of direct sunlight, *Standard Source B,* used a liquid chemical filter and an incandescent lamp to produce a correlated color temperature of 4874 K. Note that before passing through the earth's atmosphere, the solar-correlated color temperature is about 5800 K.

The sun is essentially a blackbody, or Planckian, radiator as previously discussed. It deviates from the true blackbody by many narrow-band emission and absorption "lines" in its spectrum caused primarily by interactions of the light with the atoms of the solar *photosphere,* or light-producing region. One significant emission line, hydrogen alpha (Hα), occurs in the visible portion of the solar spectrum at 656.3 nm.[9] Further modification of the solar spectrum occurs, especially in the near infrared, due to selective absorption by water vapor in the earth's atmosphere (Figure 2.3).[10] Ozone absorption of ultraviolet and carbon dioxide absorption in the deeper infrared are also significant.[11]

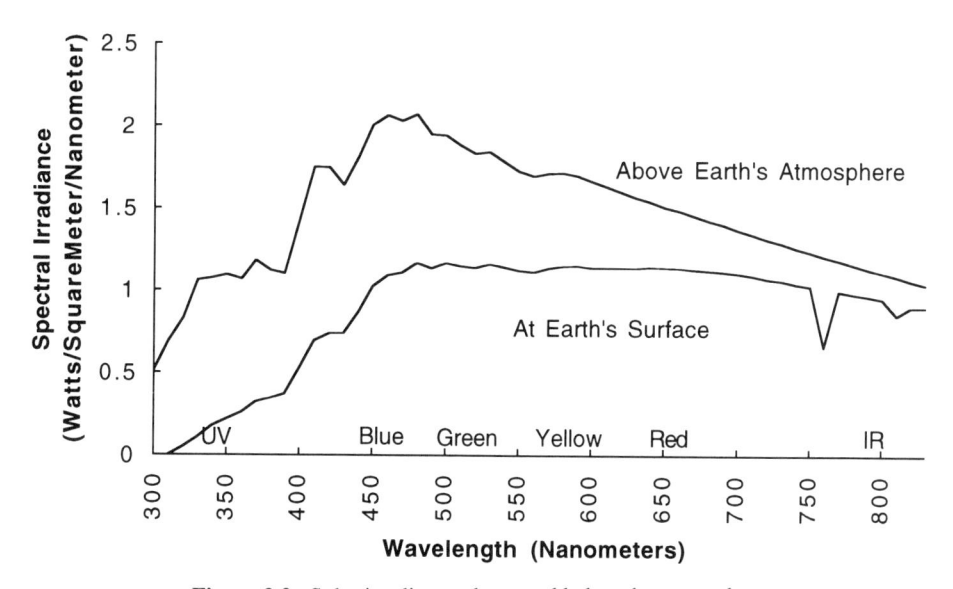

Figure 2.2. Solar irradiance above and below the atmosphere.

32 LIGHT SOURCES, FILTERS, AND DETECTORS

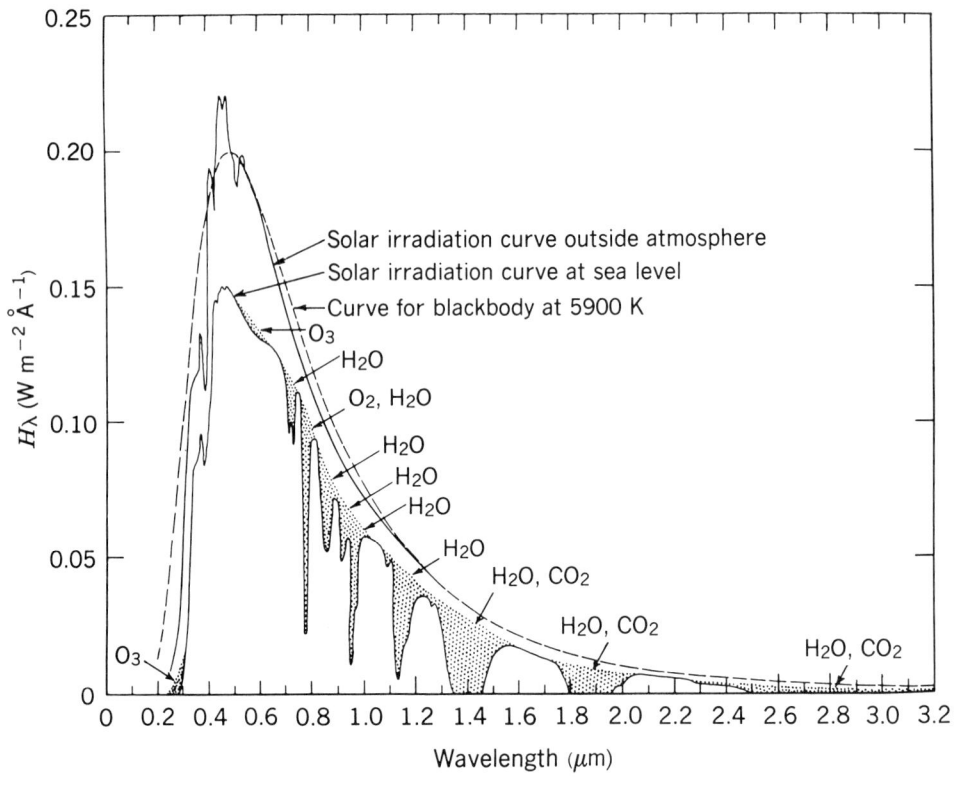

Figure 2.3. Solar irradiance showing atmospheric absorption. (From S. L. Valley, *Handbook of Geophysics and Space Environments,* McGraw-Hill, New York, 1965, p. 16-2. Reprinted by permission.)

Most stars are similar to the sun, although their temperatures and size vary considerably from one to another. These range from the so-called red giants to blue dwarfs, with our sun located in just about the middle of the range. Starlight is probably of interest from the standpoint of positioning the sun in the scheme of things and for the collective effect of all the stars for low-light-level (LLL) imaging systems.

Moon and Planets

The moon and planets shine by reflected light from the sun and therefore are similar spectrally to the sun except for differences caused by surface coloring. Only the moon is of much consideration for the purposes of illumination measurement or LLL imaging systems. The average surface luminance of the moon is about 2500 cd/m^2 and at the full phase will provide about 0.1 lx of illumination on the earth's surface.

Sky

The atmosphere contributes about 10–20% of daytime illumination.[12] It acts as a large absorber, diffuser, and scatterer of direct and reflected sunlight, especially at shorter wavelengths, hence, the blue appearance of the sky. The average luminance of a clear sky is about 1000 cd/m² but can vary considerably with atmospheric conditions. Luminance of 3000 cd/m² is typical on an overcast day but a heavy overcast can result in far less.

The correlated color temperature of the average mixture of sunlight plus skylight is about 6774 K and is approximated by CIE Illuminant C. As with Illuminant B, it may be approximated by an incandescent lamp and liquid chemical filter. Probably the most important representation of daylight related to displays is D_{65}. This is 6500 K and simulates the light from the whole sky on a horizontal surface. The reference D_{65} is the white used for balancing television monitors and computer displays to ensure uniform appearance between displays and will be discussed further in Chapter 6.

Incandescent Lamps

The incandescent lamp is a thermal light source; that is, it depends on the heating effect of a tungsten filament by passage of electrical current through it. Its behavior is in accordance with that described previously under thermal sources. The tungsten filament has an emissivity of about 0.24–0.25 at normal operating temperatures. A typical spectral emission curve in the visual portion of the spectrum is shown in Figure 2.4. Most of its energy is radiated in the infrared region and only about 2–3% is converted to visible light. This is an important consideration when measuring incandescent lighting with detectors having high near-infrared sensitivity, such as silicon photodiodes. Correction filters with good infrared blocking are necessary to avoid errors. This will be discussed in section 4.1 under Photometers.

Incandescent lamps are not a major consideration for display viewing environments since they are not used often for office illumination or for LCD backlight purposes due to their poor efficiency. Some incandescent lamps are still encountered for instrument panel lighting, status indicators, and alphanumeric readouts. The home office, which is

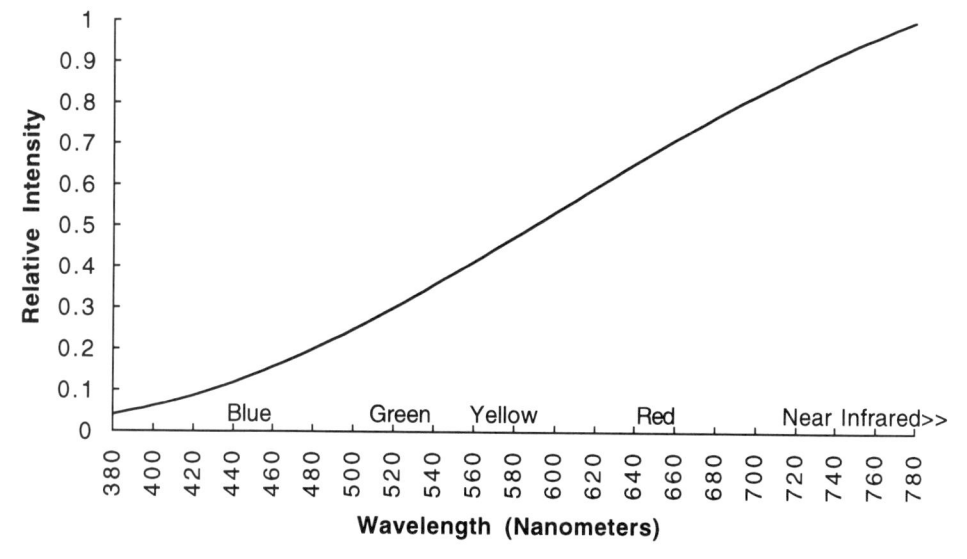

Figure 2.4. Visible spectrum of an incandescent lamp at 2854 K.

experiencing considerable growth in popularity, is probably the principal location at which incandescent illumination is encountered. The warm color of incandescent lamps is preferred in the environment of the home and their low efficiency is not necessarily a drawback. Winter is the time of year when the most light is required and the large amount of infrared energy produced by incandescent lamps heats the living space, which reduces the heating required by the furnace. And it is concentrated only in the occupied areas of the home (unless you have children who leave lights on behind them!). It is only a small part of the total heating requirement but is not necessarily wasted energy, as so often thought.

The average color temperature of incandescent lamps is in the range of 2750–2900 K depending on the type of lamp and operating voltage relative to rated voltage. The nominal value is often considered to be 2854 K, and calibration lamps are often operated at that color temperature. Localized areas of the filament may have higher temperatures, especially as the lamp ages and thinning of the filament occurs due to tungsten evaporation. The limitation of operating temperature is the melting point of tungsten, which is at 3643 K (3370°C), at which point failure of the filament will occur.

The optimum filament temperature is a compromise between two primary factors: total light output and operating life. The light output of a blackbody source increases steeply with temperature, as described earlier under Thermal Sources. The increase is exponential, with the increase in operating voltage raised to the 3.5 power. The life *decreases* at an even steeper rate by the 13.5 power of the increase in operating voltage. Thus, an increase in operating voltage of only 10% will result in a 39% increase in light output but life will decrease drastically to only 28% of the original value.[13] See Appendix C for formulas related to incandescent lamp performance. The typical efficacy for a 100-W incandescent lamp is 17.5 lm/W and is generally higher for larger lamps than smaller ones. It can range from about 7 lm/W for the lowly 1-W No. 47 radio pilot lamp to 23.7 lm/W for a 1000-W lamp and 33.5 lumens/watt for a 10,000 watt lamp.

Long-life lamps trade off light output for life and 50,000-h lamps are easily made but at a sacrifice of efficacy that may be as low as 0.5 lm/W, such as for the General Electric type 1869-D miniature lamp.[14] Any incandescent lamp may be used as a long-life lamp just by reducing the operating voltage. Some homeowners make use of this fact by purchasing 130-V lamps rather than the more common 120-V lamps although light output will be reduced. Higher efficacy lamps are also available at a premium cost. These use krypton fill gas instead of commonly used argon gas to retard tungsten evaporation. This allows operation of the filament at a higher temperature for more light output.

The light output from most incandescent lamps, especially in larger wattages, is near DC with only 10–20% of 100- or 120-Hz modulation when operated from 50 or 60 Hz AC power, respectively. This is because the relatively large thermal mass of the filament cannot heat and cool rapidly enough to follow the individual cycles of the sine wave power. Slow fluctuations of normal line voltage are another matter. Because of the exponential change in light output with operating voltage, the light output changes will be much greater than the line voltage change. Well-regulated power supplies, either AC or DC, should always be used whenever an incandescent lamp is used as a light reference standard.

Halogen Lamps

The tungsten–halogen lamp, sometimes called a quartz–iodine lamp, is a means of increasing the operating temperature of an incandescent lamp filament to over 3000 K with the attendant increase in efficacy and color temperature. A quartz envelope allows the high wall temperature above the 250°C necessary for the halogen cycle to operate. At these temperatures tungsten evaporated from the filament combines with the halogen (iodine or bromine) fill gas to form tungsten iodide or tungsten bromide. The envelope wall temperature is above the temperature at which either tungsten iodide or tungsten bromide can condense on the surface. Because of this, the halogens circulate back to the region of the filament where the higher temperature dissociates the tungsten from the halogen and the tungsten is redeposited on the filament, thus compensating for the original loss of material. Increased life, while still retaining the advantages of greater light output for the same amount of input power, results from the tungsten recombination process.[15,16] Color temperatures of 3000–3400 K are typical. The emission at the shorter wavelengths is significantly greater than that of a conventional incandescent lamp, both relative to the longer wavelengths and in efficacy (lumens per watt). The additional ultraviolet emission may require some shielding to protect nearby personnel, especially those with higher than normal photosensitivity due to some medications or some forms of erythematosus. Ordinary window glass and UV absorbing plastic are quite effective in blocking most of the UV emissions.

Gas Discharge Lamps

This category of lamps includes the familiar neon panel indicator lamp, neon sign lighting, and high-intensity discharge (HID) lamps, including sodium lamps, mercury lamps, and metal-halide lamps. The common fluorescent lamp is also considered a gas discharge lamp but will be covered separately in the next section. All depend on the production of light by passing an electrical current through a gas. Each fill gas (such as argon, neon, mercury vapor, sodium, and metal halide) emits light in a characteristic spectrum of "line" emissions at particular wavelengths (Figure 2.5) when ionized by the passage of

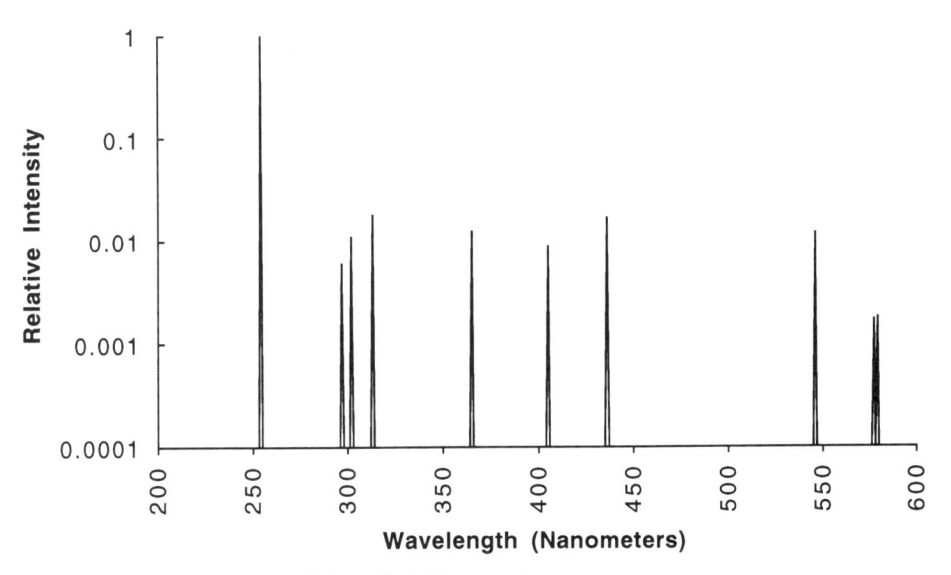

Figure 2.5. Mercury lamp spectrum.

electrical current through the gas. The gas discharge spectrum sometimes is concentrated in the ultraviolet, and the inside walls of the lamp envelope are coated with a phosphor (not to be confused with the element phosphorus). Phosphors efficiently convert ultraviolet radiation to longer wavelength visible light. Many phosphors emit broadband spectra and by mixtures of, for example, blue- and yellow-emitting phosphors will produce white light that may be optimized for either highest visual efficiency or high color-rendering index (CRI). The CRI is important for color matching of printing inks and textiles.

Gas discharge lamps operated from AC power will show a well-modulated light output at double the line power frequency since discharge current can flow in either direction and can start and stop at rates up to as much as several kilohertz. Due to the nonlinear ionization characteristics of the gas and electrical effects of the ballast resistor or inductor, the light output waveform will be anything but that of the applied sine wave. Modulation can be as much as 100% of the peak light output; in other words, no light may be emitted during part of the cycle.

Fluorescent Lamps

Fluorescent lighting is a most important consideration for many displays. Because of its widespread use and sometimes improper installation in offices, it is often a nuisance and may produce glare, reflections, and loss of display contrast for self-luminous displays such as CRT, electroluminescent (EL), and plasma displays (PDP). It may have some of the same drawbacks even for transreflective LCDs, which are dependent on ambient illumination for making their information visible. Other LCDs use cold-cathode fluorescent lamps (CCFLs), as illustrated in Figure 2.6 for backlighting with the LCD operating as a transmissive device.

The fluorescent lamp is just another gas discharge lamp that has evolved in a particular form, conventionally long and tubular but sometimes circular or U-shaped. It uses a mixture of mercury vapor and argon, krypton, and/or xenon gas to produce the characteristic spectra of mercury ultraviolet emission lines at 254, 313, and 365 nm.[17]

Phosphors applied to the inner walls of the glass tubing convert the invisible ultraviolet radiation to visible light. For general illumination applications, mixtures of two or three phosphors are commonly used. When two phosphors are used, they are usually broadband

Figure 2.6. Cold-cathode fluorescent lamp as used for liquid crystal backlighting.

blue and yellow emitters, and by controlling the ratio of blue to yellow, a range of whites can be produced with appearance from "warm" to "cool" to suit varied applications. The most common, designated *cool white* and having a color temperature of about 4200 K (Figure 2.7), has long been used in factories and offices but is considered harsh in the home environment, where it is used alongside the much warmer appearing incandescent lamp that is usually close to 2850 K. Other standard white phosphor mixes include *daylight* at 6500 K, *white* at 3500 K, *warm white* at 3000 K, and specialized merchandising whites used to enhance the appearance of freshness of vegetables or meats or make clothing more attractive to the shopper.

More efficient blends of three phosphors are becoming widely used for fluorescent lamps due to recent federal energy conservation legislation that outlawed many of the heretofore most common incandescent and fluorescent lamps in the United States. Three color phosphor lamps are known generically as triphosphor or trichromatic lamps and are available under a number of trade names. They usually contain red, green, and blue phosphors of the more efficient and more expensive rare-earth types that emit a number of narrow-band spectral lines (Figure 2.8) and are closely related to color CRT phosphors. Correlated color temperatures of 2500–6500 K are available by controlling the ratios of the three components. The color temperature is often a part of the lamp type designation rather than the older color names previously used as a descriptor.

Many other single-component phosphors are available for specialized applications with colors ranging from ultraviolet to red. These have limited uses related to displays.

Fluorescent lamp efficacy increases as the lamp length increases (Figure 2.9).[18] The miniature 4-watt fluorescent lamp produces about 25 lm/W while the 8-ft, 96-W lamps used for industrial applications produce over 70 lm/W.

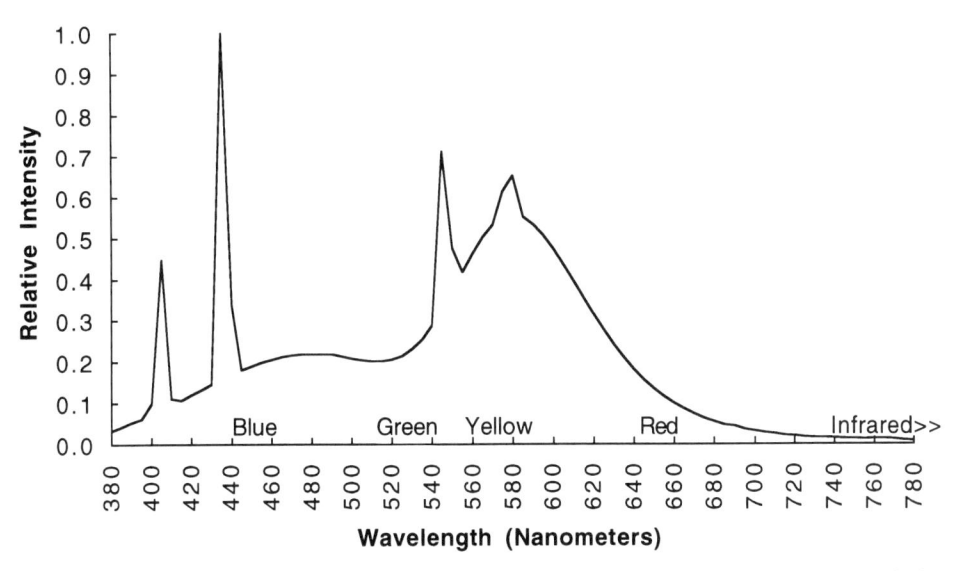

Figure 2.7. Cool-white (CW) fluorescent lamp spectrum. Note the narrow-band mercury emission lines superimposed on the more continuous phosphor emission.

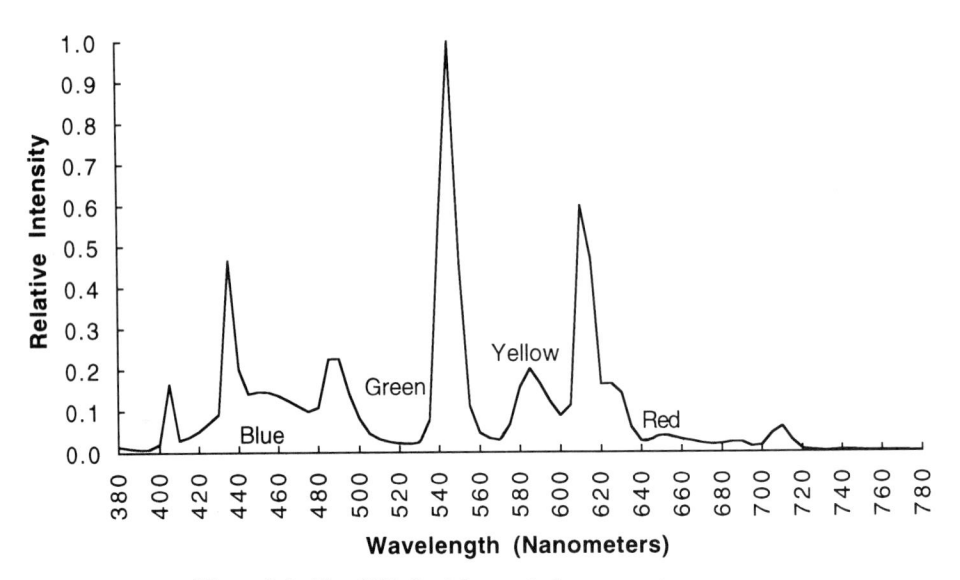

Figure 2.8. The 5000-K trichromatic fluorescent lamp spectra.

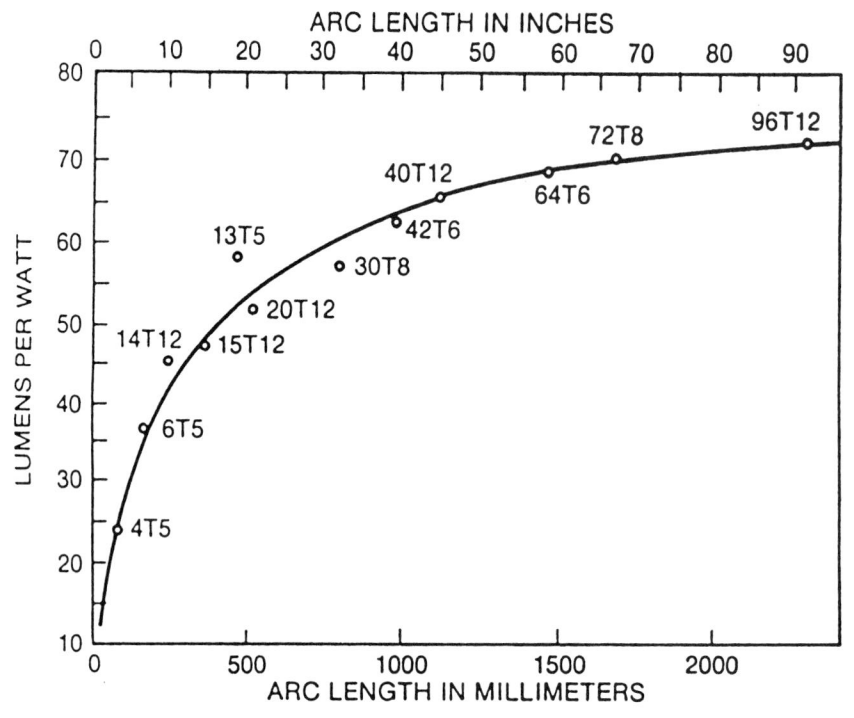

Figure 2.9. Fluorescent lamp efficacy versus arc length for common fluorescent lamp types. [From Illuminating Engineering Society of North America (IESNA), *Lighting Handbook,* 8th ed., IESNA, New York, 1993, p. 205. Reprinted by permission.]

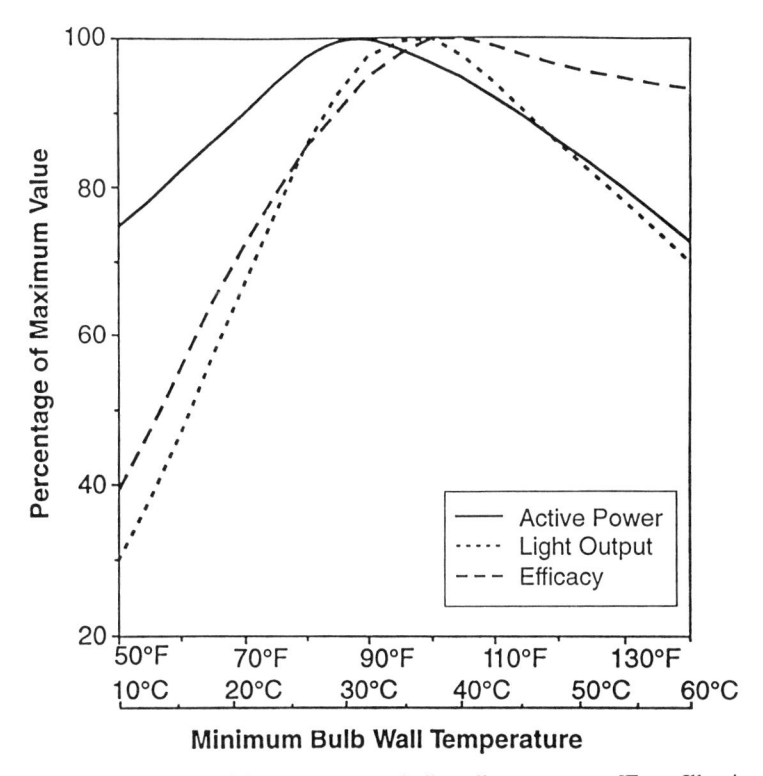

Figure 2.10. Fluorescent lamp light output versus bulb wall temperature. [From Illuminating Engineering Society of North America (IESNA), *Lighting Handbook,* 8th ed., IESNA, New York, 1993, p. 208. Reprinted by permission.]

As with other gas discharge lamps, fluorescent lamp light output will be modulated at double the AC power frequency and will have a complex waveform. Total 120 Hz modulation of 50% of the peak light output is common for a 40-W fluorescent lamp operated at 60 Hz. The light output also is very dependent upon the wall temperature of the lamp (Figure 2.10).[19] This is characteristic of the phosphors used, which are optimized for the anticipated operating ambient temperature. Usually the light output will increase after turn-on until the operating temperature is stabilized. Fluorescent lamps having two or more phosphors in particular may also show a pronounced color shift during warmup since the different phosphors may have differently shaped curves of light output versus temperature even though they are all optimized for efficiency at their anticipated operating temperature.

Electroluminescent Light Sources

Electroluminescent lamps find wide usage in aviation, automotive, and LCD backlight applications. Because they are a "cold" light source with almost no infrared radiation produced, they are particularly useful for compatability with night vision imaging systems (NVISs). The EL lamp consists of a phosphor layer sandwiched as a dielectric between two conductive electrodes (Figure 2.11). The AC power is applied between the electrodes,

40 LIGHT SOURCES, FILTERS, AND DETECTORS

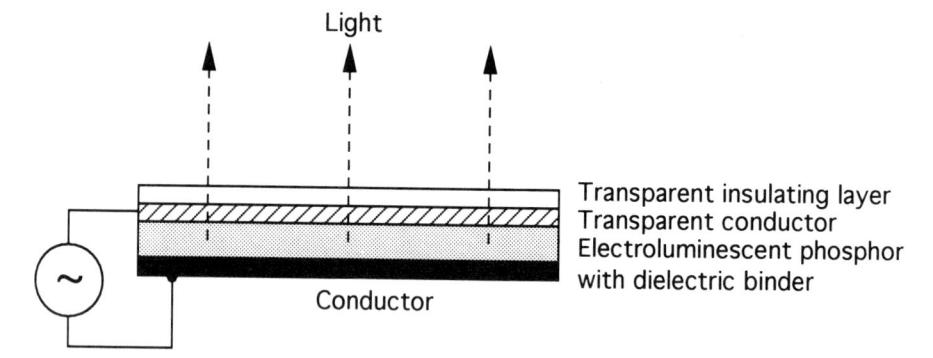

Figure 2.11. Physical construction of an electroluminescent device.

and because of the capacitor-like construction, current flows through the phosphor only in response to the rate of change of voltage. Just think of it as a luminous capacitor[20] or a lossy, light-emitting capacitor.[21]

The spectral output of an EL lamp is usually in the form of a relatively narrow band, somewhat Gaussian shaped curve (Figure 2.12) with a predominant color such as green, blue, yellow, or orange. A red color filter or fluorescent overlay may be used to produce red but at relatively low efficiency. As with fluorescent lamps, two phosphors may be mixed to produce a near-white color. The light emission color may also be modified to red or white by a method known as dye conversion or internal conversion, where a colored or fluorescent dye is added to the resin used as a phosphor binder.

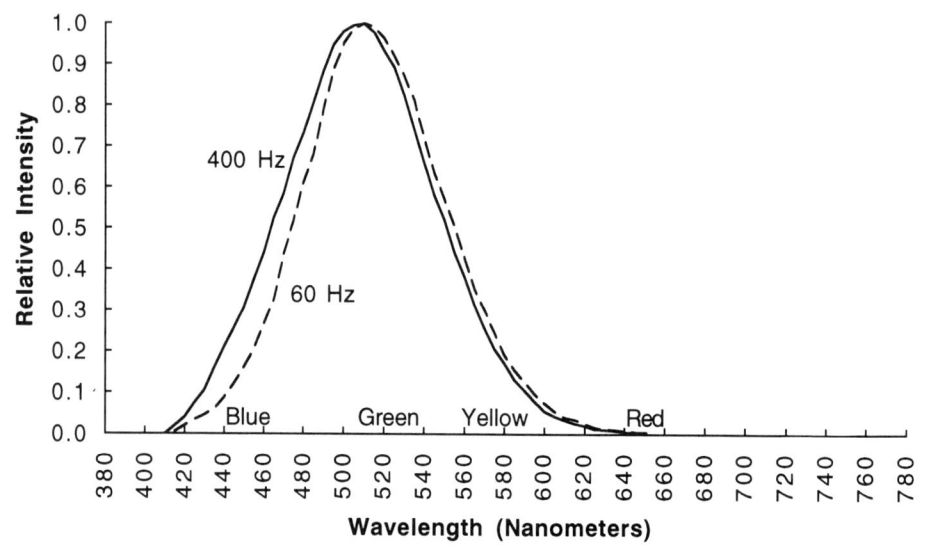

Figure 2.12. Emission spectra of a blue-green electroluminescent device showing the spectral shift that occurs with higher excitation frequency.

The color produced by an EL device may shift with applied frequency, especially for mixed phosphors and green or blue-green phosphors. The shift is caused by different saturation characteristics for the individual phosphor components. In the case of green or blue-green phosphors, it appears to be a similar effect within different crystals of the same phosphor that have slightly different emission wavelengths. The color may be distinctly green at 60 Hz, become blue-green at 400 Hz, and shift to a deep blue at several kilohertz. Unfortunately, blue is quite inefficient due to the decreased sensitivity of the human eye to that color. This has been a major obstacle in the development of full-color EL displays. Red presents somewhat the same difficulty. Yellow-emitting zinc sulfide with manganese activator is the most efficient EL phosphor.

The light output from EL devices is somewhat superlinear with change in applied voltage and increases by about the three halves power. Doubling the voltage may produce three or more times the light output. One hundred twenty to 240 volts AC or so is commonly used for excitation. Light output and power dissipation also increase with frequency and usually peak at about 5 kHz. Above that frequency, the light output again begins to fall off as series resistance losses and parallel capacitance losses increase. Light uniformity may also degrade, with highest luminance occurring nearest the voltage feed points. The light output will increase roughly with the one-half power of the applied frequency up to about 5 kHz. See Figure 2.13 for an example of the voltage and frequency effects of a typical EL lamp.

Electroluminescent lamps exhibit a near-Lambertian or cosine distribution, thus making them well suited for viewing through a range of angles up to nearly 90° from perpendicular. They are available in custom shapes and sizes up to about 0.15 m^2 area. Many are flexible to some degree and may be applied to a curved surface. Plastic encapsulation is commonly used for EL lamps (Figure 2.14), but ceramic construction with a rigid metal rear electrode is sometimes employed for greater ruggedness and improved life.

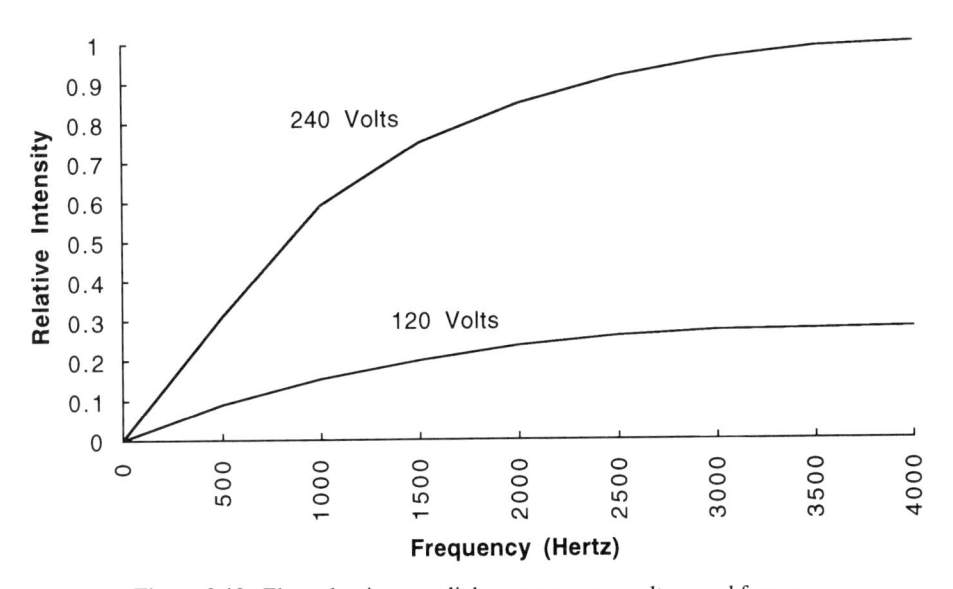

Figure 2.13. Electroluminescent light output versus voltage and frequency.

Figure 2.14. Electroluminescent backlight for LCDs.

The light output of an EL device is a periodic wave shape, with major peaks occurring at a frequency double that of the applied AC voltage. The preferred excitation waveform is a sine wave, but square, triangular, or more complex wave shapes may be used in many applications.

The life of EL lamps is sacrificed for greater light output. Because of this, they are usually best suited for low illumination level applications. Light output decreases at a faster rate during initial operation and settles down to a more gradual rate over the remaining life. Failure is gradual rather than abrupt, as with an incandescent lamp. Arcs due to transient voltages are usually self-healing although a small dark spot may remain at the point of the arc.

Electroluminescent lamps are properly measured in units of luminance since they are a self-luminous surface. Either candelas per square meter (nits) or footlamberts may be used, with the former preferred.

Light-Emitting Diodes

The LED, shown schematically in Figure 2.15, finds wide application as discrete indicator lights for instruments and appliances. Common semiconductor materials used for LEDs are GaAsP for red and yellow devices, GaP for green devices, and GaAs, GaAlAs, and InGaAsP for infrared devices. See Table 2.1 for typical emission wavelengths of each material. Recent improvements in efficiency of LEDs by use of AlGaAs along with their inherent ruggedness have resulted in the use of multiple LEDs for automobile taillights, traffic signals, and emergency exit lighting. The LED has excellent legibility in all but direct sunlight ambient light conditions, is low in cost, has long life without the sudden burn-out failures common to the incandescent lamp, produces little heat, and is compatible with the low voltages available in today's solid-state circuits. Infrared LEDs have been highly successful in wireless remote-control units for consumer products and in fiber-optic communications and are now finding application in wireless links between

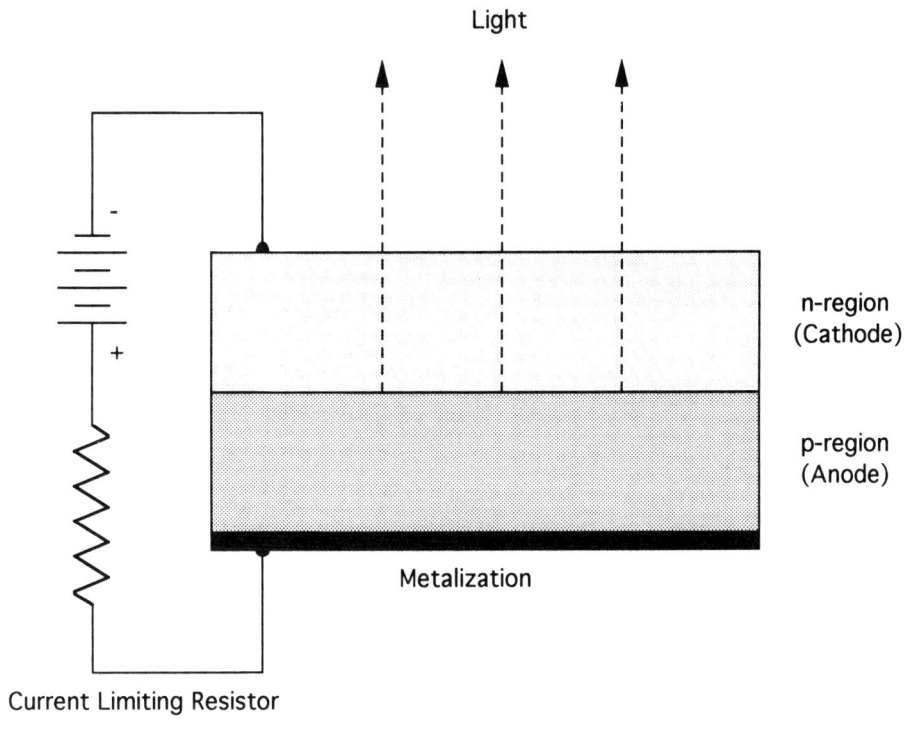

Figure 2.15. Physical construction of a LED.

TABLE 2.1 Typical Light-Emitting Diode Colors

Material	Color	Peak Wavelength (nm)
InGaN	Deep blue	450
SiC	Blue	470–480
GaN	Blue-green	490
GaP	Emerald green	558
GaP	Green	565
GaAsP	Yellow	583
GaAsP	Yellow	590
AlInGap	Amber	592
GaAsP	Amber	610
AlInGaP	Reddish-orange	615
AlInGaP	Reddish-orange	622
GaAsP	Orange	630
GaAsP	High-intensity red	635
AlGaAs	Red	645–654
GaAsP	Red	660
GaAlAs	Bright red	660
GaP	Deep red	700
GaAlAs	Infrared	880
GaAs	Infrared	950
InGaAsP	Infrared	1300
InGaAsP	Infrared	1550
GaSb-InAs	Infrared	1800–4600

44   LIGHT SOURCES, FILTERS, AND DETECTORS

electronic instruments such as printers and for some computer networking. Multisegment LED displays, once popular for digital watches, calculators, and electronic instruments, have been largely supplanted for battery-operated applications by the LCD due to the latter's extremely low power requirements.

Spectrally, LEDs are commonly available in a range of wavelengths from infrared to blue (Figure 2.16). The half-power spectral bandwidth is usually on the order of 20–50 nm. Two or more LED chips of different colors, usually red and green, may be combined in the same package to give a range of colors, between those of either chip alone, by controlling the ratio of current to each chip.

Blue LEDs are just now becoming commercially practical after years of development, albeit at a premium price and relatively low visual efficiency. These use silicon carbide, which emits at about 470–490 nm or gallium nitride, which emits at 450 nm. The latter is more efficient and is a much deeper blue. Zinc selenide is another material being investigated for blue LEDs. It has taken a long time for bright blue LEDs to evolve, and their cost is still more than 10 times higher than for comparable LEDs of other colors. The fundamental problems have been that the efficiency of LED materials decreases at shorter wavelengths and is combined with the human eye sensitivity, which decreases steeply in the blue region. Any shift toward a purer blue is met by the need for greater light output to produce the same visual intensity. Intensities of GaN blue LEDs of as much as 1000 mcd have been reported. Forward voltage drop is also higher for blue LEDs, typically 5–9 V, which makes interface with 3- and 5-V logic circuits more difficult.[22–24]

The LEDs are directional sources of light with viewing angles typically in the range of 5°–30° off-axis depending on whether they are manufactured with transparent or diffusing plastic packages (Figures 2.17 and 2.18). The package is often made from plastic colored the same color as the LED emission to enhance viewability in high-ambient-light environments.

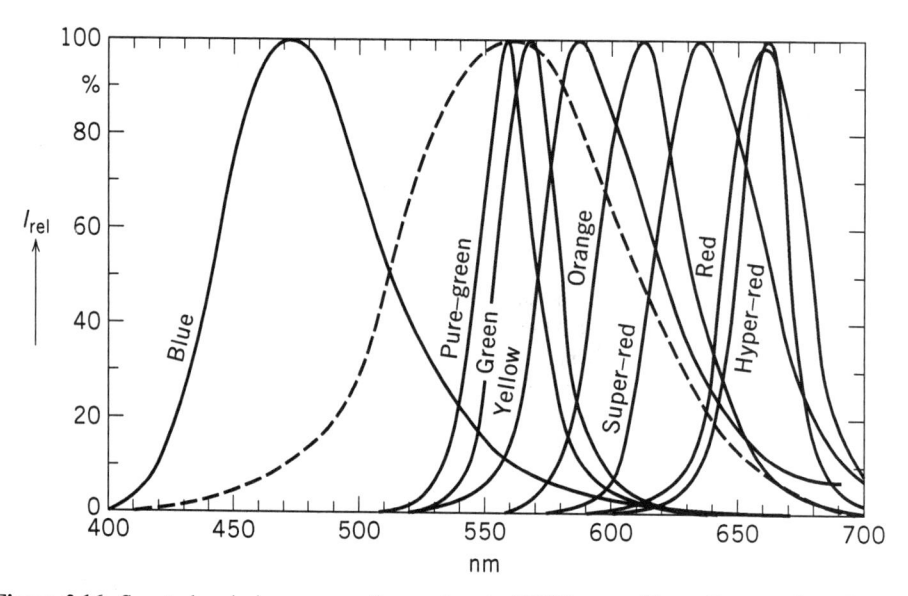

Figure 2.16. Spectral emission curves of several typical LED types. (From Siemens Optoelectronics Data Book 1995–1996, Siemens Components, Inc., Cuperino, CA. Reprinted by permission.)

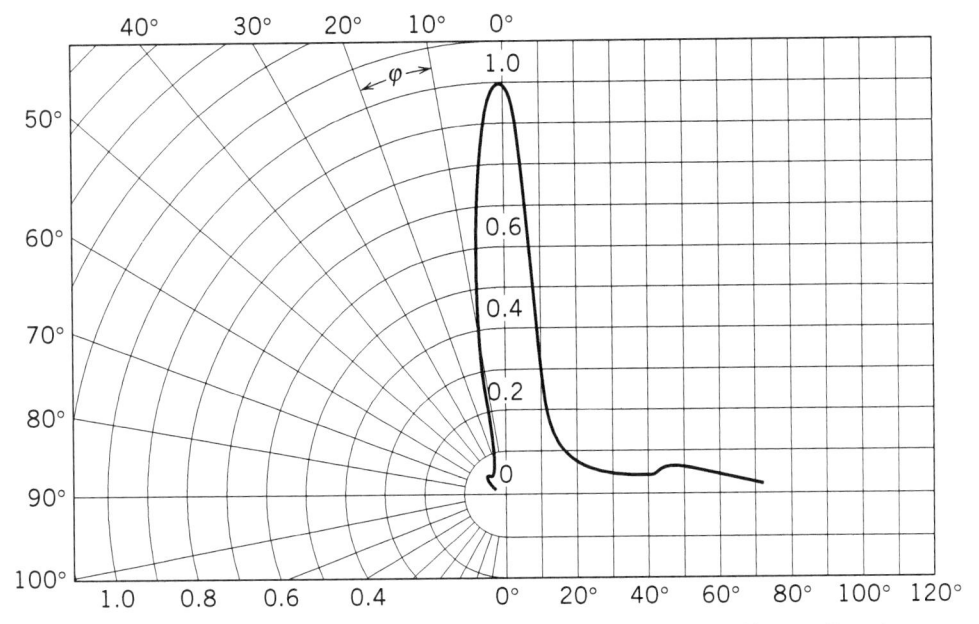

Figure 2.17. Angular distribution of light output from a clear LED. (From Siemens Optoelectronics Data Book 1995–1996, Siemens Components, Inc., Cuperino, CA. Reprinted by permission.)

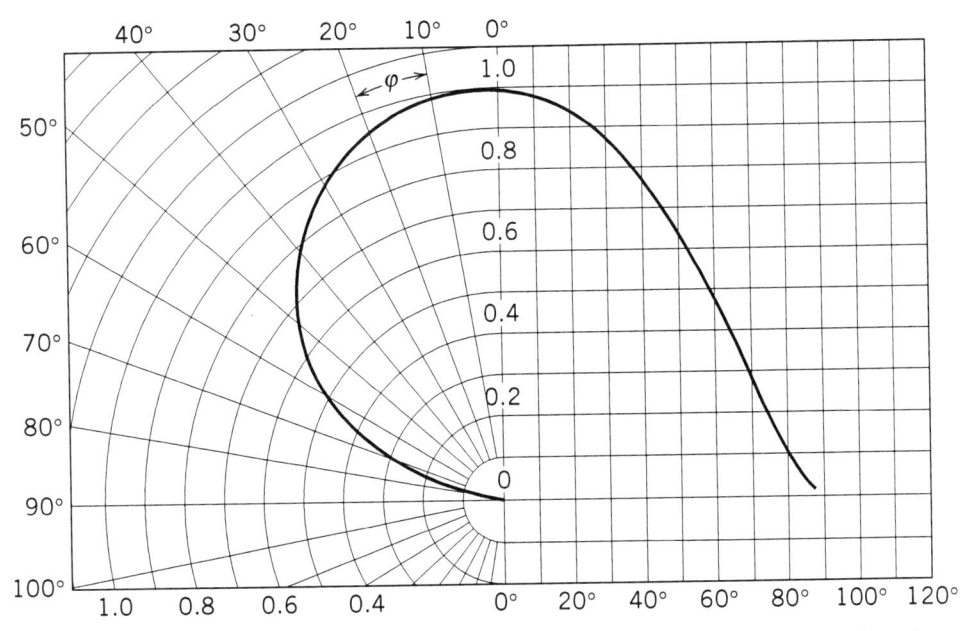

Figure 2.18. Angular distribution of light output from a diffuse LED. (From Siemens Optoelectronics Data Book 1995–1996, Siemens Components, Inc., Cuperino, CA. Reprinted by permission.)

The LEDs may be operated from low-voltage DC through a series current-limiting resistor that limits the current to the range of 10–100 mA depending on the LED type and viewing requirements. Alternating current and pulsed (strobed) operation is also used with simple driving circuitry. Typically, LEDs have response times of less than 1 μs under pulsed operation. Infrared LEDs optimized for response times in the nanosecond range are used for fiber-optic communications applications.

The efficacy of LEDs is now on the order of 10 lm/W and more, surpassing small incandescent lamps. Further gains appear to be feasible in the future. This will lead to still more applications, although lack of a white LED restricts it mostly to colored indicators and signals. White has been demonstrated using multiple red, green, and blue LEDs, but the lower visual efficiency of blue LEDs is still somewhat limiting.

Several units of measure have been used to quantify LED light output. Infrared LEDs are usually measured for total power output in milliwatts or radiant intensity in milliwatts per steradian. Visible LEDs were first measured for luminance, which was misleading, since luminance is a measure of light per unit area. Two LEDs having chips with different emitting areas might measure the same luminance value, but the one with the larger area would produce more total light and thus be more visible to the human eye. Luminous Intensity in millicandelas was soon adopted, as it was a better measure of total useful light in a forward direction. Still, this does not convey the whole picture since beam sizes vary considerably. The angular light distribution pattern must be taken into consideration for the anticipated application.

Phosphors

Phosphors are not light sources in themselves but act to convert invisible electrons, ultraviolet radiation, or X-rays to visible light. Phosphors find widespread usage in fluorescent lamps, HID lamps, vacuum fluorescent displays (VFDs), and display CRT screens. Most phosphors are chemically pure, inorganic, crystalline materials containing small controlled amounts of impurity atoms as *dopants* or *activators*. They are applied as thin coatings to the glass walls of the lamp or CRT screen. The CRT phosphors are characterized by several properties, including color, persistence, brightness, and resistance to burning or aging. Generic CRT phosphors are identified by a two- or three-letter designation (Appendix I) under the Worldwide Type Designation System (WTDS) administered by the Electronic Industries Association (EIA). The EIA publication TEP-116C, *Optical Characteristics of Cathode-Ray Tube Screens,* contains detailed spectral, persistence (decay), chemical composition, and other specifications of each of the currently registered phosphors.[25] The older EIA JEDEC designation system used in the United States is still often encountered. This system used the identifier P followed by one or two digits from P1 to P58. A cross reference between the two systems is located in Appendix J. A number of special phosphors with PC designations beginning at the number 100 were catalogued by Clinton Electronics. These were never official EIA designations but were merely "house" numbers. This sometimes caused confusion among users who overlooked the "C" and thought they were registered "P" numbers. Similar house numbers were used by Thomas Electronics (designated PT) and Westinghouse (designated PX).

Phosphors are selected and optimized for the intended excitation process. Excitation may be by electrons in the case of CRT phosphors, X-rays for radiology applications, and ultraviolet radiation for fluorescent lamps. The phosphors act as a wavelength converter by converting higher energy atomic particles to lower energy visible light. The relatively high energy of

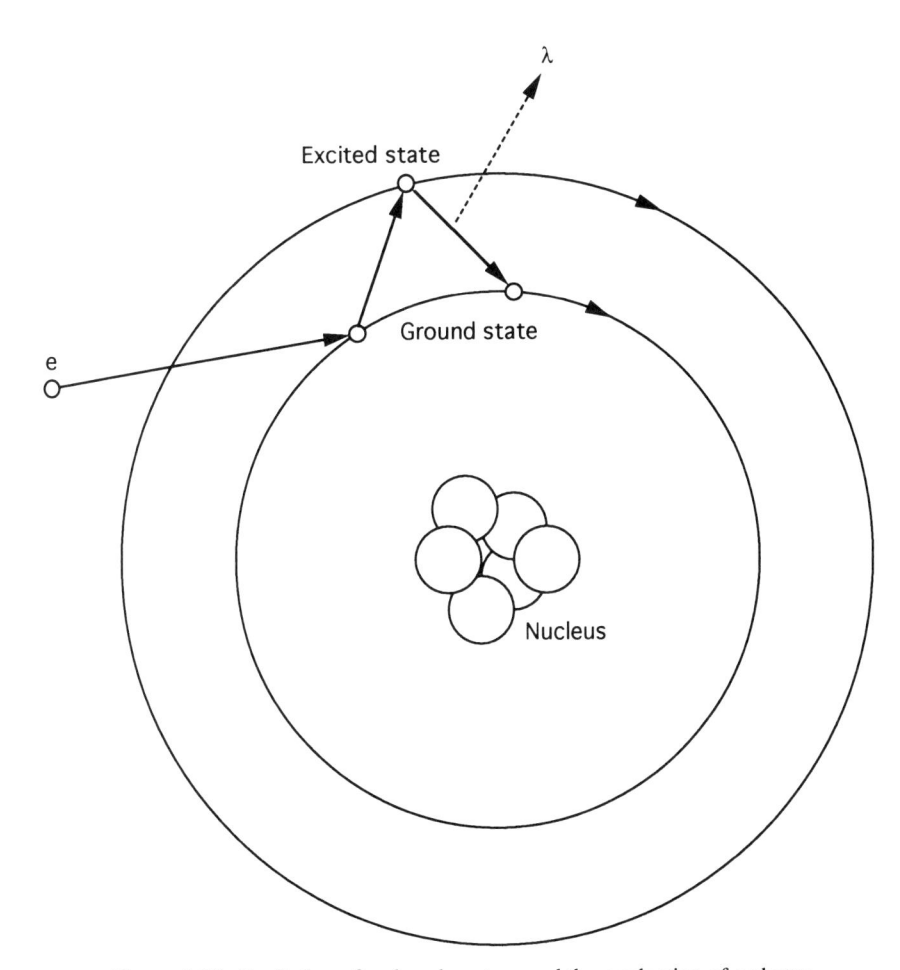

Figure 2.19. Excitation of a phosphor atom and the production of a photon.

the excitation causes the *valence* electrons in the phosphor atom's outer shell of electrons to be driven from the stability of their normal low-energy orbit or *ground* state to a less stable higher energy level called the *conduction band* (Figure 2.19). The electron gains energy in the process. After a finite amount of time, the electron returns to its normal or an intermediate orbit and gives off some of the previously gained energy as a photon of light. The wavelength and hence the color of the emitted light depend on the difference in energy level states of the particular atoms in which the transitions occurred. The length of time from excitation to emission varies greatly with different materials and is known as persistence or decay.

There are, however, a few phosphors that can emit shorter wavelengths than that of the excitation. This phenomena is known as *anti-Stokes* fluorescence. Stokes's law states that the light emitted will always be at a lower energy and longer wavelength than the excitation energy. Anti-Stokes fluorescence depends on additional thermal energy to raise the electrons to a higher energy state than would normally be expected from the excitation energy.[26] The emitted photon energy will then be higher than the excitation energy. These materials find application in viewing screens used to make the invisible beam of an infrared laser visible for study.

GJA (P1) PHOSPHOR

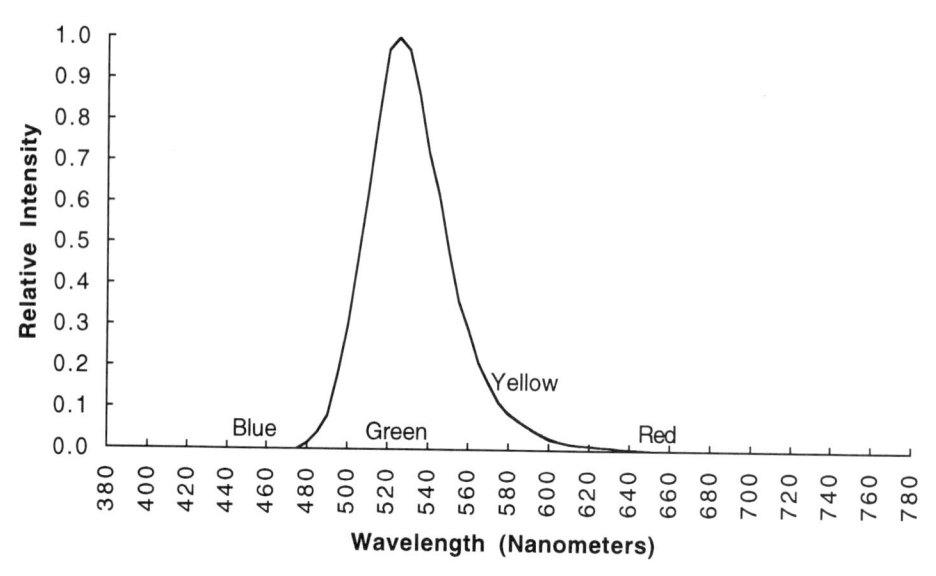

Figure 2.20. Zinc orthosilicate type GJA (formerly designated P1) phosphor spectra.

Phosphors may emit virtually any color. Their spectra are usually somewhat narrow in bandwidth (Figures 2.20 and 2.21), with the rare-earth phosphors having very narrow "line" emission spectra (Figure 2.22). The use of mixtures of two or three phosphors to obtain white are common (Figure 2.23) for CRTs, LCD backlights, fluorescent lamps, and HID lamps. Dot- or stripe-patterned phosphor screens are employed in color CRTs. These use three phosphors, red, green, and blue, to produce white (Figure 2.24).

Spatially, phosphors approximate a Lambertian, or cosine, emitter and are usually in the form of an emitting surface for displays. Measurement of surface luminance in candelas per square meter (nits) or the deprecated footlambert is the means most often employed to characterize display phosphor light output. Directionality of the light output is altered slightly for CRTs at extreme angles due to total internal absorption within the glass faceplate or to a much greater extent by the polarizers employed for LCDs.

As previously mentioned, phosphor persistence (decay) is dependent on the atoms of the phosphor and their energy transition states. Decay is analogous to fall time in electronic circuits. The decay of certain phosphors may be in the nanosecond realm while others may continue to emit light at very low light levels for many hours. Short decay phosphors are used for photographic recording and flying spot scanning applications where image "smearing" due to image or CRT electron beam movement must be minimized. Long-decay phosphors have been traditionally used for slow-scan radar displays where images must be retained between scans. Inexpensive electronic memory is displacing the need for long-decay CRTs. Decay times are usually specified to the 10% level, but this is not always valid for a given use. Where a CRT display is viewed in subdued ambient illumination, the range of 0.1–1% is more representative of what the eye will actually see due to its wide dynamic range.

GHA (P31) Phosphor

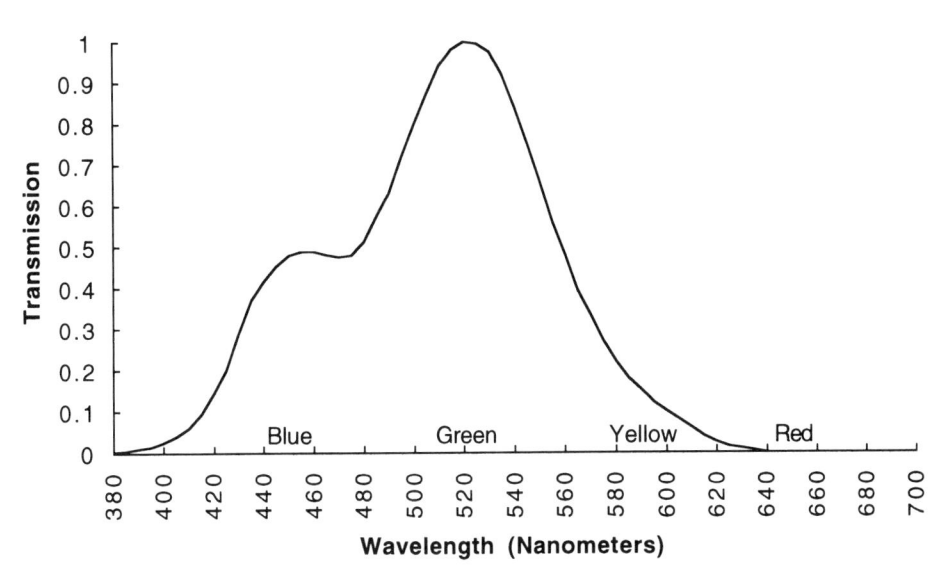

Figure 2.21. Zinc sulfide, copper-activated-type GHA (formerly designated P31) phosphor spectra.

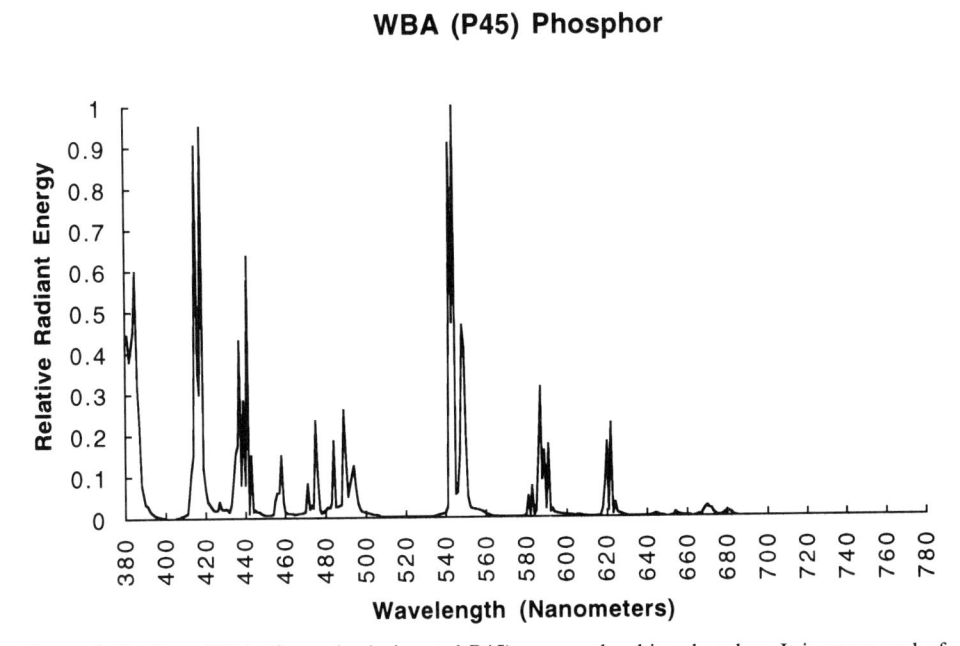

WBA (P45) Phosphor

Figure 2.22. Type WBA (formerly designated P45) rare-earth white phosphor. It is composed of yttrium oxysulfide with terbium activation.

50   LIGHT SOURCES, FILTERS, AND DETECTORS

WWA (P4) Phosphor

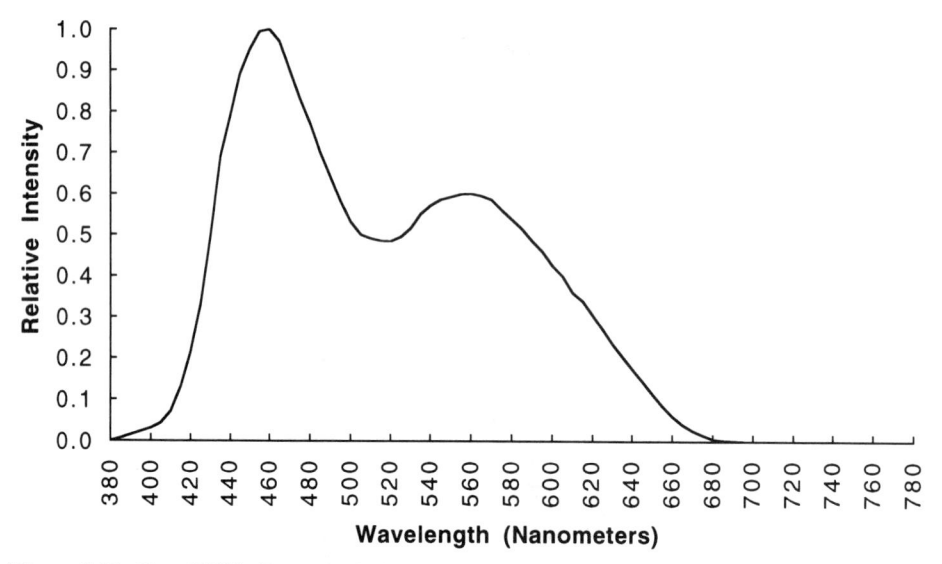

Figure 2.23. Type WWA (formerly designated P4) phosphor spectra. This is composed of a blend of blue-emitting zinc sulfide with silver activator and yellow-emitting zinc–cadmium sulfide with silver activator.

XXD (P22 Sulfide/Oxysulfide) Phosphors

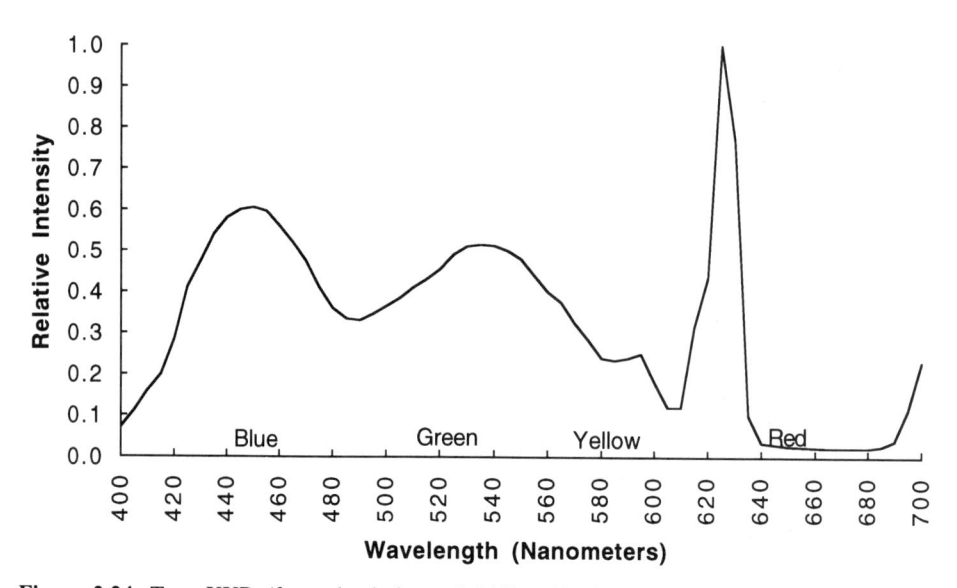

Figure 2.24. Type XXD (formerly designated P22) sulfide/oxysulfide phosphor screen spectra. Note that the red component is a rare-earth narrow-band "line emitter" at about 624 nm. It is actually producing several line emissions in that region but the resolution limit of the spectroradiometer used prevents resolving the individual lines.

Decay time may vary not only with materials but also with excitation conditions. Figure 2.25 shows a phosphor decay curve for zinc orthosilicate that is exponential. Zinc and cadmium sulfide phosphors exhibit decay curves (Figure 2.26) that follow the power law and are particularly dependent on excitation duration and intensity. High-current and short-duration excitation pulses will result in shorter decay times. Published decay curves for these materials[27] should be used with caution. The curves apply only to the excitation conditions specified. Sulfide phosphors also have another little-known decay characteristic, that of dependence of decay upon emission wavelength. These phosphors emit light over a range of wavelengths approximately 200 nm wide. By selecting individual wavelengths within that range using narrow-band optical filters, the decay times will be observed to decrease with decreasing wavelength.[28] Within that range of wavelengths, decay to 10% may vary from less than 5 μs in the blue to about 90 μs in the green region for some materials. Therefore, decay measurements will be found to vary according to the spectral response of the detector, and for most measurements a photopically corrected detector should be used. Color filters on the display make use of this effect depending on whether short or long decay is desired for a particular application.

Phosphors also exhibit risetime and buildup characteristics. Risetime is similar to that in electronics and is measured to the 10% level. The term *buildup* has usually been used to denote the increase in light output with repeated excitation pulses for screens used for radar.

Phosphor screens are subject to permanent damage, or "burning," at high excitation levels, with slow-moving intense electron beams, and prolonged display of the same image. The first two conditions are due to thermal destruction of the phosphor by exceeding the temperature at which the heat can be conducted away from the phosphor crystals

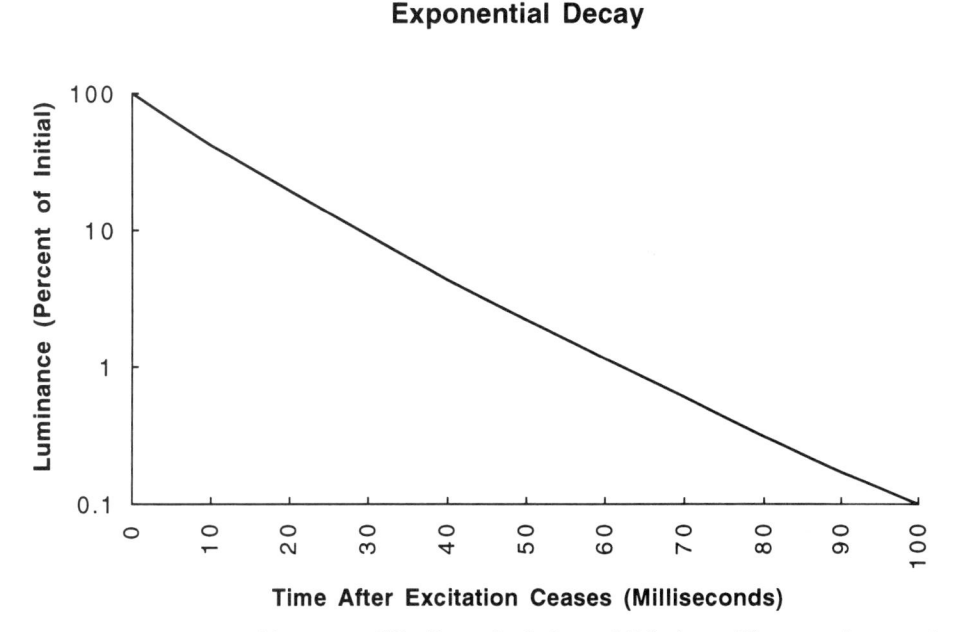

Figure 2.25. Zinc-orthosilicate-type GJA (formerly designated P1) decay. The curve is approximately exponential.

52 LIGHT SOURCES, FILTERS, AND DETECTORS

Power Law Decay

Figure 2.26. Zinc sulfide, copper-activated-type GHA (formerly designated P31) decay. The curve approximately follows the power law.

quickly enough to prevent changes in chemical composition or even melting. The individual crystals are not in good thermal contact with other crystals or the glass faceplate due to their irregular shapes. The sensitivity of a phosphor to burning depends primarily on its chemical composition and secondarily on the deposition techniques used for screening. Zinc oxide is extremely resistant to burning while magnesium fluoride phosphors are intolerant of even the slightest abuse.

Repetitive scanning of the same areas of a screen by the electron beam causes electron and X-ray browning of both the phosphor crystals and the glass faceplate. The result is that some light emitted by the phosphor will be absorbed within these materials instead of reaching the viewer.[29] This aging is often referred to as *Coulomb aging* since it is dependent primarily upon the beam current multiplied by time. It is specified as the number of Coulombs required to reduce the screen light output to a given percentage of initial light output. From a screen aging standpoint, it is far better to use 10 kV at 100 μA of beam current than trying to obtain the same brightness with 1 kV at 1 mA. The power input to the screen is the same in both cases, but the number of coulombs will be 10 times less in the former case. This is further aggravated by the fact that beam penetration into the phosphor crystals is much less at low accelerating voltages, thus causing the damage to be concentrated near the crystal surface rather than spread deeper into the phosphor crystals.

2.2 FILTERS

Light filters are widely used for displays to improve viewing contrast. They are also used extensively for display measurement where they may be used to control input intensity or the spectral response of measurement instruments and by themselves to test for measure-

Figure 2.27. One-inch-diameter thin-film filters.

ment linearity errors or infrared leakage. Two types of filters are common: the absorbing filter and the reflective, or thin-film, filter (Figure 2.27). Within these categories, the filters may exhibit light attenuation (Figures 2.28–2.30), narrow bandpass (Figure 2.31), wide bandpass (Figure 2.32), long-wave pass (Figure 2.33), short-wave pass (Figure 2.34), spectral shaping (Figure 2.35), and wavelength separation or dichroism (Figure 2.36) characteristics either separately or in combination. Each is defined by its spectral transmission curve.

Smoke Gray Plastic Filter

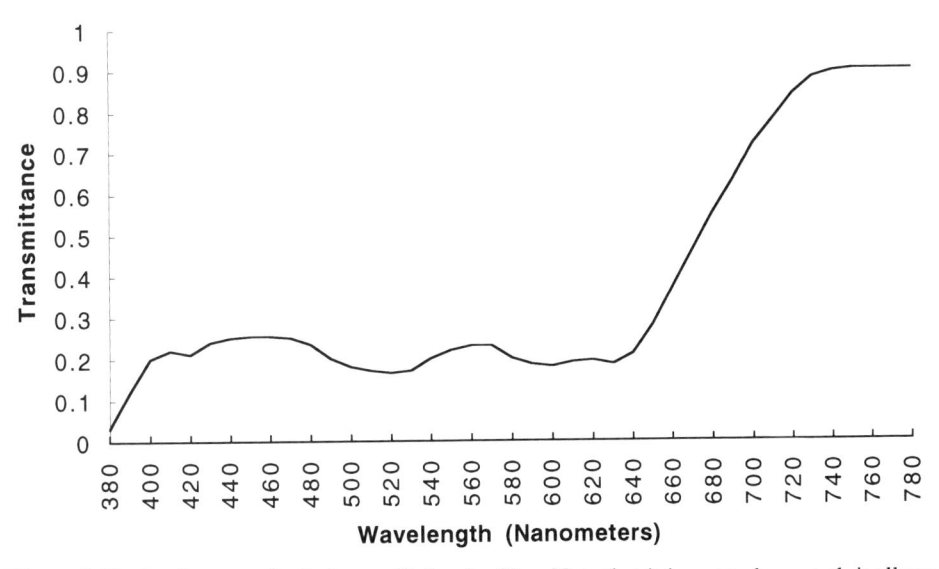

Figure 2.28. Smoke gray plastic "neutral"-density filter. Note that it is not truly neutral, it allows longer wavelengths to be transmitted.

54 LIGHT SOURCES, FILTERS, AND DETECTORS

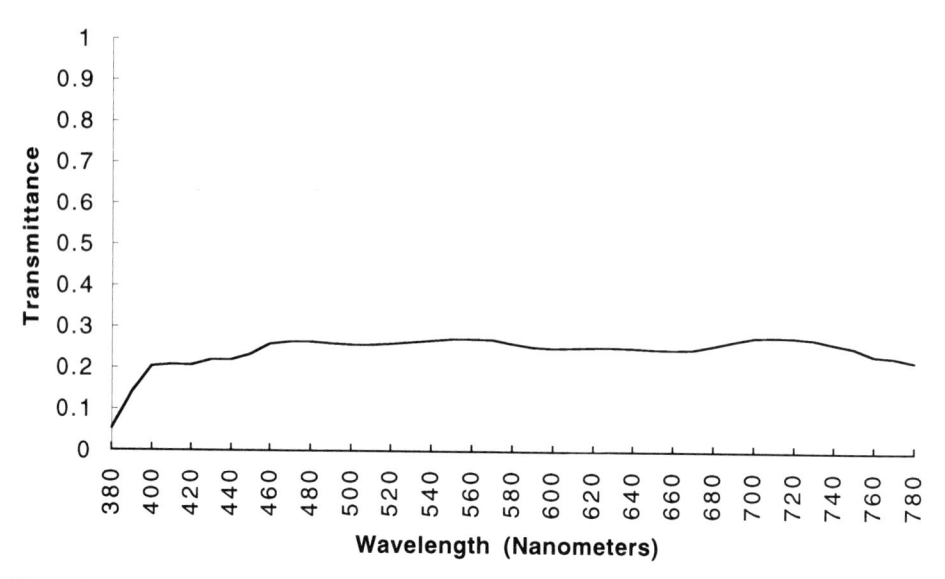

Figure 2.29. Glass neutral-density filter. These are considerably more neutral than the previously illustrated plastic filter.

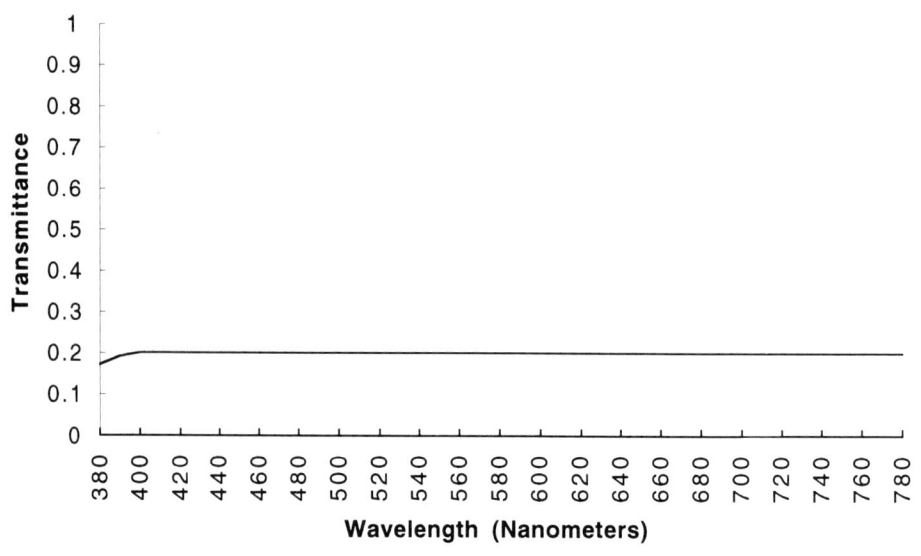

Figure 2.30. Thin-film neutral-density filter. These are the most neutral but attenuate by reflection rather than absorption and thus have a mirrorlike appearance.

2.2 FILTERS 55

Narrow Pass Filter

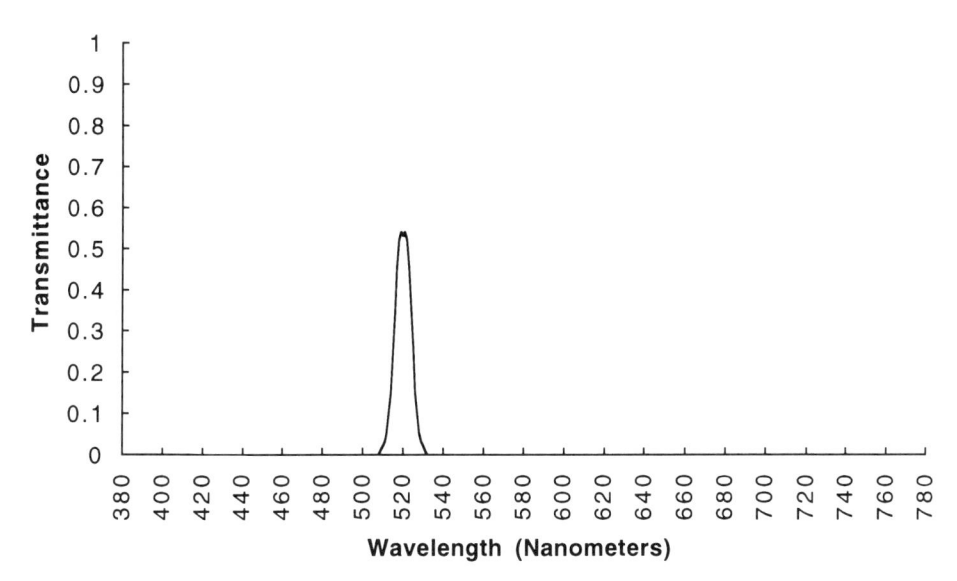

Figure 2.31. Thin-film narrow bandpass filter.

Wide Pass Filter

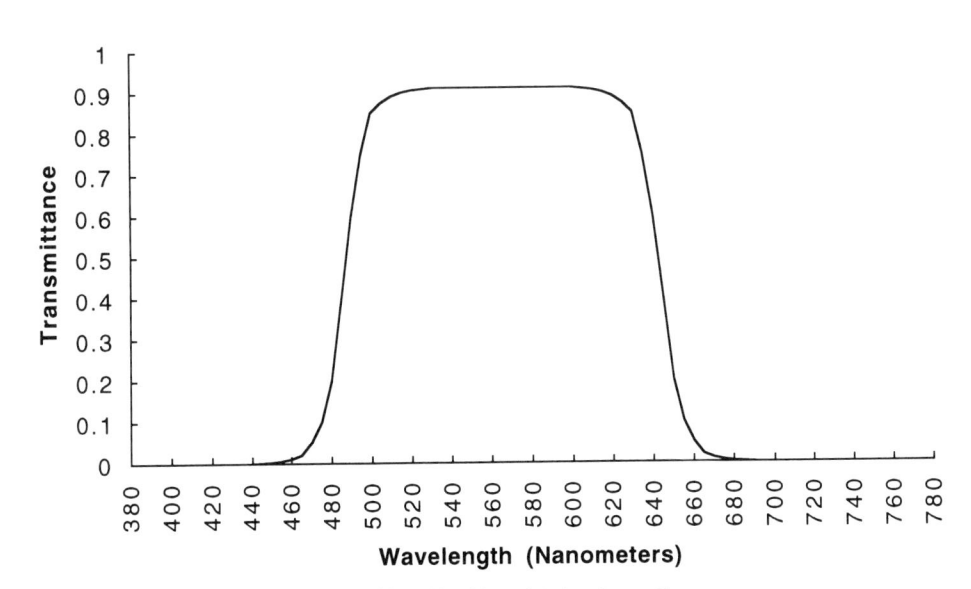

Figure 2.32. Thin-film wide bandpass filter.

Long Wave Pass Filter

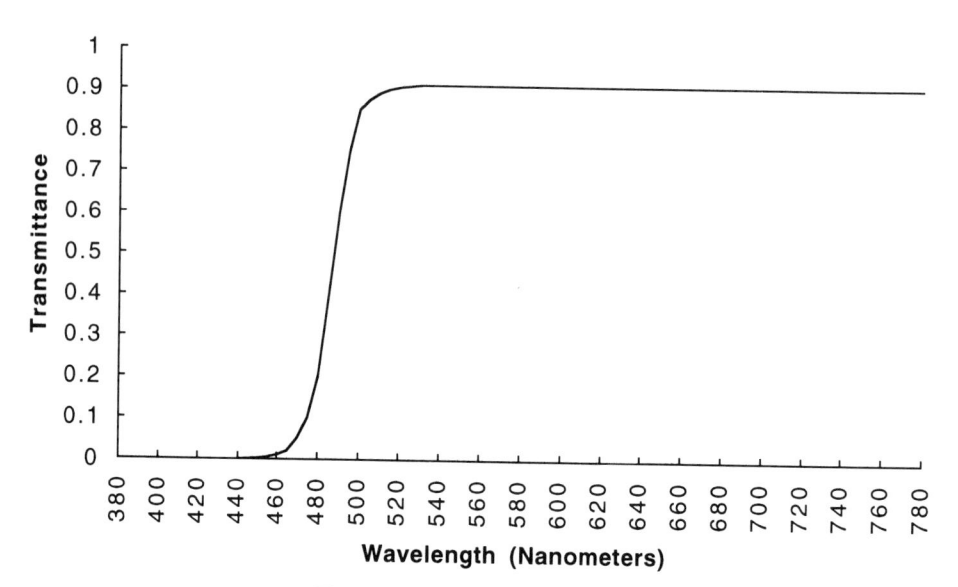

Figure 2.33. Long-wave pass filter.

Short Wave Pass Filter

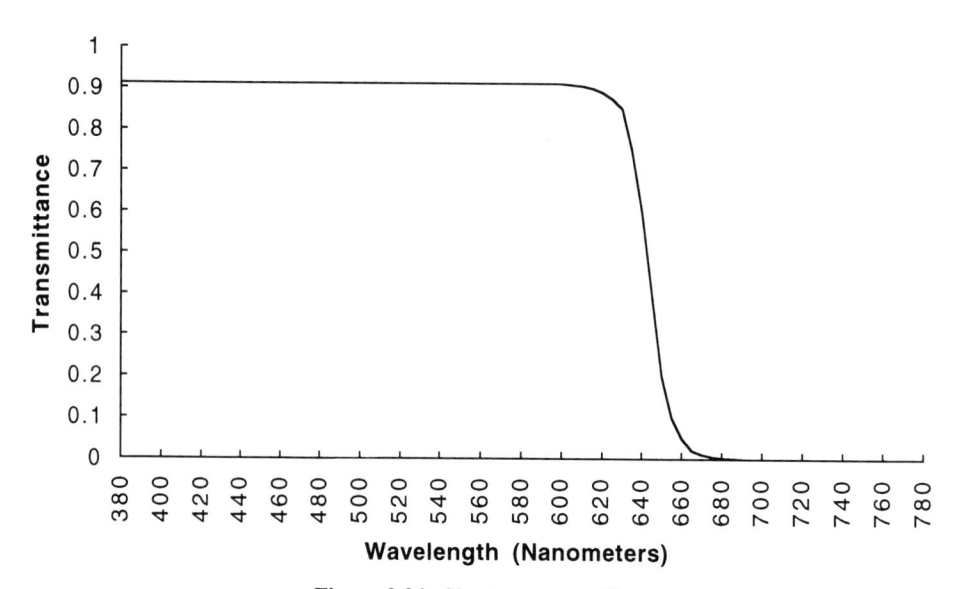

Figure 2.34. Short-wave pass filter.

Typical Glass Color Filters

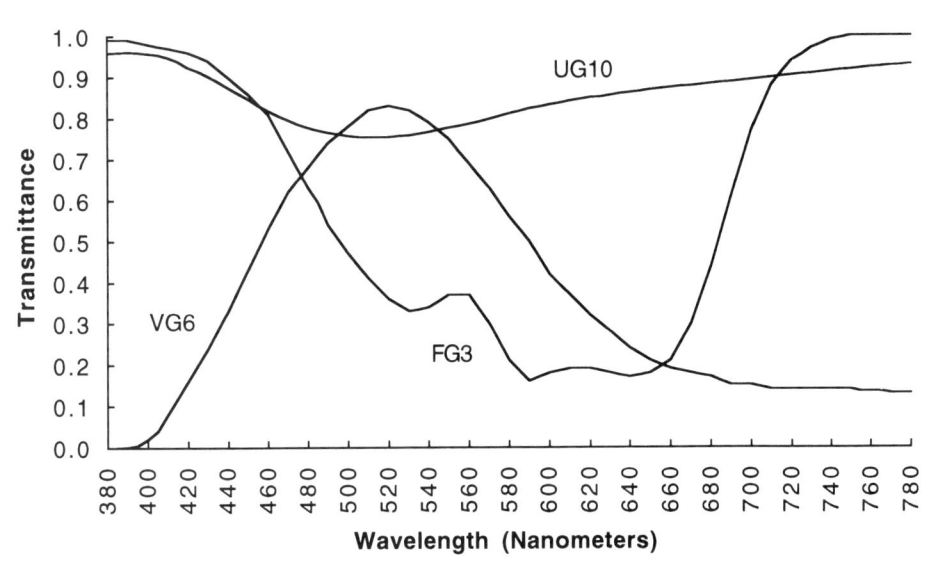

Figure 2.35. Typical glass color filters for spectral shaping.

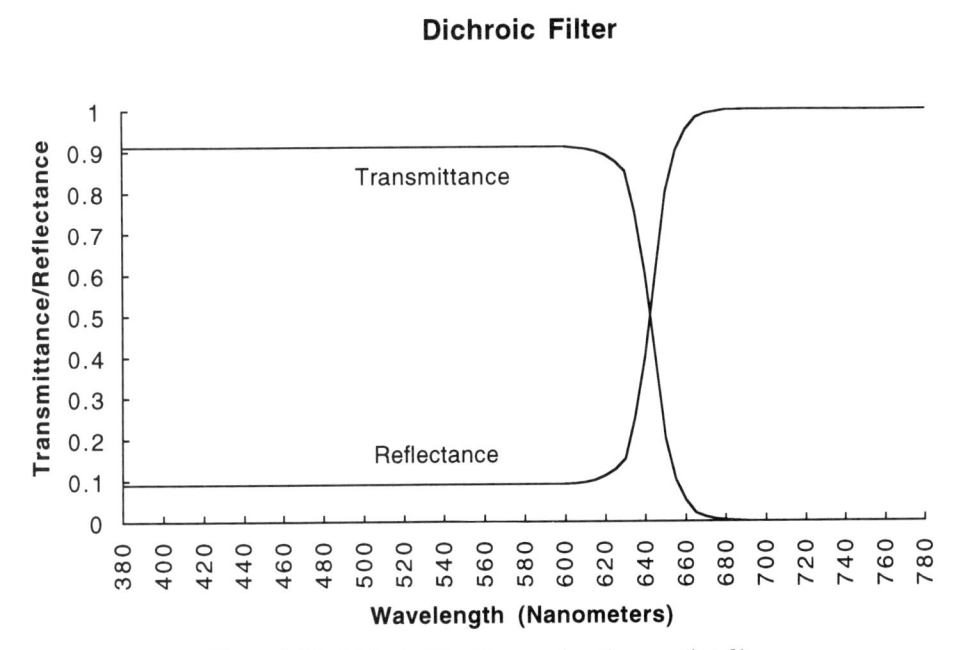

Dichroic Filter

Figure 2.36. Dichroic thin-film wavelength separation filter.

Light filters may be specified in percent transmission, transmittance, internal transmittance, transmissivity, absorptance, absorptivity, optical density, or reflectance. Some of these include *specular,* or mirrorlike, reflection of light from the air-to-filter interface; others represent just the internal properties of the filter. Except for some neutral density filters, all filter types are usually specified in one or more of these terms for a range of wavelengths in either tabular or graphical form. Figure 2.37 illustrates the fundamental concepts of incident light, reflectance, transmittance, and absorptance.

Because of the similarity of names of some of the terms and because each has a precise meaning, they will be defined before discussing the various types of filters in more detail. The definitions given below are derived from three authorative and widely used sources: CIE 17.4/IEC 50 *International Lighting Vocabulary,*[30] the IESNA *Lighting Handbook,*[31] and the *Photonics Dictionary.*[32]

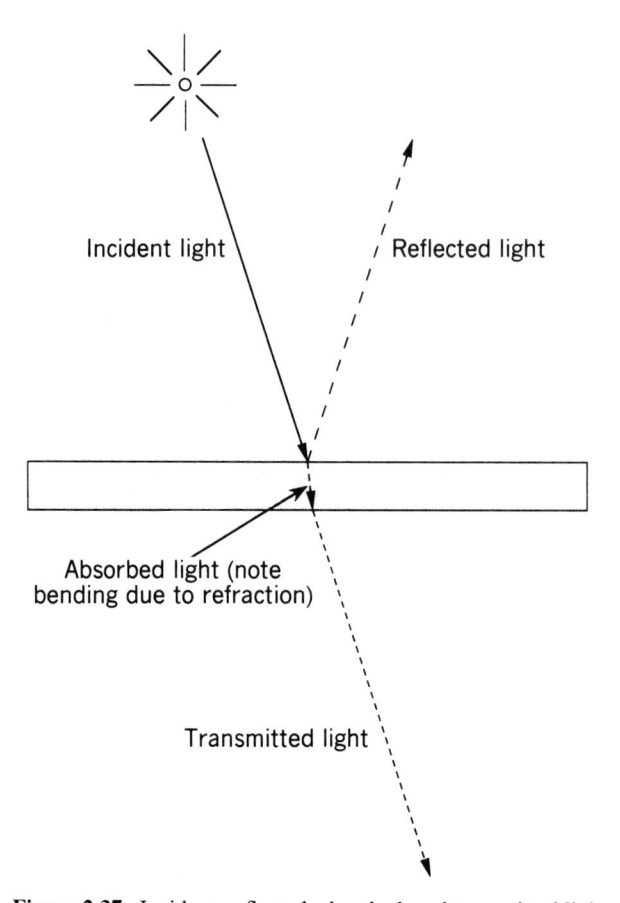

Figure 2.37. Incident, reflected, absorbed, and transmitted light.

Percent transmission is perhaps the easiest to understand. It is the percentage of light passing through a filter compared to the total light entering it. Percent transmission would normally include surface reflection losses unless expressed as percent internal transmission.

Transmittance (τ) is the ratio of the light passing through a filter divided by the light incident on it. The transmittance will always be less than 1.00 and includes surface reflectance losses unless designated as internal transmittance (τ_i).

Transmissivity is the internal transmittance per unit of thickness and is expressed, for example, as 0.3 mm^{-1}. Surface reflectance is excluded from transmissivity. Transmissivity is useful for filters such as glass that may be specified in thickness for a particular application. For calculating the transmittance for any other specific thickness, figure the ratio of that thickness to the thickness specified in the transmissivity and use that ratio as an exponent for the transmittance value specified. Using the example above, if we wanted to calculate the transmittance for a material 2 mm thick, we would use $0.3^{(2/1)} = 0.3^2 = 0.09$.

Absorptance (α) is similar in nature to transmittance except that it is the ratio of the light absorbed by a filter divided by the light incident on it. Surface reflectance is excluded. For absorbing filters such as glass and plastic it is the reciprocal of internal transmittance.

Optical density (D) is a logarithmic scale used in photography and is often used to specify neutral density filters. It is the log to the base 10 of the reciprocal of the transmissivity. A neutral density filter that transmits 10% of the light incident on it (transmittance = 0.1) will have an optical density of 1.0; a filter that transmits 1% (transmittance = 0.01) will have an optical density of 2.0, and so on.

Reflectance (ρ) is the ratio of the light reflected by a filter divided by the light incident on it. If all of the light was reflected, the reflectance would be 1.00. Actual filters will always be less than 1.00.

Surface reflection results in about 4% light loss per surface at each air-to-filter interface for typical polished glass and plastic surfaces. Therefore, a single filter would have about 8% loss (92% transmission or 0.92 transmittance) even if it had 100% internal transmission. Stacking an additional similar filter would reduce the remaining light by another 8% so the result would be a little less than 85% of the original light being transmitted ($0.92 \times 0.92 = 84.6$). Cementing the two filters together with an optical cement having an index of refraction similar to that of the filters being laminated would eliminate two air-to-filter interfaces of 4% each and would reduce the total surface reflectance loss to only 8% again. The use of antireflection coatings on the remaining surfaces can reduce surface reflections to less than 1%.

The transmittance at a particular wavelength of a stack of filters may be calculated by multiplying the transmittances of each ($T_\lambda \times T_{1\lambda} \times T_{2\lambda} \times T_{3\lambda} \times \cdots$).[33]

Glass Filters

Colored glass filter materials have many applications in electro-optics and displays. They are most commonly used for the glass faceplate of display CRTs for contrast enhancement. Faceplates have undergone a number of contrast enhancement improvements over the years to allow viewing under bright ambient illumination. Neutral gray glass faceplates were introduced in 1950 for television picture tubes. The contrast enhancement resulted because ambient light passes twice through the absorbing gray glass as it is

60   LIGHT SOURCES, FILTERS, AND DETECTORS

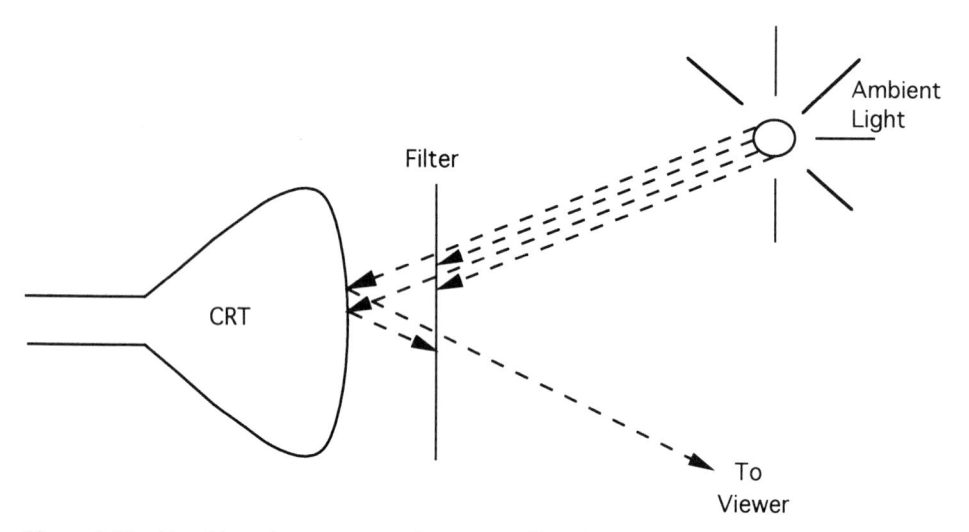

Figure 2.38. Absorbing glass contrast enhancement filter for a CRT display. Ambient light must pass through the filter twice while light from the display only passes through it once before reaching the viewer.

reflected from the phosphor while the light from the image produced by the phosphor only passes through it once (Figure 2.38). A related method used in monochrome avionics displays that employ green phosphor is the use of a narrow-spectral-transmission dark-green glass faceplate that passes only the color of the phosphor. External gray or green filters have a similar effect but at the expense of adding two surfaces, each with 4% surface reflectance unless treated with an antireflection coating.

External spectrally selective filters for the enhancement of display contrast were first described in a German patent application filed in 1936 and in U.S. patent applications filed in 1937 and 1958.

Neodymium, praseodymium, and a mixture of the two, didymium, glasses were used because of their unique spectral transmission characteristics. Didymium glass along with neodymium glass for color television picture tube filters was investigated by Zenith in 1969. These glasses have a series of peaks and valleys across the spectrum (Figure 2.39) that selectively pass the colors of the three-color phosphor primaries with little loss while ambient light of other wavelengths is heavily absorbed. Under ambient fluorescent light, they have a pleasant pale blue coloration. This approach was later used for color CRT panels by at least Mitsubishi and General Electric, the latter using the name Neovision. Nippon Electric Glass is a supplier of neodymium-doped barium–strontium glass CRT panels. Higher cost appears to be the main reason that it has not gained widespread acceptance.

Glass color filters are used for spectral correction of sensors for tristimulus filter colorimeters and higher quality photometric instruments. These filters are used to match the spectral response of a detector, usually a silicon photodiode, to that of the CIE tristimulus functions or the average human eye (CIE photopic response). Two or more glass filters are employed to shape the detector's response in different spectral regions. Glass filters have a strong advantage over the low-cost gelatin filters used by some manufacturers.

Didymium Glass

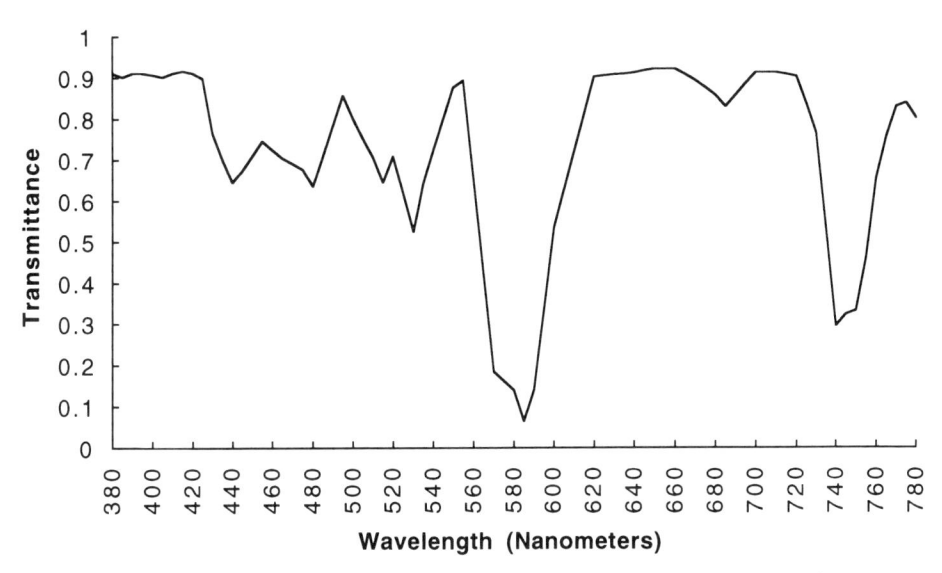

Figure 2.39. Didymium glass filter suitable for tricolor phosphor display contrast enhancement.

That advantage lies in long-term stability since the proper selection of glass types will withstand exposure to heat, humidity, and high light levels far better than any gelatin filter. The trade-off is cost. Each element of the filter must be ground to a precise thickness for the individual sensor to optimize the spectral match to the desired curve. Lamination of the filter stack with an optical cement further improves durability and ruggedness and eliminates internal reflection losses.

Two construction techniques are used for spectral correction filters.[34] The first is the stacked, or *homogeneous,* filter (Figure 2.40), where each element successively removes a portion of the spectrum.[35–38] The disadvantages of the stacked filter are heavy absorption of light and the resultant sensitivity loss if an accurate spectral correction is to be achieved. Photopic correction accuracies (f_1') of 2–3% are attainable with four-element stacked filters. Stacked filters are usually incorporated in moderate-cost portable and laboratory instruments.

The second construction method is that of the *mosaic* filter, based on the sensor developed by A. Dresler in Germany in 1933[39] and marketed today by LMT in Berlin. The mosaic filter consists of a glass filter substrate that roughly approximates the desired spectral response.[40] Smaller pieces of different colored glasses are cemented to the surface of the substrate to selectively absorb light in specific portions of the spectrum (Figure 2.41). Photopic correction accuracies with f_1' values of less than 1% are attainable with mosaic filters. Drawbacks include the expense of individually adjusting thickness, area, and position of a number of pieces of glass and the variation of spectral response that occurs with differing angles of incoming light. The latter effect is lessened when a cosine correction diffuser is employed or where all light comes from only one direction. Mosaic filters are usually limited to higher quality laboratory instruments.

62 LIGHT SOURCES, FILTERS, AND DETECTORS

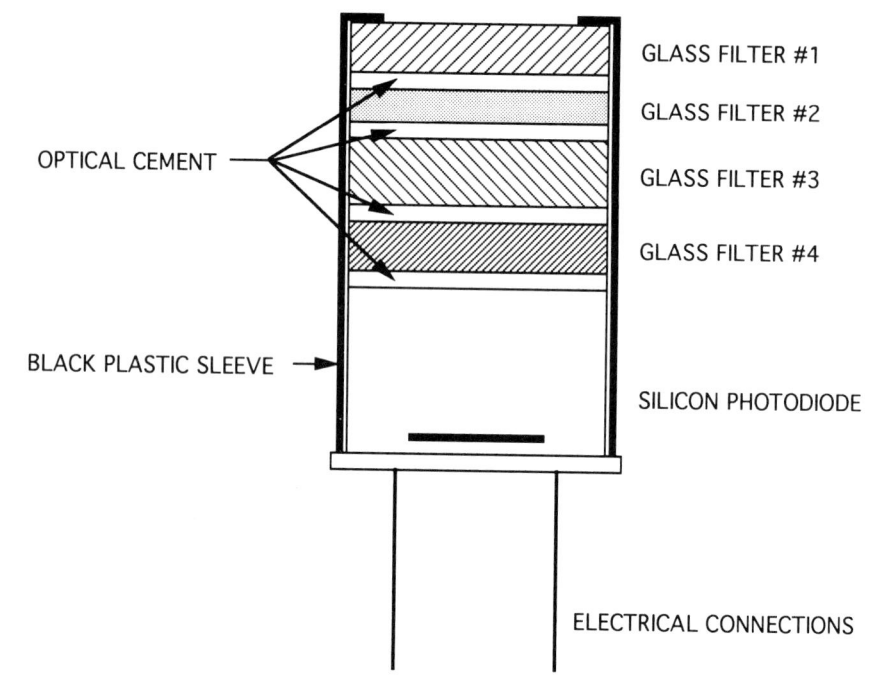

Figure 2.40. Stacked or homogeneous spectral correction filter and photodiode assembly as used to match CIE Standard Observer (photopic response) or one of the CIE tristimulus functions.

Plastic Filters

This category of filters includes gelatin filters and sheet plastic filters. Gelatin filters such as the familiar Wratten photographic filters are significant from the display standpoint only as spectral correction filters for low-cost photometers. Lack of long-term stability with exposure to humidity and light is a drawback that precludes their use for serious dis-

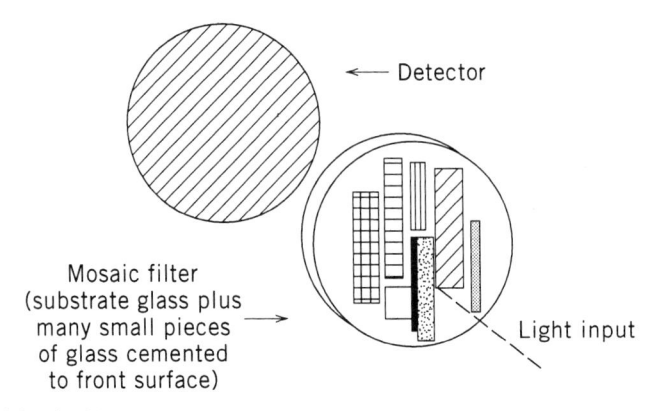

Figure 2.41. Mosaic filter as used to correct the response of a photodetector to that of the CIE Standard Observer (phototopic response) or one of the CIE tristimulus functions. (From *Mosaic Filters, What Are They and How Do they Work?* LMT Application Note AN2-1187. Reprinted by permission.)

play measurement work. One other gelatin filter, the Wratten neutral density filter, should be mentioned here. While perfectly suited to photographic applications, it should never be used for photometric work due to its lack of a flat transmission spectral response curve across the visible spectrum, which can cause measurement errors with many types of light sources.

Plastic filters, on the other hand, are in widespread use for contrast enhancement of many types of displays. Plastic color filters are often similar to glass color filters in appearance and spectral transmission curves. The principal advantages of plastic filters are their low cost, which is especially important for large filters; their resiliency, which provides physical protection of the display device behind them; and their ease of molding, cutting, and shaping, even as far as curving the surface to conform to a CRT faceplate. Their principal shortcoming is their lack of resistance to scratching.

Thin-Film Filters

Thin-film filters consist of one or more films of evaporated, sputtered, or plated metal and transparent dielectric coatings deposited in a vacuum onto a heated glass or quartz substrate. As many as 50 or more layers may be used for more complex filters. Film thicknesses are on the order of one wavelength of light, approximately 0.1–5 μm. The substrate provides physical support for the films. Most thin-film filters may be identified by their mirrorlike coating(s). Two comprehensive books on thin-film filters from a user's perspective are J. D. Rancourt's *Optical Thin Films*[41] and OCLI's *Interference Filter Handbook*.[42] Additional information may be found in the Optical Society of America's *Handbook of Optics*.[43]

Thin-film filters are available in many forms, ranging from neutral density, cutoff, and bandpass to dichroics, which selectively transmit and reflect different wavelengths. Another form of thin-film coating, the *antireflective coating,* is used to treat display screens to minimize reflections. It will be discussed further in the section on display contrast enhancement filters. Finally, transparent conductive coatings of mixtures of indium and tin oxides (ITO) may be applied to glass substrates by themselves or in combination with antireflective coatings. Conductive coatings are used in the display industry to prevent charging of CRT display screens by the high voltage applied to the inside of the CRT faceplate, to reduce the emission of electromagnetic interference (EMI) from a display, or to provide internal electrical connections for rows, columns, characters, or individual pixels in many flat-panel displays.

Neutral-Density Filters The simplest thin-film filter is the neutral-density (ND) filter. These may be as simple as a single metallic film on one piece of glass. Two metal films may be applied to opposite sides of a single piece of glass to reduce the effect of "pinholes" that may be present in the films. To improve durability, two pieces of substrate glass with the film or films on the inner surfaces may be laminated together.

Thin-film ND filters reduce the light levels with metal films that are so thin as to be semitransparent. The light that is not transmitted is reflected back toward the source, as opposed to the absorption that occurs in glass and plastic ND filters. The metal film may be aluminum, chromium, or the alloy *Inconel,* the latter two having higher abrasion resistance. Aluminum is easiest to work with but is easily damaged and prone to oxidation. Oxidation may be reduced by "overcoating" it with a transparent dielectric film. The filter transmittance, which may be as little as 0.00001 (ND 5.0), is controlled by the thickness of the metal films. Glass substrates are commonly used, but fused quartz substrates allow the filter characteristics to be extended into the ultraviolet in addition to the visible and infrared ranges common to glass substrates. The cost is substantially higher for fused quartz substrates and they are usually confined to the smaller filter sizes.

Multilayer Filters Bandpass, cutoff, and dichroic filters composed of alternating thin metallic and insulating dielectric films or alternating low- and high-refractive-index dielectric films are often referred to as *interference filters* due to the effects of light interference between layers. The thicknesses and refractive indices are selected to shape the filter passband characteristics.

Thin-film *narrow-bandpass* filters transmit only a single, narrow-band portion of the spectrum while reflecting most other wavelengths. They may be constructed for any wavelength from the ultraviolet to the infrared. The shape of the passband is somewhat squarish but will often have minor peaks and dips in the flattened peak. Bandwidth (BW) is specified at the full-width, half-maximum points (FWHM) and may be as narrow as 0.1 nm or as wide as tens of nanometers. Wavelength in nanometers or micrometers is specified for the center of the FWHM bandwidth for light perpendicular to the filter (0° angle of incidence). Filter transmission in percent is specified for the average peak transmission of the flattened top. Narrow bandpass filters are commonly used to isolate light of specific wavelengths of interest, such as when using a radiometer to measure 632.8 nm emission from a helium–neon laser in the presence of broadband ambient light. Excellent rejection of unwanted wavelengths (blocking) is a characteristic of these filters. In some cases, they may incorporate a colored-glass, sharp-cutoff, low-pass filter (Figure 2.42) in the design for complete blocking of all wavelengths shorter than that of the bandpass filter.

Glass Cutoff Filter

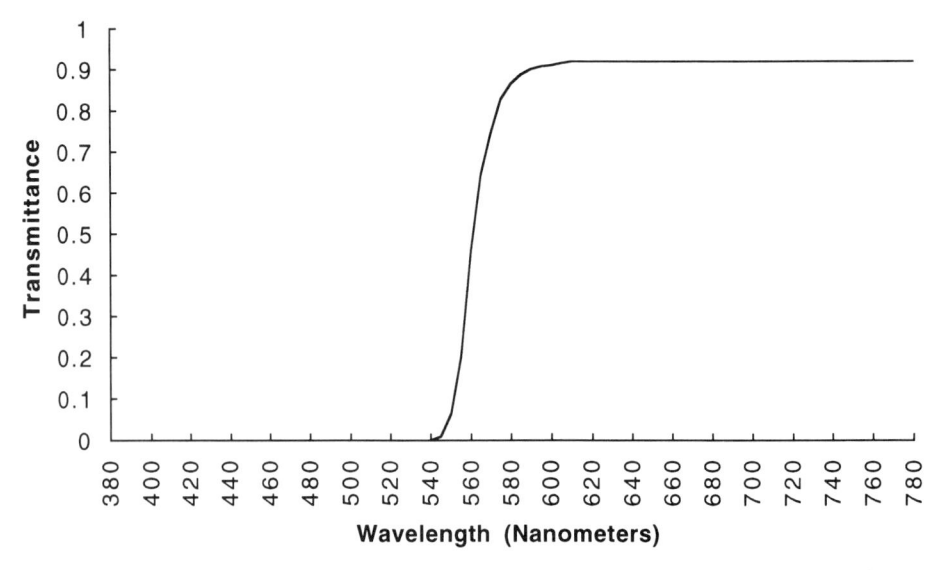

Figure 2.42. Colored-glass, sharp-cutoff, low-pass filter as used for blocking unwanted wavelengths as part of or in conjunction with thin-film filters.

The wavelength of peak transmission of interference filters shifts to a shorter wavelength as the angle of incidence of light upon the filter increases. Care should be taken to ensure that the filter is used in a configuration where light arriving at an angle is blocked either at the entrance or exit side of the filter. If light passes through the filter over a range of angles, the effect will be a wider apparent passband combined with a variation of wavelength dependent on the angle of arrival of light.

Filters that block shorter wavelengths while transmitting longer wavelengths are termed *long-wave pass* filters. Many glass and plastic filters exhibit this characteristic and provide a low-cost source of such filters. In these filters, the wavelengths not transmitted are absorbed within the filter. Thin-film long-wave pass filters can provide sharper cut-on slopes or the ability to have cut-on wavelengths at wavelengths not available in glass or plastic filters. Also, those wavelengths not transmitted are reflected. This ability is known as *dichroism*. Some dichroic filters are optimized for use as a wavelength separation filter or *beamsplitter* at a 45° angle to split a light beam into its short- and long-wave components at 90° angles to each other. This is sometimes used for light measurement instruments and various imaging systems. Where used to transmit infrared energy and reflect visible light, the dichroic filter is called a *cold mirror* since it reflects only the "cooler" visible light. An example is used in the common overhead transparency projector. The cold mirror is often applied to the projection lamp reflector.

The *short-wave pass* filter is just the opposite. It transmits the shorter wavelengths while reflecting longer wavelengths. Here, there is no glass or plastic filter that has the same function. Some "heat-absorbing" glasses are available, but they have a very shallow cutoff characteristic and merely tail off gradually in the infrared. Again, they block by absorption rather than reflection. This means that the level of infrared energy must be kept moderately low to prevent filter damage by infrared absorption and consequent heating. Thin-film short-wave pass filters, on the other hand, reflect the unwanted infrared energy in a direction where it may be safely dissipated. These are also considered dichroic filters and may be termed *hot mirrors* as well. Hot mirrors are often used with incandescent lamps to transmit the useful visible light while reflecting the infrared light, which contributes only heat, in a direction away from the object to be illuminated.

Display Contrast Enhancement Filters

Contrast enhancement for emissive displays is divided into two categories: (1) glare reduction, which deals with specular or mirrorlike reflections, and (2) reduction of diffuse reflectance, which more or less uniformly causes illumination of the entire display screen. Specular reflections tend to be especially annoying but are often easily resolved by repositioning the display screen or the offending source of light. Diffuse reflectance may sometimes be reduced by blocking light from falling directly on the screen. A simple hood around the screen may suffice. For more stubborn cases of either specular or diffuse reflections, it is necessary to employ surface treatment and/or filters.

Display glare reduction techniques use various surface treatment processes to reduce the specular reflections from a polished glass surface. These reflections occur because of the difference in refractive indices of air and glass at the air–glass interface and are on the order of about 4%. The most common surface treatment is a surface etching that breaks up direct reflections from the front surface of the glass and scatters the light in random directions. Some loss of resolution results because light from the display itself undergoes a small amount of scattering as well. The amount of diffuse reflectance is increased also.

A more sophisticated contrast enhancement technique is the evaporation on the glass surface of the display of a single dielectric layer of one-quarter wavelength thickness and having an index of refraction (N_d) intermediate to those of the glass and air. A common example is magnesium–fluoride (MgF_2) with an index of refraction of 1.38. This coating is called an "antireflection" (AR) coating and serves to improve the match of the index of refraction of the glass to that of air, thus reducing surface reflections from 4% to less than 1% for typical glasses having an index of refraction near 1.52. Optimum reduction of reflectance from the surface occurs near the wavelength for which the coating is one-quarter wavelength. These coatings give the glass surface a slight purplish appearance and are also termed "V" coat because of their V-shaped spectral reflectance curve. This is because they are optimized to prevent reflections in the center of the visible spectrum and reflectance increases toward the red and blue ends of the spectrum, resulting in a purple mixture of reflected light. Multilayer coatings have been developed to broaden the wavelength range of lowest reflectance, but this usually must be balanced against slightly increased average reflectance. Broadband high-efficiency antireflection (HEA) coatings with excellent low-reflectance characteristics across the visible spectrum are available in sizes up to 1.8 m in diameter. The only drawbacks of AR and HEA coatings are higher manufacturing cost and the reflective fingerprint spots that become evident because the oils from a finger reduce the absorption of reflections as a result of their higher index of refraction. More frequent cleaning is thus required. No resolution loss occurs with AR or HEA coatings. This is important for the higher resolution displays used for computer-aided design (CAD) workstations, military applications, and medical imaging, where the additional cost is a minor factor. These coatings are similar to those employed in high-quality photographic lenses.

Methods of diffuse reflection reduction include the use of external glass or plastic light filters, louvered films, and polarizers that may also be combined with etching, an antireflection or HEA coating, and/or conductive ITO coating. These may be either laminated directly to the display panel or spaced some distance from it. They may also provide additional protection against accidental implosion or breakage when used for CRT displays.

Neutral-density glass and plastic filters can provide excellent contrast enhancement for displays at relatively low cost. They are effective because ambient light must pass through the filter twice before reaching the viewer while light emitted by the display itself only passes through it once, as previously shown in Figure 2.38. Thus a 20% transmission neutral filter will reduce the display luminance to 20% of its original value while ambient light will be reduced to 20% of 20%, or 4%.

Colored glass and plastic filters, another low-cost approach, work best with narrowband display emissions such as LEDs and monochrome CRT phosphors. Here, the color filter will usually have a relatively narrow bandpass of high transmission matched to the emission spectrum of the display. All other wavelengths are blocked (Figure 2.43). The light from the display passes through with very little loss while ambient light that is broadband will be attenuated substantially.[44] A result is that the display background will have a color similar to the display emission. Some further improvement may be gained by use of a filter having a broader passband that gives it a different color than the display. Tektronix exploited this for many of their oscilloscope displays. Green CRT phosphors and cyan plastic filters (Figure 2.44) have long been used. The cyan filter transmits most of the green emission since it is still within the relatively wide blue bandpass of the filter but it absorbs the yellow and red portions of the ambient light spectrum. The outcome is a

2.2 FILTERS 67

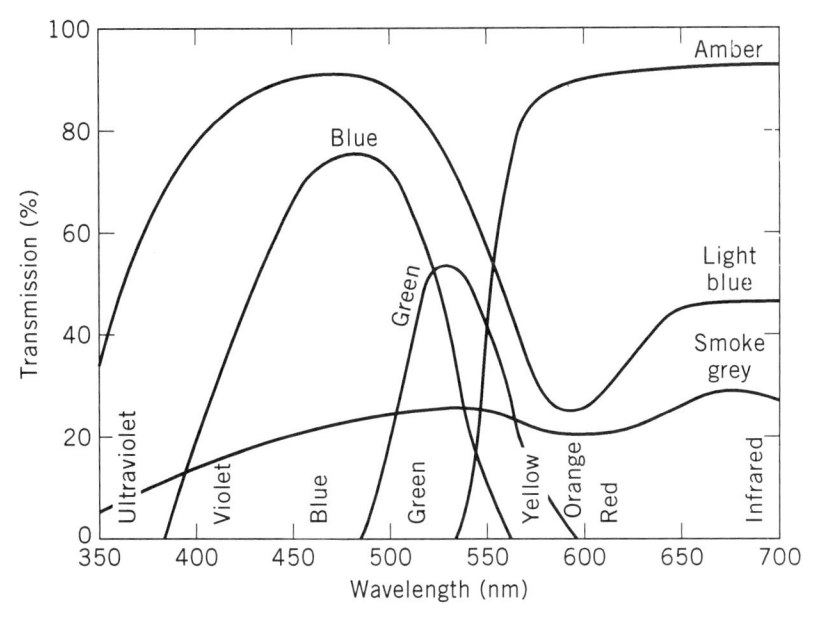

Figure 2.43. Spectral transmission of plastic light filters for contrast enhancement of CRT displays. (From *Light Filter Characteristics,* Tektronix CRT Data, Copyright 1969 Tektronix, Inc. All rights reserved. Reproduced by permission.)

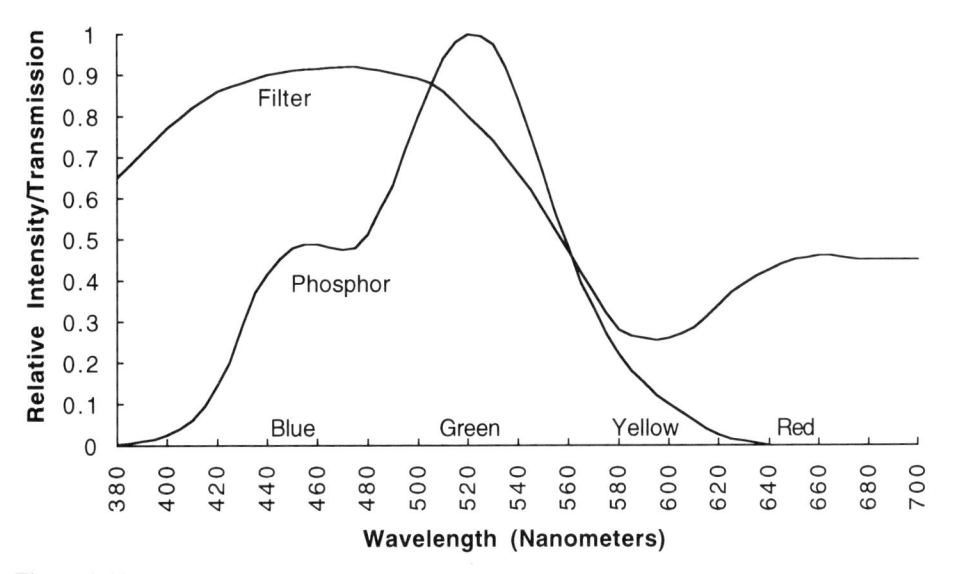

Figure 2.44. Spectral transmission of cyan plastic light filter for enhancement of color contrast of green phosphors.

green display against a blue background. Numerically, the luminance contrast is not particularly impressive. The improvement is through *color contrast,* which also has the ability to improve viewability. A display having a luminance contrast ratio of 1:1 would not be viewable if the background was the same color as the display but could be entirely satisfactory if they were of different colors.

Circularly polarized filters have an end result similar to ND filters. Light from the display becomes circularly polarized but is passed without substantial loss while ambient light is also circularly polarized by passage through the filter, but when reflected from the display, it is of circular polarization opposite to that required for passage back through the filter and is largely blocked. This may be demonstrated by viewing a circular polarizer in a mirror. The effect on a display is not as pronounced, however. The diffusing characteristics of phosphor screens tend to depolarize the light to some degree, which limits the effectiveness of the circular polarizer.[45]

Louvered filters, developed by 3M, consist of a venetian blind–like louvre structure embedded within a plastic film. When used for display contrast enhancement, light from the screen passes through with little loss to the viewer who is viewing from a direction essentially perpendicular to the screen. Ambient light from other angles above and below the centerline of the display is prevented from reaching the screen and is mostly absorbed by the dark color of the louvres (Figure 2.45). The depth of the louvres and their angle may be selected to permit optimum viewing from certain angles or narrowing of the viewing angle for greater contrast enhancement in high ambient environments. Privacy of the displayed information may also be gained at the same time. This is an important consideration for automatic teller machines (ATMs).

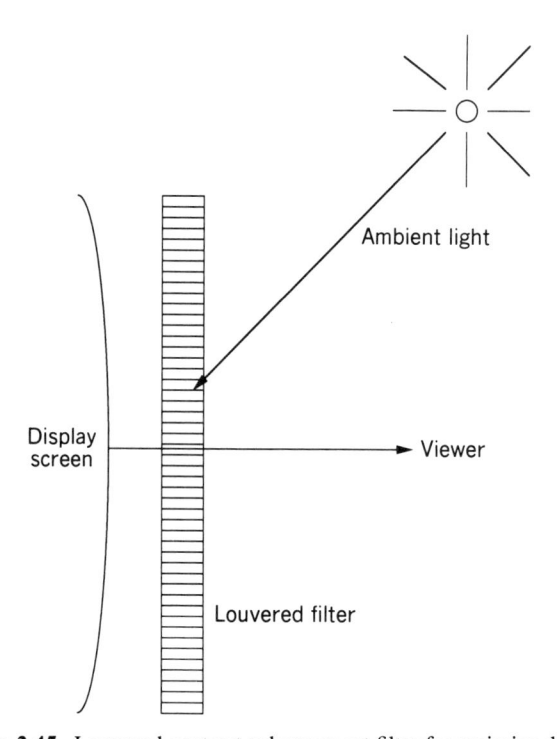

Figure 2.45. Louvered contrast enhancement filter for emissive displays.

2.3 LIGHT DETECTORS

Thermal Detectors

One of the earliest and simplest of all sensors of light is the thermal junction or radiant thermocouple. The thermocouple consists of a junction of two dissimilar metals that generates a small voltage (approximately 25 μV/°C) when heated. This is known as the *Seebeck effect* and is the reverse of the *Peltier effect* utilized in thermoelectric cooling devices. An absorber of radiant energy is thermally coupled to the junction to convert the light to be measured into heat to which the thermocouple will respond. Use of a material such as lampblack for the absorber results in near-total absorption of the incoming radiant energy over a wide range of wavelengths including ultraviolet, visible, and infrared. The measurement of irradiance in watts per square meter was most common. The sensitivity of the thermocouple detector is very poor and the output signal correspondingly low (microvolts), but it was used for much early radiometry where sunlight and large incandescent lamps were the light sources to be measured. The time response of the thermocouple is slow (seconds) due to the thermal mass of the absorber and junction.[46–48]

The thermopile consists of several blackened thermocouple junctions connected in series to produce higher voltages that are easier to measure (Figure 2.46). The increased sensitivity of the thermopile is not without its own problems. It is very sensitive to changes in ambient temperature and to air currents. These are at least partially offset by improvements in construction, such as an enclosed housing with a quartz window and compensating reference junctions in the same housing which are not exposed to the radiation being measured.[49]

In recent years, miniature thermopiles have become commercially available. These are hermetically sealed in a conventional TO-5 semiconductor package and have given the thermopile a new lease on life. The prime advantage of the thermopile is its flat response over a broad spectral range for applications in radiometry. No other detector can match this characteristic. The small junction size used in the TO-5 package improves response time into the millisecond range and allows easier control of their ambient temperature for more stable measurements.[50]

Photoconductive Sensors

The term *photoconductive* is applied to sensors that exhibit a change in the bulk resistance of their resistive material with incident light. Cadmium sulfide and cadmium selenide are the most common materials employed. The photoconductive cell is similar to a negative-temperature-coefficient thermistor in operation except that it responds to light rather than temperature. Increasing the light level on a photoconductive cell decreases its resistance. The photoconductive cell is quite sensitive to small changes in illumination and has found many applications where a low-cost sensor is required. Examples are shown in Figure 2.47. These applications include camera exposure meters, sensors to turn on lights at dusk, and digital sensing such as card readers. Other advantages include simple circuitry, ruggedness, direct operation on up to 240 V AC or DC, and wide dynamic range. Disadvantages of the photoconductive cell for measurement applications include a spectral response (Figure 2.48) that is neither flat nor easily matched to the CIE photopic response curve, nonlinear resistance versus illumination level, temperature errors and slow response time at low illumination levels, and "fatigue" or light memory effects.[51–53]

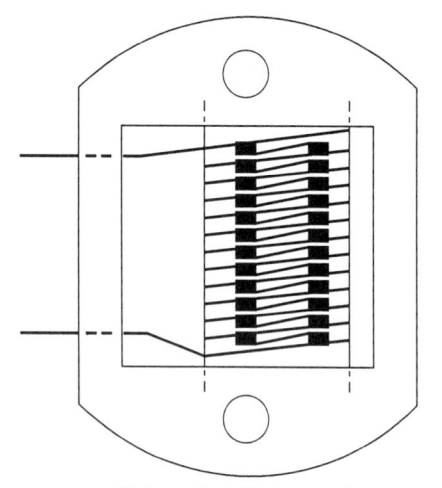

12 Junction linear surface type thermopile

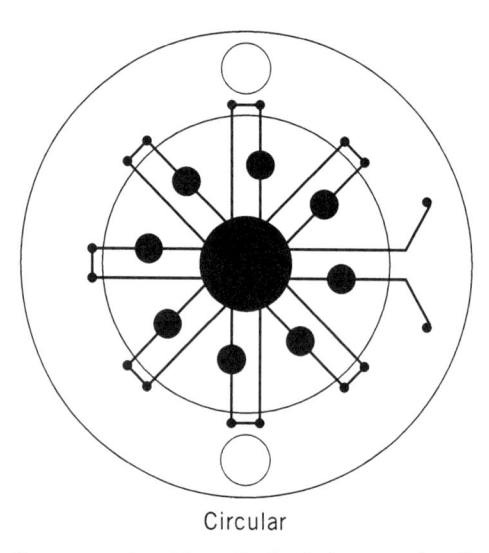

Circular

Figure 2.46. Thermopile construction. (From Eppley Laboratory, Inc. Reprinted by permission.)

Figure 2.47. Typical photoconductive sensors.

CdS Photoconductive Cell

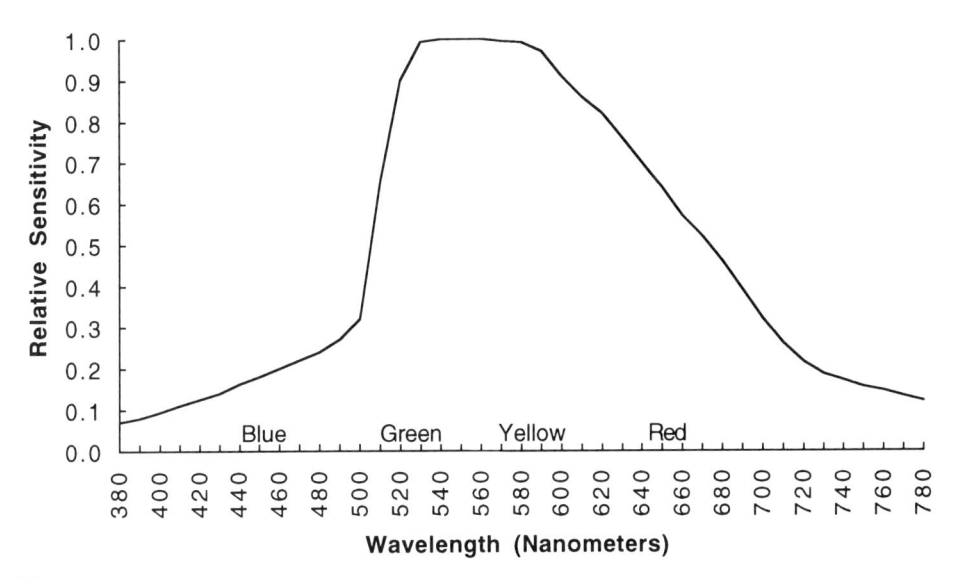

Figure 2.48. Typical photoconductive sensor spectral sensitivity curve. While it is peaked near the wavelength of maximum sensitivity of the average human eye (555 nm), it is not easily matched to the eye at other wavelengths. This limits its applications in photometry to those requiring only approximations of the light level.

Vacuum and Gas Photodiodes

The vacuum photodiode (Figure 2.49) is a photoemissive device; that is, electrons are emitted from a photocathode when it is struck by photons. These electrons are attracted to the electrically more positive anode where they are collected. The current through the anode is proportional to the light incident on the photocathode. A number of materials are photoemitters and may be evaporated onto a metal or glass substrate to form the photocathode. These materials are based on combinations of cesium, antimony, potassium, silver, oxygen, rubidium, sodium, telluride, and bismuth. Each combination has a particular spectral sensitivity (Figure 2.50). Operating voltages for phototubes range from about 15 V to 250 V. Advantages of the vacuum photodiode include good linearity, fast response time (nanoseconds), simple circuitry requiring only an anode power supply, and spectral response that may be easily made to approximate the CIE photopic curve by the addition of a Wratten type 106 filter. Disadvantages of the vacuum phototube include greater fragility, spectral response that may change with time and light exposure, and comparatively large physical size.[54]

The gas photodiode is of similar construction, except for the use of an inert fill gas such as argon instead of a vacuum. The electrons emitted by the photocathode collide with gas atoms on the way to the anode, with additional electrons being ejected by the collision as a result of ionization. This *gas amplification* or *multiplication* results in 5–10 times the number of starting electrons being collected at the anode depending on operating voltage. The trade-off for the sensitivity gain is response time limited to about 100 μs.[54] At one time widely used for motion picture soundtrack pickup, the gas photodiode is almost never used today.

72 LIGHT SOURCES, FILTERS, AND DETECTORS

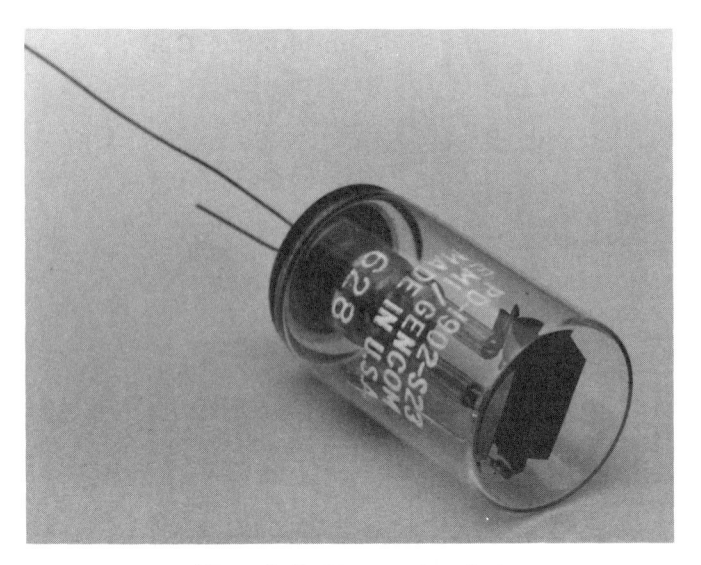

Figure 2.49. Vacuum photodiode.

Although obsolete for most applications, the vacuum photodiode is still used in highly refined configurations to provide extremely fast risetimes (picoseconds) or for "solar blind" ultraviolet sensors that have virtually no response to visible light. The vacuum photodiode is also the basis for the photomultiplier tube.

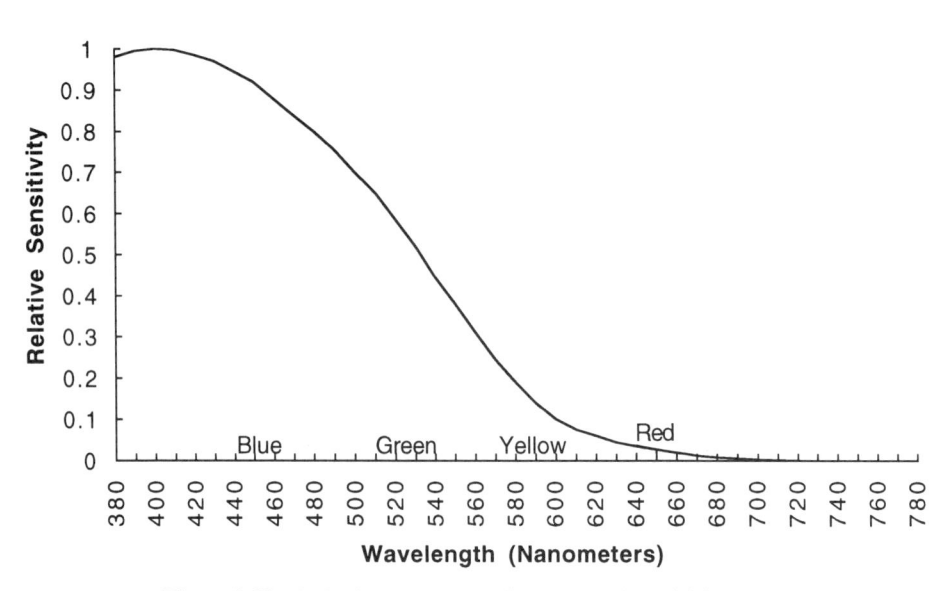

S-4 Photocathode

Figure 2.50. Typical vacuum photodiode spectral sensitivity curve.

Figure 2.51. Side-window PMT employing photocathode deposited on a metal substrate. This design dates from the RCA 931 introduced in 1940, which has been highly successful. Many PMTs of the same mechanical configuration and type number are still being manufactured today incorporating material and process improvements.

Photomultiplier Tubes

The photomultiplier tube (PMT) employs either an opaque photocathode on a metal substrate (Figure 2.51) or a semitransparent photocathode deposited on the end of the glass envelope (Figure 2.52). The photocathode materials are similar to those previously described for vacuum phototubes. The spectral sensitivity curves for common photocathode materials are defined by S designators by the Electronic Industries Association in a manner similar to that of the earlier P designations for phosphors.[55] Some examples of the most common photocathode spectral sensitivity curves are shown in Figure 2.53.

Between the photocathode and the anode is a series of *dynodes* (Figures 2.54–2.57), each coated with a material that emits several electrons for each electron that strikes it. This is called *secondary emission*. The dynodes are operated with each successive dynode at a higher positive voltage. The dynodes are configured such that the electron-optical fields cause the photoelectrons emitted by the photocathode to strike the first dynode, secondary electrons emitted by the first dynode to strike the second dynode, and so on, until the ever-increasing quantity of electrons reach the most positive element, the anode.

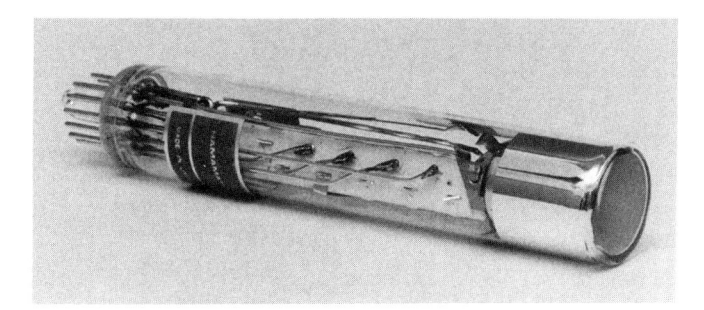

Figure 2.52. End-window PMT having a semitransparent photocathode deposited on its glass window.

Photocathodes

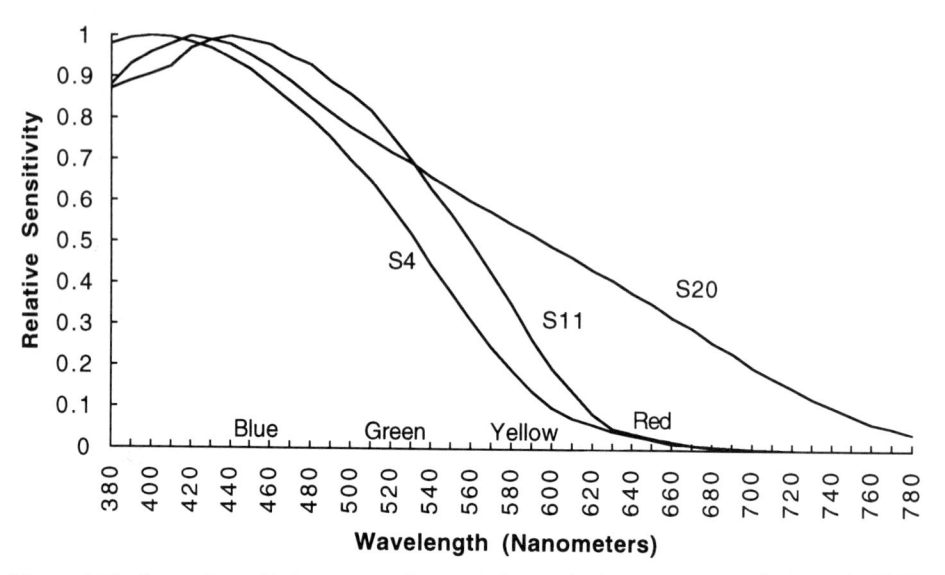

Figure 2.53. Spectral sensitivity curves of several photocathode types commonly found in PMTs. The S20 photocathode is also often called "trialkali" or "multialkali" due to its chemical composition of $Na_2KSb:Cs$.

0 = Opaque photocathode
1–9 = Dynode = electron multiplier
10 = Anode

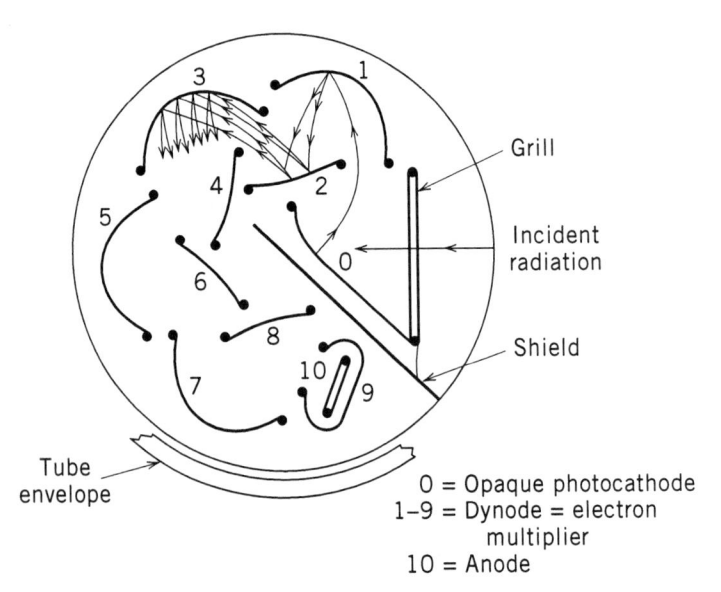

Figure 2.54. Side-window PMT electrode configuration. This arrangement is popularly known as a "squirrel cage" multiplier. (From Burle Industries, Inc., *Photomultiplier Handbook,* Publication TP136, Oct. 1989, p. 28. Reprinted by permission of Burle Technologies, a subsidiary of Burle Industries, Inc.)

2.3 LIGHT DETECTORS   75

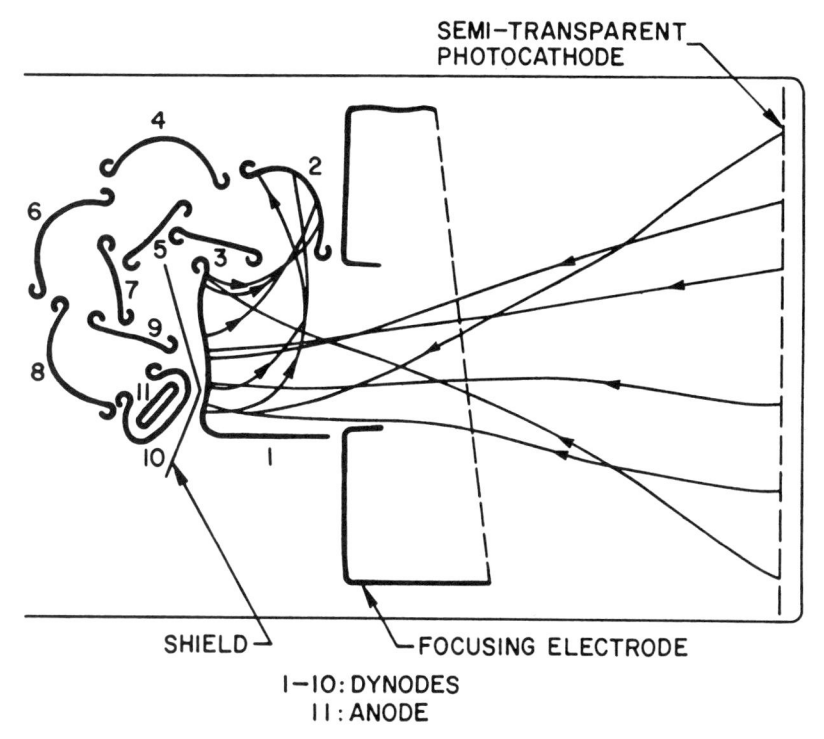

Figure 2.55. End-window PMT electrode configuration employing the same squirrel cage multiplier structure as the side-window type. (From Burle Industries, Inc., *Photomultiplier Handbook*, Publication TP136, Oct. 1989, p. 28. Reprinted by permission of Burle Technologies, a subsidiary of Burle Industries, Inc.)

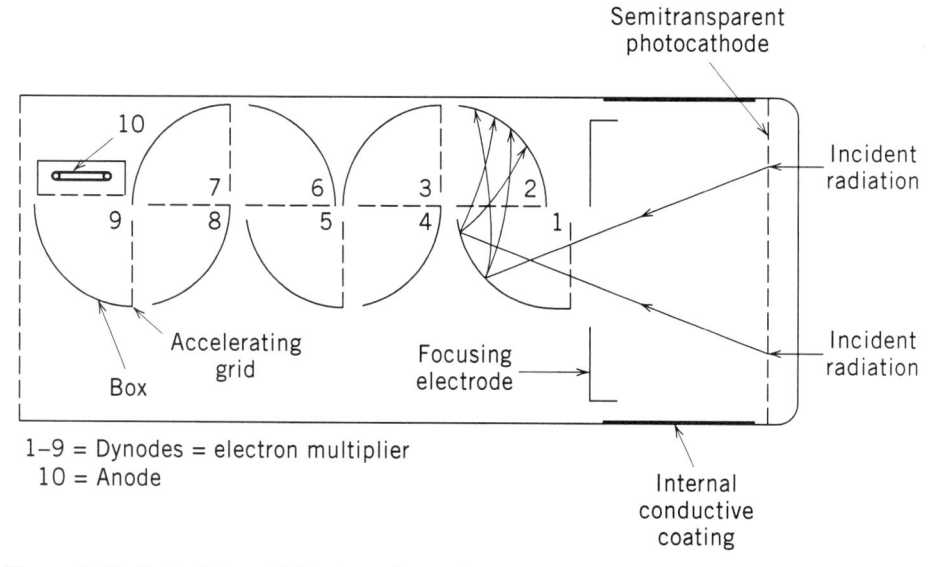

Figure 2.56. End-window PMT electrode configuration employing a "box-and-grid" multiplier structure. (From Burle Industries, Inc., *Photomultiplier Handbook*, Publication TP136, Oct. 1989, p. 28. Reprinted by permission of Burle Technologies, a subsidiary of Burle Industries, Inc.)

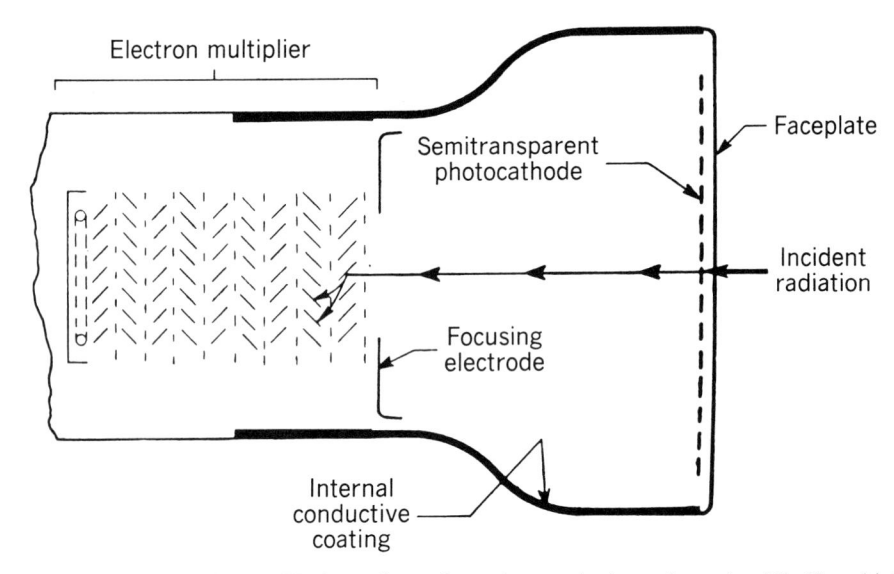

Figure 2.57. End-window PMT electrode configuration employing a "venetian blind" multiplier structure. (From Burle Industries, Inc., *Photomultiplier Handbook,* Publication TP136, Oct. 1989, p. 28. Reprinted by permission of Burle Technologies, a subsidiary of Burle Industries, Inc.)

Total amplification of from 100,000 to 10,000,000 times occurs, which makes it possible to measure low light levels that even the human eye cannot sense. Risetimes in the low-nanosecond realm make the PMT well suited for measurement of phosphor or display risetime and falltime and other temporal effects. As with the vacuum phototube, photopic correction is easily approximated with a Wratten type 106 filter.

The PMT is especially well suited for astronomical, spectrographic, and other research applications. It is also employed for photometric measurement of extremely small areas on displays such as single pixels where the total amount of light is too small to permit the use of simpler sensors. Disadvantages of the PMT include cost, fragility, need for a power supply of 800–1200 V, risk of permanent damage from exposure to ambient light with voltage applied, and changes in sensitivity and spectral response with time.[56] Frequent calibration checks using a stable light source can help overcome the latter changes.

Selenium Cells

For many years the selenium cell was used in most instrumentation for measurement of illumination, photographic exposure, and television luminance. It is still found in a few very low cost imported incident light meters. The selenium cell is a photovoltaic device; that is, it generates a voltage when exposed to light. In actual use for photometry, it is more correctly termed *photoamperic* operation since current is measured for best linearity rather than voltage. Selenium cells up to several square inches in area are practical and provide high sensitivity at far lower cost than an equivalent area of silicon. Approximation of photopic response may be accomplished by the addition of a Wratten type 102 filter.

On the down side, selenium cells exhibit "fatigue" with exposure to light and take a while after exposure to bright light before low light levels may be measured. Ruggedness, dynamic range, and long-term stability are all inferior to the silicon photodiode.[57]

Silicon Detectors

The silicon detector in one form or another is the most widely used light sensor today. Small size, rugged construction, low to moderate cost, good sensitivity, wide linear dynamic range, and excellent long-term stability are all in its favor. Silicon detectors rely on the same silicon technology and materials that have been highly refined for the semiconductor, integrated circuit, and solar cell industries.

The patent application for the basic silicon light sensor was filed by Russell Ohl of Bell Telephone Laboratories in 1941 and the patent was issued in 1946.[58] The germanium phototransistor, another Bell Labs first, was described in 1950.[59] During the development of the silicon transistor in the late 1950s and early 1960s it was found that it was an excellent light sensor due to the responsiveness of silicon to light combined with the gain of the transistor itself. The phototransistor is most suited to digital applications such as card readers rather than light measurement since its collector current is somewhat nonlinear with light level. The simpler silicon photodiode, first developed as a solar cell, overcame this problem and was soon used for most light measurement applications. The favorable sensitivity of the phototransistor could be met by increasing the surface area of the silicon since sensitivity is proportional to surface area.

In its simplest form (Figure 2-58), the silicon photodiode consists of a thin, high-purity, silicon wafer with electrical connections to the front and rear surfaces. Small amounts of impurities, or "dopants," are diffused into the surface that is intended to accept incoming radiation. The type of impurity, quantity, and diffusion depth are all controlled to obtain the desired electrical characteristics as well as spectral sensitivity. Electrically, the silicon photodiode has an equivalent circuit as represented in Figure 2.59.

Four modes of operation are used for the silicon photodiode. Each mode has its own particular advantages, and the silicon is usually processed in a manner designed to enhance the particular mode for which it is intended. Although they are optimized for a particular mode of operation, almost any silicon photodiode may be operated in any of the first three modes with varying degrees of performance compromises.

Photovoltaic The first and simplest mode is *photovoltaic* operation. In this mode, voltage is generated by light impinging on the silicon junction and is electrically coupled into an external load resistance. This is the basis for the operation of a solar cell. Photovoltaic operation results in approximately logarithmic output voltage with light level at higher load resistances. Solar cells are optimized for efficiency of power transfer to the load resistance. Low internal series resistance is an important characteristic for maximum power transfer. Efficiency is good due to its wide spectral response in the range of maximum solar output.

Photoamperic For light measurement applications, the *photoamperic* mode is used. Photodiodes intended for this mode are usually called photovoltaic by manufacturers to differentiate from photoconductive operation, but photoamperic more correctly describes the actual usage. The photoamperic mode is similar to the photovoltaic mode except that the load is essentially a short circuit through which the current is measured. Excellent linearity results in this configuration. The linear operating range may be as many as 8–10 decades for a high-quality photodiode.[60] Photodiodes optimized for this mode have both minimum internal series resistance for operation at high light levels and extremely high silicon bulk resistivity for high shunt resistance to aid detection at very low light levels.

78 LIGHT SOURCES, FILTERS, AND DETECTORS

Planar Diffused Silicon Photodiode

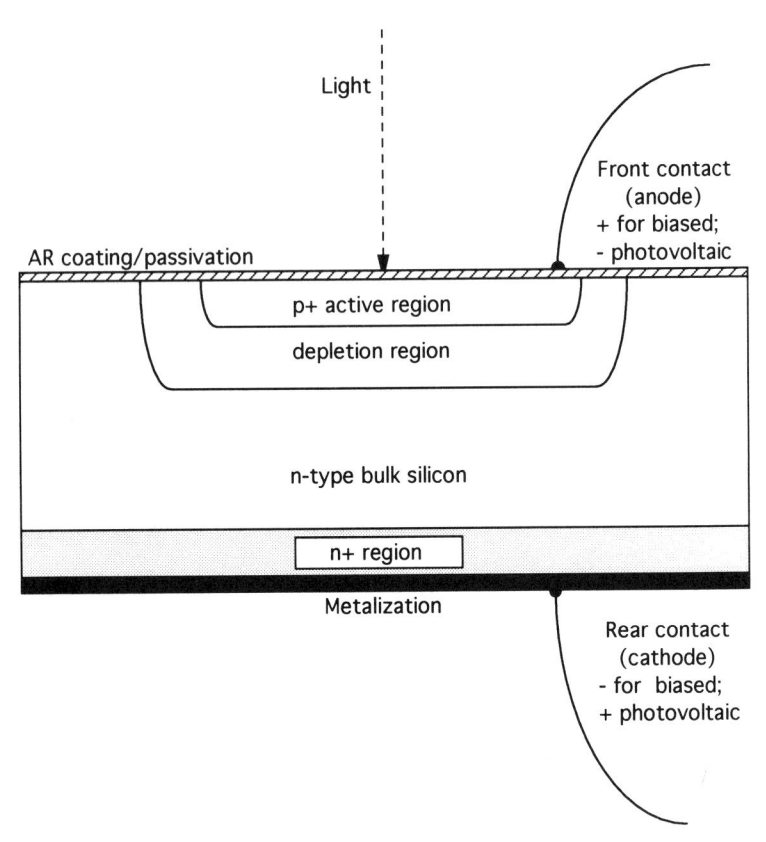

Figure 2.58. Silicon photodiode cross section.

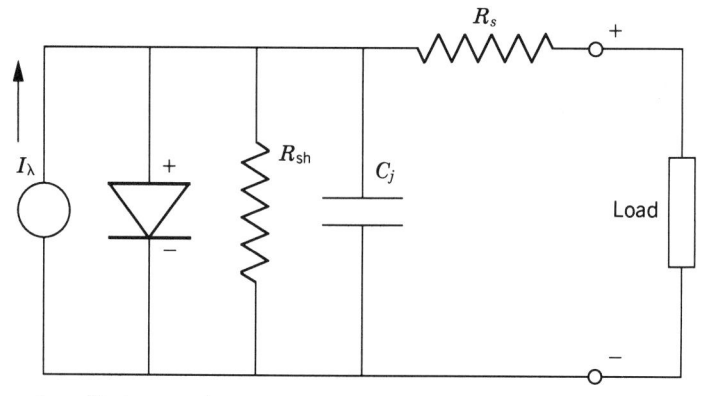

I_λ = Photocurrent
D = Ideal diode
R_{sh} = Shunt resistance due to bulk resistivity of silicon
C_j = Junction capacitance
R_s = Series resistance

Figure 2.59. Equivalent circuit of a silicon photodiode.

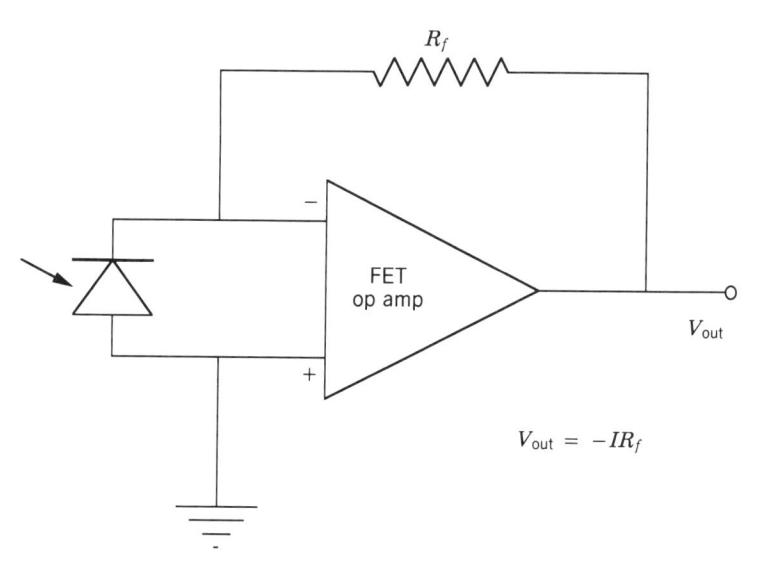

Figure 2.60. Transimpedance amplifier circuit for photoamperic operation of silicon photodiodes.

The latter is important, not from the standpoint of affecting the operation of the photodiode itself since it is operating with a near-zero ohm load anyway, but because lower shunt resistances cause greater offset errors due to minute bias currents from within the operational amplifier used to convert the photodiode current to a voltage. This configuration is termed a *transimpedance* amplifier. See Figure 2.60 for the basic circuit of a photoamperic photodiode and transimpedance amplifier. Risetimes of 1 μs or less are common for silicon photodiodes operated in the photoamperic mode excluding the time response characteristics of the transimpedance amplifier.

Photoconductive The third operational mode is the biased or photoconductive mode. In this mode, a reverse-bias voltage, usually 100 V or less, is applied in series with the photodiode (Figure 2.61). The photodiode is connected to a transimpedance amplifier as for photoamperic operation. The advantage of reverse-biased operation is that risetime is considerably reduced by the applied bias voltage, which increases the speed at which the photoelectrons produced by the photons of absorbed radiation are collected. A frequency response of 10–100 MHz is typical of reverse-biased silicon photodiodes. The trade-off for biased operation is a residual "dark current" that is present at all times and doubles for each 10°C increase in temperature.

Avalanche Mode The last operating mode for silicon photodiodes is the avalanche mode. The silicon avalanche photodiode (APD) is a device operated similarly to the reversed-biased photodiode. The difference is that the APD is operated with a high-voltage field in the depletion region of the silicon. Photoelectrons produced by photons are accelerated into this region where they collide with atoms within the field, thereby producing a larger quantity of secondary electrons, which in turn collide with other atoms, producing still more electrons. This is known as the avalanche effect and amplifies the original current considerably. The optimum bias voltage for highest signal-to-noise ratio varies as a

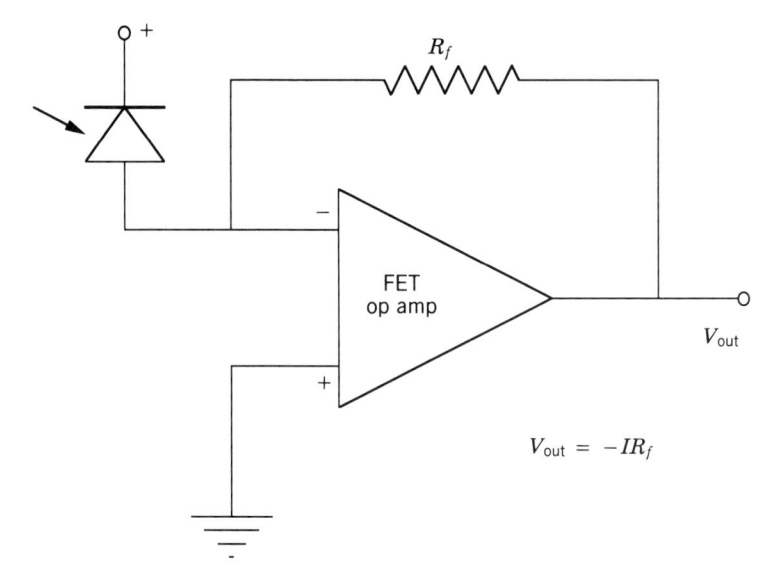

Figure 2.61. Reverse-bias and transimpedance amplifier circuit for photoconductive operation of silicon photodiodes.

$V_{out} = -IR_f$

function of temperature, and the power supply should be compensated to allow it to track the optimum bias voltage with temperature. Avalanche photodiodes are optimized for relatively high reverse-bias voltage. The APD is characterized by gains of 10–100 or more and high-speed operation that can exceed 1 GHz for small-area devices.

Silicon photodiodes are available in many sizes and package configurations, some of which are shown in Figure 2.62. They may be unmounted for custom packaging, mounted on plastic or ceramic substrates, or hermetically sealed in standard sized semiconductor "cans." Active sensing areas are usually between 1 and 100 mm^2. Special shapes and configurations are available, such as four-quadrant position sensors and strip sensors. Integrated detector/transimpedance amplifiers are commonly available that simplify design and packaging of products requiring light-sensing capability and to improve the electrical noise immunity of the connections between the photodiode and amplifier by minimizing their lead length. Others provide light intensity to frequency conversion to aid digital processing of the light intensity or to directly interface to a microprocessor.

Figure 2.62. Typical silicon photodiode packaging styles.

Blue Enhanced Silicon Photodiode

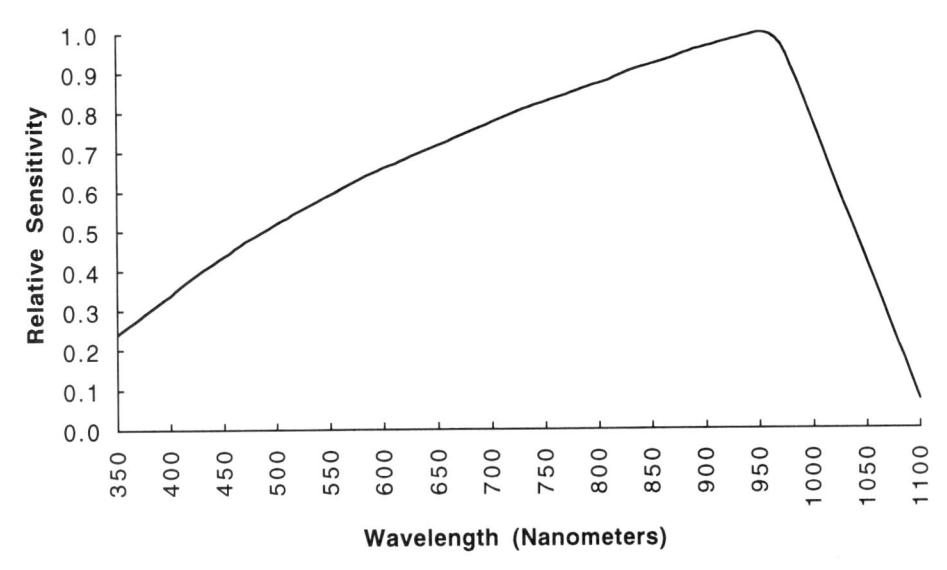

Figure 2.63. Silicon photodiode spectral sensitivity curve. The shape of the curve for particular types will vary somewhat with the processing used to optimize its electrical and optical characteristics for a given application.

The typical spectral response of silicon photodiodes is shown in Figure 2.63. Maximum sensitivity occurs in the near-infrared region around 900–1000 nm and cuts off sharply around 1100–1200 nm. It slopes downward throughout the entire visible spectrum and has very low response in the ultraviolet. Photodiode processing may be optimized to improve response in the short-wavelength region, and these sensors are termed "UV enhanced" or "blue enhanced" depending on their response curve. Even with enhancement, sensitivity is always considerably lower than at the infrared end of the spectrum. Short-wavelength absorption occurs near the surface while longer wavelengths are absorbed deep within the silicon wafer. Below the long-wavelength cutoff point, the silicon is near transparent. For extended spectral response, an additional sensor having infrared sensitivity may be mounted behind the silicon sensor and connected in parallel with it.

Glass or thin-film filters are sometimes built into the photodiode package to meet particular needs for spectral response other than that of the silicon itself. More often, the filters are added externally. This is especially true for photopic correction filters for photometry. Higher quality photopic filters consist of several colored glass filters laminated into a stack with the silicon photodiode for maximum long-term stability.

Charge-Coupled Devices There is a family of silicon detectors that is becoming increasingly important for photometry, radiometry, and especially colorimetry. These devices are based on an array of photodiodes and are known variously as the charge-coupled photodiode array (CCPD), charge-transfer device (CTD), or, simply and most commonly, charge-coupled device (CCD) depending on its form and manufacturer's preference. The photodiode array may consist of a single row, or *linear array,* of as few as 128 individual sensors to two-dimensional arrays, or *imagers,* containing as many as 2048 × 2048 elements. Figure 2.64 illustrates a typical linear CCD array in a standard DIP package configuration.

The CCD was originally demonstrated by Boyle and Smith of Bell Telephone Laboratories in the early 1970s and was intended as a computer memory device.[61] Its intended purpose has been long since discarded, but its use for scanning and imaging applications is expanding greatly. Medical imaging, television cameras, video photometers, spectroradiometers, astronomy, automated inspection, image scanners for computers, and barcode scanners are just a few of the many important uses for CCDs.

CCDs consist of an array of metal–oxide–silicon (MOS) capacitor elements on or just below the surface of a silicon strip or wafer. Charges generated within the silicon by photons absorbed by it are stored in the capacitors, which may be thought of as "buckets." These charges are proportional to the amount of light incident on each element and are accumulated in each bucket for a controlled period of time. To read out the series of accumulated charges, voltage pulses are applied sequentially to the top metal *gate* electrodes of the capacitors to cause each accumulated charge to shift to the next element by capacitive coupling between elements or *pixels.* This is called scanning or *charge transfer.* The last element in the row is an amplifier used to convert the charges to an output voltage. This voltage varies in amplitude with time as each bucket is transferred to it in a manner similar to a "bucket brigade." The individual elements are emptied during the process and readied to resume charge accumulation for the next scan. Most CCDs have the circuitry for the charge transfer process and output built into the CCD chip itself. At a minimum, external clock pulses, start pulses, and power are required for operation of these devices.[62,63]

By combining the linear photodiode array with a thin-film linear variable filter (LVF) having narrow-band spectral transmission that varies along its length (Figures 2.65 and 2.66), a relatively simple spectroradiometer can be made that will measure the amount of light at each wavelength. These data can be used to graph the spectral curve of the source being measured or further processed to obtain chromaticity coordinates for the source. Such a device is well suited for interface to a computer or microprocessor to further process the information.

Figure 2.64. Typical linear CCD array in a standard "DIP" package configuration.

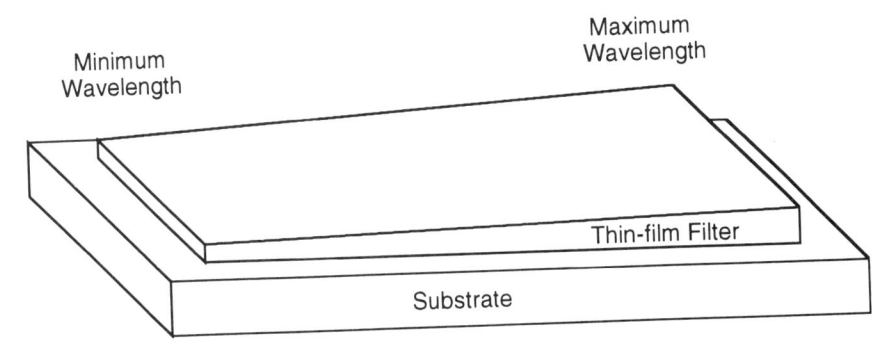

Figure 2.65. Linear variable-filter construction. The wedge-shaped thin-film interference filter passes different wavelengths at each point along its length.

The dynamic range of CCDs is less than that of silicon photodiodes, and the sensitivity of each pixel is also lower due to their small size. Charge-coupled devices may be easily cooled with thermoelectric coolers to extend their sensitivity to low light levels by reduction of dark current.

Other Detectors

There are many types of thermal, semiconductor, and photoconductive sensors intended for the infrared region. These are outside the scope of this text since they have little application to displays. There are several excellent texts on the subject for anyone interested in infrared detectors or detectors in general.[64–66]

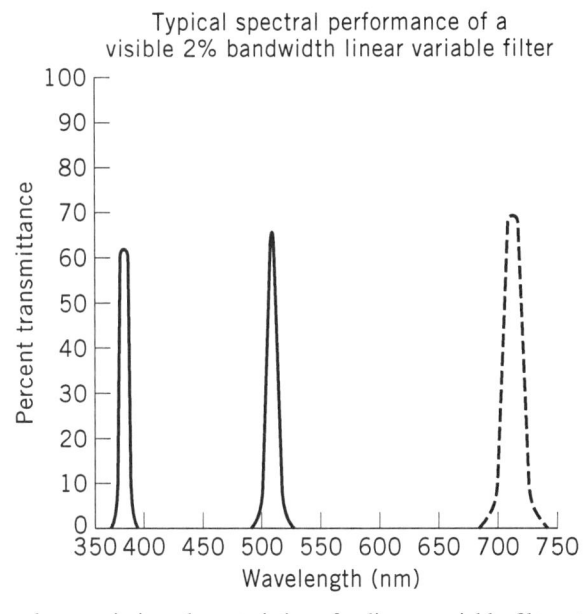

Figure 2.66. Spectral transmission characteristics of a linear variable filter at selected positions along its length. (From Optical Coating Laboratory, Inc.)

REFERENCES

1. L. Levi, "Blackbody Temperature for Threshold Visibility," *Appl. Opt.*, Vol. 13, No. 2, p. 221, 1974.
2. Optical Society of America, *Handbook of Optics*, McGraw-Hill, New York, 1978, pp. 1–15.
3. J. R. Meyer/Arendt, *Introduction to Classical and Modern Optics*, Prentice-Hall, Englewood Cliffs, NJ, 1972, pp. 458–459.
4. J. R. Meyer/Arendt, *Introduction to Classical and Modern Optics*, Prentice-Hall, Englewood Cliffs, NJ, 1972, pp. 469–470.
5. J. R. Meyer/Arendt, *Introduction to Classical and Modern Optics*, Prentice-Hall, Englewood Cliffs, NJ, 1972, p. 460.
6. H. Brandhorst et al., "Interim Solar Cell Testing Procedures for Terrestial Applications," NASA Lewis Research Center, Report TM-X71771, 1975.
7. M. P. Thekaekara, "Solar Energy Outside the Earth's Atmosphere," *Solar Energy*, Vol. 14, p. 109, 1973.
8. Commission Internationale de l'Eclairage (CIE), *Colorimetry*, 2nd ed., CIE Publication, No. 15.2, Vienna, 1986.
9. J. Audouze and G. Israel, *Cambridge Atlas of Astronomy*, 3rd ed., Cambridge Univ. Press, Cambridge, UK, 1994, p. 28.
10. S. L. Valley, *Handbook of Geophysics and Space Environments*, McGraw-Hill, New York, 1965, p. 16-2.
11. RCA, *Electro-Optics Handbook*, Harrison, NJ, 1974, pp. 61–62.
12. RCA, *Electro-Optics Handbook*, Harrison, NJ, 1974, p. 68.
13. General Electric Co., *Incandescent Lamps*, GE, Cleveland, 1977.
14. General Electric Co., "Miniature and Subminiature Lamp Catalog," Publication No. 208-9165R.
15. Sylvania, "Tungsten Halogen Miniature Lamps," Catalog No. 308A, undated.
16. J. A. Cox, *A Century of Light*, The Benjamin Co., New York, 1979, p. 101.
17. Illuminating Engineering Society of North America (IESNA), *Lighting Handbook*, 8th ed., New York, 1993, p. 197.
18. Illuminating Engineering Society of North America (IESNA), *Lighting Handbook*, 8th ed., New York, 1993, p. 205.
19. Illuminating Engineering Society of North America (IESNA), *Lighting Handbook*, 8th ed., New York, 1993, p. 208.
20. E. C. Payne, E. L. Mager, and C. W. Jerome, "Electroluminescence: A New Method of Producing Light," *Illumin. Eng.*, pp. 688–693, Nov. 1950.
21. R. S. Rakowski, "Inverter Matrix Simplifies Matching Power Sources: EL Lamps," *Inform. Display*, pp. 14–15, Oct. 1985.
22. Sharp Optoelectronics, "Developers Continue to Refine Blue LED Technologies for Display Use," Opto Application Note, pp. 4-56, undated.
23. D. Maliniak, "LEDs Quit Singing the Blues," *Electron. Design*, Nov. 21, 1994.
24. Y. Hara, "Bright Blue LED Could Enable Color Displays," *Electron. Eng. Times*, p. 4, Mar. 21, 1994.
25. Electronic Industries Association, "Optical Characteristics of Cathode-Ray Tube Screens," EIA TEPAC Publication No. TEP116-C, Feb. 1993.
26. G. Wyszecki and W. S. Stiles, *Color Science*, Wiley, New York, 1982, p. 236.
27. Electronic Industries Association, "Optical Characteristics of Cathode-Ray Tubes," EIA Publication No. TEP-116B, Nov. 1987.

REFERENCES 85

28. P. A. Keller, "Phosphor Decay vs. Wavelength," Tektronix internal technical bulletin, July 26, 1967.
29. P. A. Keller, *The Cathode-Ray Tube,* Palisades, New York, 1991, p. 21.
30. Commission Internationale de l'Eclairage (CIE)/International Electrotechnical Comission (IEC), *International Lighting Vocabulary,* CIE Publication 17.4/IEC, Publication 50, Geneva, 1987.
31. Illuminating Engineering Society of North America (IESNA), *Lighting Handbook,* 8th ed., IESNA, New York, 1993, pp. 909–941.
32. *Photonics Dictionary,* 40th ed., Laurin, Pittsfield, MA, 1994.
33. W. G. Driscoll, *Handbook of Optics,* McGraw-Hill, New York, 1978, Chap. 8.
34. W. E. R. Davies and G. Wysecki, "Physical Approximation of Color-Mixture Functions," *J. Opt. Soc. Am.,* Vol. 52, No. 6, pp. 679–685, June 1962.
35. H. Wright, C. L. Sanders, and D. Gignac, "Design of Glass Filter Combinations for Photometers," *Appl. Opt.,* Vol. 8, No. 12, pp. 2449–2455, Dec. 1969.
36. G. Czibula, "Producing a Detector with Predetermined Spectral Responsivity," IMEKO 10th Intl. Symp. Tech. Comm. on Photon Detectors, pp. 189–199, Sept. 1982.
37. J. P. Ritzel and S. Sojourner, "Silicon Photodiodes Matched to the CIE Photometric Curve Using Color Filter Glass," *Optics and Photonics News,* pp. 16–19, April 1993.
38. K. Muray, "Photometry with Filtered Detectors," Proc. of Measurement Science Conference, Anaheim, CA, 1992.
39. B. Merik and V. H. Gittery, "A New Detection System for Automotive Headlamp Photometry," *J. IES,* pp. 77–82, Oct. 1973.
40. LMT, "Mosaic Filters, What They Are and How They Work," Application Note AN2-1287, undated.
41. J. D. Rancourt, *Optical Thin Films,* McGraw-Hill, New York, 1987.
42. OCLI, *Interference Filter Handbook,* Optical Coating Laboratory, Santa Rosa, CA, 1991.
43. W. G. Driscoll, *Handbook of Optics,* McGraw-Hill, New York, 1978, Chap. 8.
44. Hewlett Packard, "Contrast Enhancement Techniques for LED Displays," Application Note 1015, Nov. 1982.
45. R. C. Jones, "Use of Circularly Polarizing Filters to Increase the Contrast of Kinescope Displays," Polaroid Report No. 1582-6/65, June 1978.
46. A. Stimson, *Photometry and Radiometry for Engineers,* Wiley, New York, 1974, pp. 48–49.
47. F. Grum and R. J. Becherer, *Optical Radiation Measurements,* Vol. 1: *Radiometry,* Academic, New York, 1979, pp. 204–206.
48. R. W. Boyd, *Radiometry and the Detection of Optical Radiation,* Wiley, 1983, pp. 106–112 and 211–216.
49. "Thermopiles," Eppley Laboratories, undated.
50. R. McCluney, *Introduction to Radiometry and Photometry,* Artech House, Boston, 1994, pp. 188–190.
51. A. Stimson, *Photometry and Radiometry for Engineers,* Wiley, New York, 1974, pp. 19–27.
52. *Optoelectronics Designers Handbook,* Clairex Electronics, Mount Vernon, NY, 1988.
53. *Optoelectronics Data Book: Photoconductive Cells,* EG&G Vactec, St. Louis, MO, 1990.
54. "Phototubes and Photocells," RCA Technical Manual PT-60, Lancaster, PA, Oct. 1963.
55. Electronic Industries Association, "Relative Spectral Responsc Data for Photosensitive Devices ("S" Curves)," EIA JEDEC Publication No. 50, Oct. 1964.
56. *Photomultiplier Handbook,* Publication TP136, Burle Industries, Lancaster, PA, Oct. 1989.

57. *Solar Cell and Photocell Handbook,* 8th ed., International Rectifier Corp., El Segundo, CA, 1966.
58. R. S. Ohl, "Light-Sensitive Electric Device," U.S. Patent No. 2,402,662, filed May 27, 1941, issued June 25, 1946.
59. J. N. Shive, "The Phototransistor," *Bell Lab. Record,* Vol. 28, No. 8, pp. 337–342 1950.
60. G. Eppeldauer, "High Sensitivity Absolute Radiometer," *IMEKO 10th Intl. Symp. Tech. Comm. Photon-Detectors,* Vol. 1, pp. 145–156, Sept. 1982.
61. G. F. Amelio, M. F. Tompsett, and G. E. Smith, "Experimental Verification of the Charge Coupled Diode Concept," *Bell System Tech. J.,* Vol. 49, p. 593, 1975.
62. J. Janesick and M. Blouke, "Sky on a Chip," *Sky and Telescope,* pp. 238–242, Sept. 1987.
63. Scientific Imaging Technologies, Inc., "An Introduction to Scientific Imaging Charge-Coupled Devices," Application Note, July 15, 1994.
64. R. W. Boyd, *Radiometry and the Detection of Optical Radiation,* Wiley, New York, 1983.
65. R. McCluney, *Introduction to Radiometry and Photometry,* Boston, Artech House, 1994.
66. F. Grum and R. J. Becherer, *Optical Radiation Measurements,* Vol. 1: *Radiometry,* Academic, New York, 1979.

CHAPTER 3

Displays

Some of today's many competing display classes include the CRT, liquid crystal, EL, plasma, vacuum fluorescent, and LED display. Each class has further and often competing distinctions, thus creating subclassifications. Displays are also divided into two primary categories: emissive and transmissive. The CRT, EL, plasma, and LED displays are all examples of emissive displays: They all produce light internally. The LCD is the principal example of a transmissive display: It selectively absorbs light passing through it in order to display information. The backlit LCD, while still considered a transmissive display, can appear similar in appearance to some emissive displays.

This chapter will describe the principal display types, their basic operation, and pros and cons of each type as they pertain to display performance and measurement. No attempt will be made to discuss the construction variations of each type as this is primarily of concern to the device manufacturer. What counts in the end is how well the device performs for a given application.

3.1 DISPLAY FORMATS AND CHARACTERISTICS

Format

Three display formats are available to match the format of images or data to be displayed. These are page or portrait mode, landscape mode, and square format.

The first consists of the display mounted with the long axis of the screen in a vertical direction. This is known as page mode to the electronic publishing market due to its usefulness for displaying a complete page. The same configuration is usually referred to as portrait mode for medical imaging applications and is especially useful for chest X-ray presentations. For CRT displays, the raster scan lines may be in either the vertical or horizontal direction depending on performance requirements. Aligning them vertically can give about a 10% gain in performance since there will be fewer scanning lines in that orientation. Fewer lines means less time lost during retrace of each line, which in turn reduces the video bandwidth needed to display the same resolution. The scan orientation needs to be understood mutually between the manufacturer and user since digital image data needs to be translated for the appropriate orientation. Also, the origin of the scan must be specified. Top-right origin is the most common where the scan lines are aligned vertically.

Landscape mode is the conventional television scan format. In this mode, two complete pages of text may be presented side by side or multiple medical images may be displayed. This is a relatively straightforward display from the specification standpoint. The origin of the raster scan is always in the upper left corner of the screen.

Square-format displays may use special square-screen CRTs. They are specialized displays and not yet very common. They find application in the display of square-format imaging sensors such as 1024 × 1024 and 2048 × 2048 pixel CCD arrays. Either portrait or landscape CRT displays may be used for square format since a monitor can easily be adjusted to display a square raster. Some screen area will be wasted, however. The raster scan origin for square displays is in the upper left, as with the landscape format.

Aspect Ratio

Aspect ratio refers to the ratio of width to height of the raster. A ratio of 4:3 is most common due to its long history of television use and the fact that most CRT glassware is in that configuration. The ratio 5:4, which is close to the aspect ratio of X-ray and some photographic film and paper, is also used and is a fairly good fit on standard 4:3 ratio CRTs. The ratio 1:1 may also be used as described for square-format displays. Displays with 16:9 aspect ratios for high-definition television (HDTV) are now becoming available, especially in Japan. It remains to be seen what impact HDTV displays will have on computer and imaging displays.

Gray Scale

Medical imaging in particular requires a wide range of light intensities, or "gray scale," to be accurately reproducible with as many discernable levels as possible. "Halftones" is another term for gray scale. Gray scale is frequently expressed exponentially in terms of binary "bits." An 8-bit gray scale is 2^8, or 256, steps. Some displays used for early computer displays were 1-bit displays and were either black or white with no intermediate steps. As with newspaper printing, it is possible to reproduce halftones by using many small dots of different sizes or spacings.

Higher resolution CRT displays require wider bandwidth video amplifiers to display fine image detail. This works counter to gray scale since amplifier noise tends to increase with wider bandwidths. For a 2048 × 2048 pixel display, the required bandwidth is about 200 MHz, with a resultant noise level nearly equal to 1 bit out of 256, thus limiting gray scale to 8 bits.

Gamma

It is a characteristic of all CRTs for the intensity of light from the screen to respond nonlinearly to a linear increase in video drive voltage. The effect is analogous to "gamma" in photography and bears the same name. Gamma is an exponent of the voltage input required to produce a given light output. A gamma of 1 results in a 1-to-1 relationship between the input and output. A gamma of 2 gives a 10-to-1 relationship. A typical value is about 2.3. Gamma may be corrected in the software for driving a digital display through use of a "look-up table" that converts the exponential output to a linear output.

Display Size

Directly viewed displays are available in a wide range of sizes from less than 25 to 1100 mm (1–43 in.). For CRTs, this dimension often refers to the overall CRT glass faceplate diagonal dimension from corner to corner, and the usable screen area may be somewhat less. This is often further reduced as a CRT may be "underscanned," particularly for computer displays. Underscanning is desirable for two reasons. The first is to avoid possible loss of information at the extreme edges of the image, and the other is to avoid the image distortion and defocusing that increases rapidly near the edges and especially the corners of the screens. Curvature of CRT faceplates for strength against implosion also makes it somewhat difficult to specify their size. The increase in usage of flat panel displays which are specified by their actual display dimensions has led to similar specifications being adopted for CRT computer displays.

Projection displays, CRT or otherwise, on the other hand, are more truthfully designated by their actual viewable dimensions. These may range up to theater sized and are used for simulators, entertainment, and systems monitoring.

Resolution Versus Addressability

The difference between resolution and addressability is one of the most frequently misunderstood parameters for information display specifications. Monochrome flat-panel displays are made up of discrete elements in a geometric pattern, and each may be addressed individually or in a matrix by electrical signals. Each element constitutes one *pixel* or picture element. For color flat panels, one pixel consists of at least three elements termed *subpixels,* one each for red, green, and blue. Addressability for flat-panel displays is defined by the number of complete pixels in each axis of the matrix; that is, 640 × 480. Resolution does not apply in the case of flat-panel displays.

Cathode-ray tubes are a more complex matter. Here, both addressability and resolution apply. Resolution of CRT displays is primarily a function of CRT spot size combined with any spreading in the fast scan axis caused by video bandwidth limitations and defines the smallest individual area of the CRT that may be individually illuminated. Commonly, this is expressed as spot size in millimeters or in thousandths of an inch (mils) at the half-intensity points, although the human eye usually can detect to much lower levels and the apparent spot size is correspondingly greater than the half-intensity values indicate. The light output from the spot is usually a near-Gaussian distribution as in Figure 3.1. Dividing the raster dimensions by the effective spot size in each axis gives the number of *resolvable* pixels.

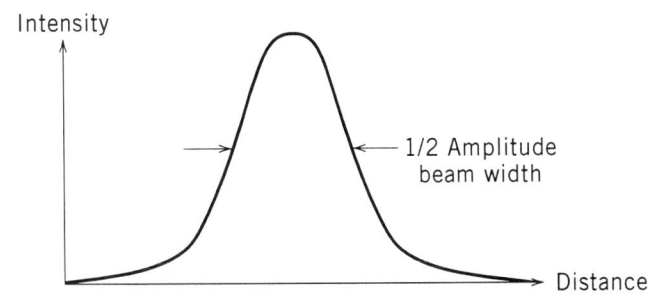

Figure 3.1. Gaussian profile of a CRT beam or spot.

Cathode-ray tube addressability is determined by the electronics used to drive the CRT and defines how many points in each axis may be individually controlled as distinct pixels. In one axis, usually vertical, it is defined by the number of active scan lines of the raster. Addressability in the other axis is determined by the bandwidth of the video amplifier.

Ideally, the addressability and resolution should be fairly closely matched to avoid wasting the capabilities of either the CRT or the electronics. Both resolution and addressability may be expressed in pixels, such as a 1280 × 1024 pixel display, although addressability is the more commonly used specification, at least for digital displays. It is important to verify that a display described thusly is indeed 1280 × 1024 resolvable pixels since it is of little use to be able to address pixels that cannot be resolved. There are two cases where lower resolution than addressability can be of some benefit. The first is to reduce "Moire" patterns in shadow-mask color displays where the mask pattern and a displayed pattern "beat" together to produce a closely spaced series of dark and light lines that are not present in the original image. The other case is for diagonal lines that produce a staircase effect as the lines cross the horizontal scanning lines. This effect is called "jaggies."

3.2 CATHODE-RAY TUBE DISPLAYS

A brief discussion of CRT fundamentals is included to provide an understanding of the single most important component of the CRT display, which largely determines the ultimate performance attainable from a CRT display. In these days of solid-state electronics, electron tube fundamentals are usually overlooked in electronics courses. For that reason, a discussion of the basics is in order at this point. As a side note, it has become all too common for uninformed writers to refer to a complete CRT display monitor as a "CRT." Correct usage of the term CRT means the cathode-ray tube by itself.

Cathode-Ray Tube Fundamentals

The heart of a CRT display and its single most costly component, the CRT, is crucial to the overall performance of the display. Brightness, resolution, color, contrast, viewer comfort, life, and cost are all strongly influenced by the selection of a CRT by the display designer. These factors are all the more important for displays used for medical diagnostics where patient safety and comfort hinge on the ability of the display to present easily readable, high-resolution images accurately and rapidly.

The CRT in its present form dates back to 1897. In that year, Karl Ferdinand Braun, a German physicist and Nobel Prize recipient, demonstrated a tube intended for the measurement of electrical waveforms. It was not until 1929 that the CRT was applied to the display of actual images for television by Vladimir K. Zworykin of Westinghouse Electric. Subsequently, television and radar provided much of the impetus for further refinements of the CRT, with imaging and computer display applications finally emerging rapidly during the 1980s with increased performance demands placed on the CRT and associated electronics. Today, CRT displays are used in almost every business, and much of society is dependent on them for low-cost entertainment.[1]

For the reader to better understand the capabilities and limitations of the present-day CRT, we will first discuss the principles and operation. The CRT may be divided into six functional groups (Figure 3.2):

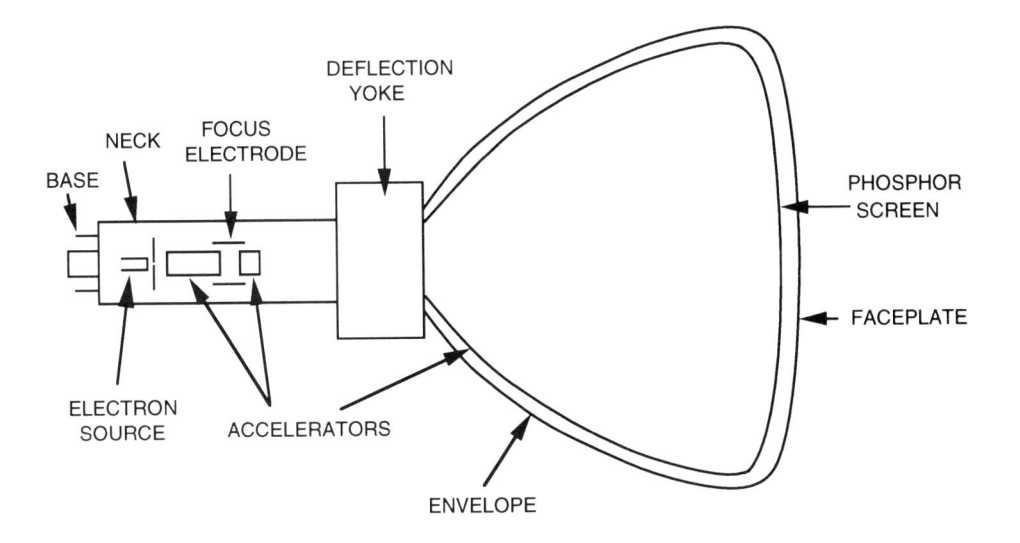

CRT FUNCTIONAL PARTS

Figure 3.2. Basic functional parts of a CRT.

1. a mechanical structure known as the envelope to maintain a vacuum, provide electrical connections to the internal electrodes, and insulate them from each other;
2. the electron source and beam intensity control section;
3. one or more acceleration electrodes to increase the velocity of the electron beam for increased light output from the screen;
4. the focusing section to bring the electron beam to a sharp focus at the screen;
5. a deflection system to position the beam to a desired location on the screen or scan the beam in a repetitive pattern; and
6. a phosphor screen to convert the invisible electron beam to visible light.

The assembly of electrodes or elements mounted within the neck of the CRT is commonly known as the "electron gun" (Figure 3.3). The term electron gun is a good analogy since it is the function of the electron gun to "shoot" a beam of electrons toward the screen or target. The velocity of the electron beam is a function of the overall accelerating voltage applied to the tube. For a CRT operating at an accelerating voltage of 20,000 V the electron velocity at the screen is about 400,000,000 kmph, or about 37% of the velocity of light. While the velocity of the electrons is extremely high, the mass is very small and normally the result is to cause the phosphor screen to luminesce where struck by the beam. However, if the beam power (accelerating voltage times beam current) is sufficiently high, intense localized heating may occur with a resultant phosphor burn or even glass damage.

Two or more electron guns may be combined into a single assembly. The use of three guns (Figure 3.4) is the most widely used means of producing color displays with one gun for each of the primary colors: red, green, and blue.

Each of the above functions will now be discussed briefly.

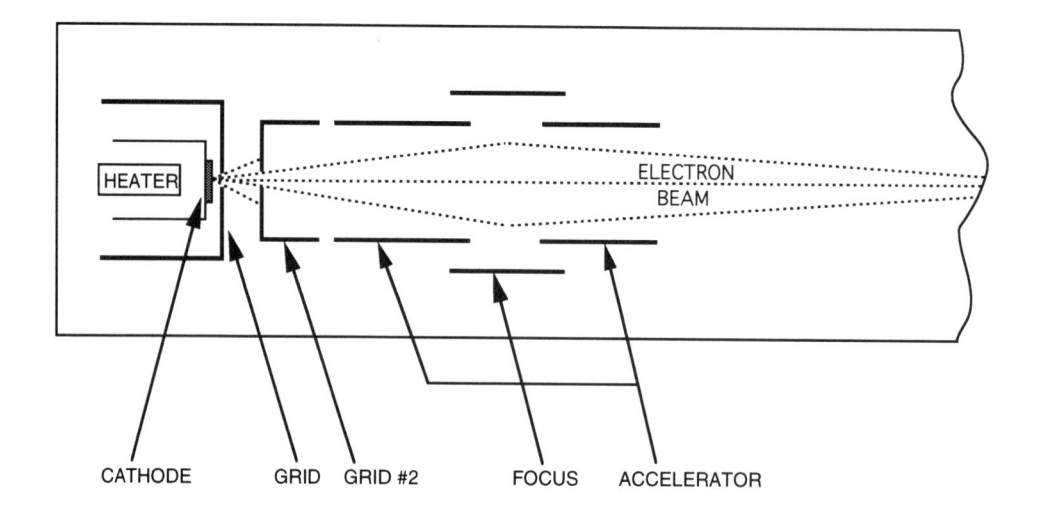

CRT ELECTRON GUN

Figure 3.3. Elements of a CRT electron gun.

Physical Structure

The glass envelope or bulb serves several purposes. These include maintaining the very high vacuum required to allow free movement of the electron beam without colliding with residual gas atoms, providing electrical connection to the electron beam forming and accelerating electrodes, and insulating these voltages, which may be as high as 30,000 V or more from one another. The shape of the envelope required to meet the CRT performance objectives largely dictates the physical configuration of the entire display. The envelope is divided into three distinct parts (Figure 3.5): the neck, which contains the electron gun and the base for making electrical connections to the gun; the funnel, which literally contains nothing except an electrically conductive coating with high voltage applied to it to accelerate the electron beam; and the faceplate, or "panel," which contains the luminescent phosphor screen.

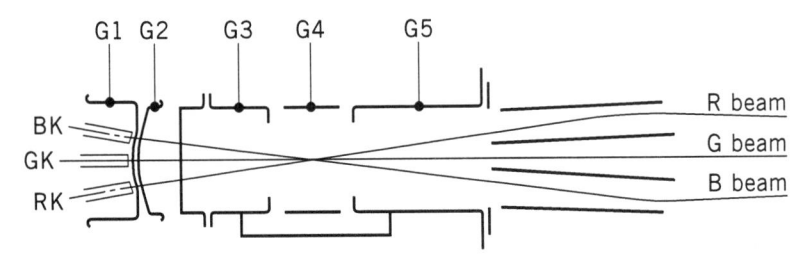

Figure 3.4. Elements of a three-beam color Trinitron electron gun. Note the commonality of a number of the elements to all three beams. (From S. Yoshida, A. Ohkoshi, and S. Miyoaka, "A Wide-Deflection Angle (114°) Trinitron Color-Picture Tube," IEEE BTR, 19 (4), p. 235, Nov. 1973. © 1973 IEEE, reprinted by permission.)

3.2 CATHODE-RAY TUBE DISPLAYS

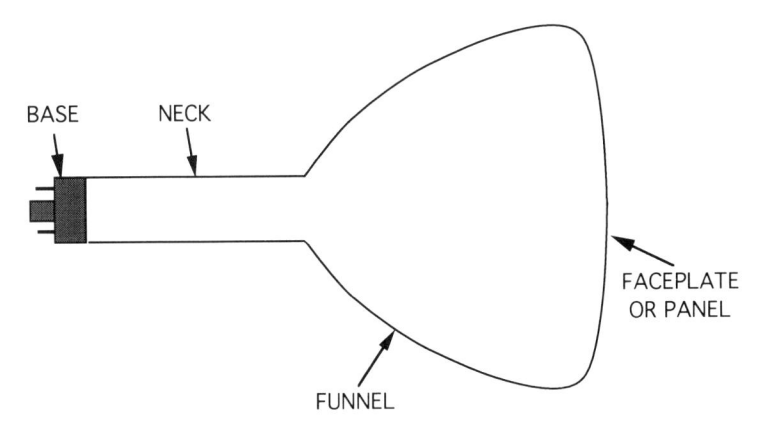

Figure 3.5. Components of the CRT envelope.

Glass has been the customary material for CRT bulbs, with the only exceptions being the use of metal or ceramic for the funnel for specialized applications. Glass has the advantages of transparency; good insulating properties; strength under compression resulting from the high vacuum; an easily cleaned, smooth, continuous surface to maintain the high vacuum integrity; and relative ease of sealing the three separate pieces together.

Electron Source

The source of the electrons to form a beam is the cathode (Figure 3.6). The cathode consists of a small metal cup containing a filament or heater to raise the temperature of an oxide cathode coating on the end of the cup to a sufficient temperature for the emission of electrons, usually around 800°C. Electrons, which are negatively charged, are drawn away from the cathode and toward the screen by positive voltages on other electrodes.

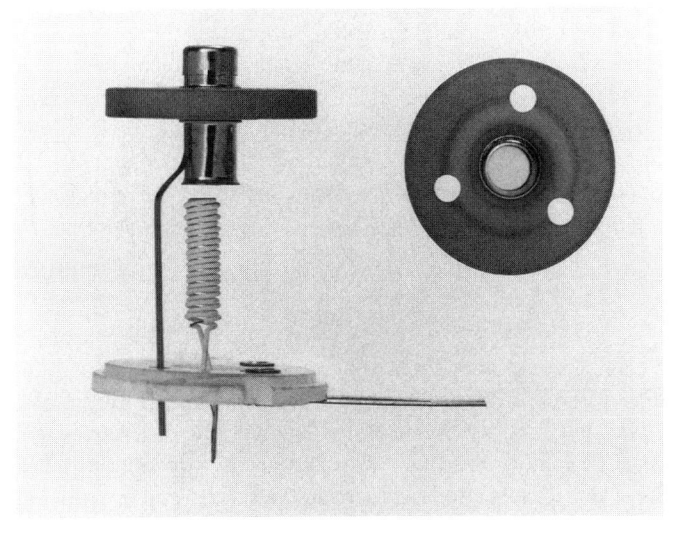

Figure 3.6. Photograph of a CRT and heater assembly. (From Tektronix, Inc. All rights reserved. Reprinted with permission.)

A second cup that surrounds the cathode cup is the control grid. The grid has a small aperture in front of the cathode coating that normally permits the electrons to flow through on their way to the screen if it is at the same voltage as the cathode. By applying a negative voltage to the grid, the number of electrons is decreased until at a sufficiently high voltage, usually around −60 V, a point is reached known as "cutoff" where all electrons are repelled back toward the cathode. By controlling the relationship of the grid to cathode voltages in synchronism to the scanning of the electron beam across the screen, the intensity of any point on the screen may be controlled to create a picture with many shades of gray, or gray scale. The number of gray levels that may be produced are primarily determined by the associated electronics used to drive the grid or cathode of the CRT. In many cases, the grid voltage is held at a fixed potential and the cathode driven positive instead. This is known as cathode drive but the effect is the same. It is the relation of grid to cathode voltage that causes intensity modulation.

Acceleration

To produce a bright display, the electron beam must be accelerated to a high velocity before impinging on the phosphor screen. This is accomplished by a two-step process.

Another cup-shaped electrode with a central aperture known as grid 2 or the screen grid immediately follows the control grid. This electrode has a positive voltage of several hundred volts and forms a controlled field to pull electrons from the cathode and provide initial acceleration prior to focusing the beam.

The final positive accelerating voltage of typically 15,000–25,000 V is applied to the entire funnel and screen as well as a cylindrical electrode at the end of the electron gun. The entire region is called the anode or *ultor*. Contact is made through a metal snap connector sealed into the glass funnel, referred to as the "anode button." The interior of the funnel is coated with a conductive carbon paint called *aquadag* or just "*dag*" to form the major portion of the anode and to connect with the screen, which usually has a conductive aluminum backing, as well as extend into the neck to make contact with the end of the gun through "snubber springs." Resistive aquadag may be used to provide current limiting to prevent damage if an internal high-voltage arc occurs.

The voltage applied to the anode is roughly related to screen size. For a given beam current (number of electrons), a tube having a larger screen will spread the available electrons "thinner" by scanning the larger area. Consequently, the brightness of the image will be lower for a large screen than for a small one. This brightness may be gained back by the use of a higher accelerating voltage that increases the beam power to the screen. The rule of thumb has been to use approximately 1 kV per inch of screen size.

Focusing

Focusing of the electron beam to a sharp spot at the screen may be done through use of either an electrostatic field or a magnetic field. Both methods form an electron optical "lens" capable of reconverging the beam, which is diverging as it leaves the cathode and goes to the first-anode region. Electrostatic lenses are most common today and usually consist of a cylindrical focus electrode inserted in the space between grid 2 and the anode cylinder. An adjustable voltage that is intermediate between those of grid 2 and the anode allows the focus to be adjusted. A number of other focusing methods are in use today.

These differ in the configuration of the electrodes in the focusing region and the magnitude of the focus voltage required. Magnetic focus coils have also been used around the same region of the CRT neck for magnetic focusing. Better focus may be attained this way but at a cost and weight penalty.

Deflection

So far, we have seen how the CRT produces a pencil-like beam of electrons. Without some means to position the beam to desired locations on the screen or scanning it past all areas of the screen, there would merely be a bright spot of light in the center of the screen. Either electrostatic or magnetic fields may be used to deflect the electron beam toward other locations on the screen. Today, all television, computer, and imaging displays use magnetic deflection since it produces bright, sharply focused displays over large screen areas with a minimum tube length. Electrostatic deflection is confined to specialized applications such as oscillography. Even these are being replaced by raster scanned magnetically deflected displays for digital storage oscilloscopes (DSOs) and LCDs for portable oscilloscopes.

Magnetic deflection consists of two sets of deflection coils, together called a deflection yoke, over the CRT neck at its juncture with the funnel. Each set of coils is mounted at right angles to the other to produce vertical and horizontal deflection when energized. By applying a fast sawtooth current (Figure 3.7) to the horizontal coil, the spot is scanned rapidly across the screen to form a thin horizontal line repetitively. The horizontal sweep rate may be anywhere between 15,750 Hz for television to 150,000 Hz for ultra-high-resolution imaging displays. Now, by applying a considerably slower sawtooth current to the vertical deflection coils, a series of horizontal lines will be formed beginning at the top of the screen and continuing to the bottom before rapidly returning to the top (retrace) to begin again (Figure 3.8). This process, known as a "raster scan," typically occurs 50–80 times per second. The persistence of vision makes the raster appear continuous, and the only visible evidence of it may be a slight flicker, and under close examination of the screen, the individual scan lines may be seen on a well-focused display. At the same time the beam is being swept around the screen, video signals are fed to the cathode or grid to produce up to one million or more individually controlled points, called pixels, making up the total picture.

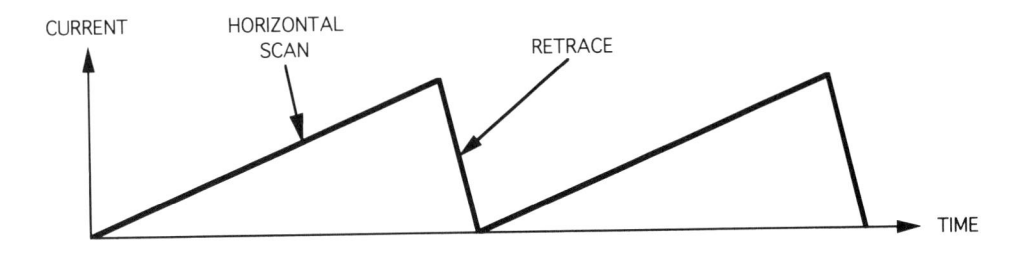

SAWTOOTH CURRENT WAVEFORM

Figure 3.7. Sawtooth current waveform used to produce a CRT raster.

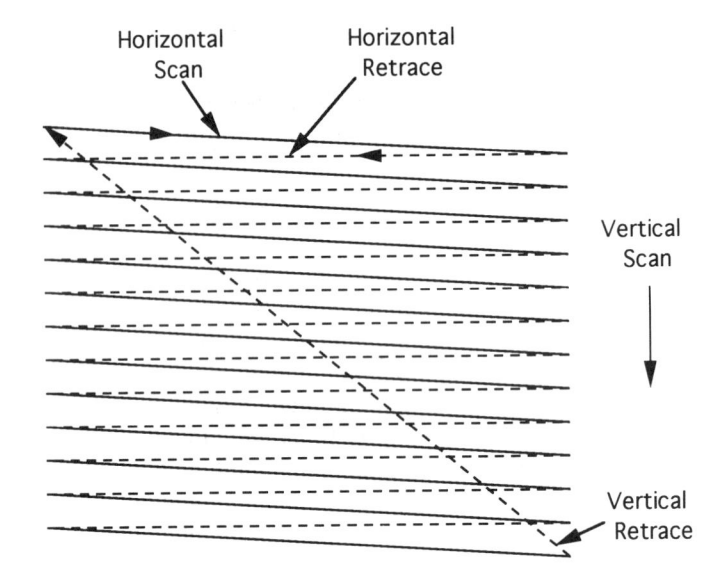

PROGRESSIVE RASTER SCANNING

Figure 3.8. Path of CRT electron beam during raster scanning.

Electrostatic deflection consists of two pairs of metal deflection plates perpendicular to each other. By applying voltages between the plates in each pair, the electron beam is deflected toward the more positive plate and away from the negative plate. Sawtooth voltage waveforms applied to the deflection plate pairs result in scanning of the beam linearly across the CRT face in a manner similar to magnetic deflection, except that voltage rather than current is used.

Screens

The screen is the portion of the tube upon which pictures or images are displayed. A crystalline material called a phosphor and having the property of emitting light (luminescence) upon excitation by an otherwise invisible electron beam is applied uniformly to the glass faceplate. The CRT phosphors are characterized by several properties, including color, persistence (decay), luminance, and resistance to burning and aging.

Color displays use a complex screen structure made up of a shadow mask mounted a short distance behind a screen consisting of a pattern of red, green, and blue phosphor dots or stripes. Three individual electron guns are mounted within the CRT neck and aligned such that one gun lines up with small holes or slots in the shadow mask to strike only the red phosphor dots or stripes (Figure 3.9 and Plates 3–5). The other two guns can only excite the green and blue phosphors, respectively. All three beams are scanned in unison, and the grids or cathodes of each are driven to produce the correct proportions of each color to reproduce the resultant blend of color for each location of the image. Because of limitations on the size that the shadow-mask holes and phosphor patterns may be manufactured, resolution of color CRTs is significantly lower than for monochrome tubes. Also, the electron beams must be aligned precisely to pass through the shadow-

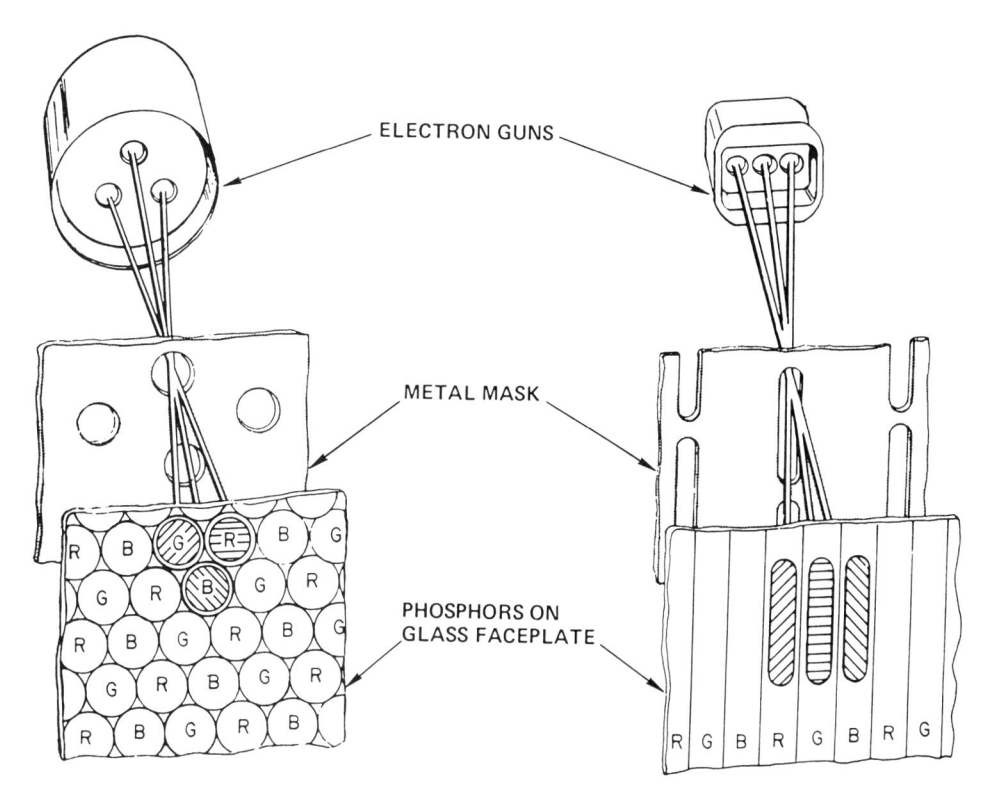

Figure 3.9. Color CRT shadow mask screen structure. (From J. J. Moscony, "Process Requirements for a New Generation of Precision In-line Shadow Masks," *RCA Picture Tubes*, 1980, p. 40.)

mask holes and strike only the desired color phosphor. Any misalignment of the three beams due to magnetic fields or improper adjustment will result in loss of color "purity" in some areas of the screen or "misconvergence" where displayed images have color fringes around any sharply defined object or character.

To maximize viewability of a display, some form of contrast enhancement is usually applied to the CRT screen. This may take several forms. The simplest is a slight etching of the glass faceplate to prevent specular or mirrorlike reflections from the screen. This prevents the images of objects behind the viewer from being visible but results in diffuse light being reflected from the glass surface. Another simple means of contrast enhancement is the use of neutral gray glass faceplates to make the entire screen background appear dark. This is effective and is based on the principle of light from the phosphor being attenuated once as it passes through to the viewer while ambient light must pass through twice in the process of being reflected from the phosphor. A more costly approach to contrast enhancement is the use of an AR coating on the glass surface to better match the index of refraction of the glass to that of air. Antireflective coatings reduce specular reflections from the glass surface without increasing diffuse reflections as etching does. These coatings are identifiable by their purplish cast as on a camera lens. Antireflective coatings are durable, but oils left by fingerprints cause noticeable reflective spots requiring more frequent cleaning. See Chapter 2 for more information on display contrast enhancement.

Dynamic Focus and Astigmatism

As the electron beam is deflected further from the screen center, the spot loses focus and becomes elliptical or astigmatic due to the electron beam having to travel a greater distance to the edges of the screen than to the center. This effect is becoming more of a factor with the continuing trend from spherical shaped faceplates to flatter, more squarely shaped screens.

The defocusing may be corrected by use of dynamic focus circuits that automatically adjust the CRT focus voltage as the beam is deflected across the screen. The simplest technique is to apply parabolic waveforms derived from each axis of the deflection circuits to the focus electrode of the CRT. The parabola approximates the variation in focus voltage required with distance from the screen center. In a more elaborate scheme, the correction voltages for each location on the screen are digitally derived and stored in programmable read-only memory (PROM) for the individual CRT. This solves the problems of slight electron gun misalignments and nonuniform deflection yoke magnetic fields that often result in three corners being in focus and the fourth fuzzy.

The PROM approach may also be applied to correct for the astigmatic spot in the corners; however, a special CRT electron gun is required that contains additional "stigmator" electrodes to control the spot ellipticity dynamically. Such extreme spot control pays large dividends in precision ultra-high-resolution displays for critical applications such as medical imaging and photoreconnaissance. Corner spot size can be controlled to be virtually the same as in the center of the screen rather than the 1.5 times larger that is typical with dynamic focus alone.

Color Displays

Color displays have become widely accepted for most display applications; however, color comes at the price of a significant penalty in resolution. The current limit of about 1280×1024 pixels is imposed by two factors, both of which are a result of the shadow-mask screen structure necessary to produce colors. These factors are the spot size attainable at the high electron beam current required to overcome efficiency losses in the shadow mask and the number of phosphor dot or line triads that can be deposited on the screen. Beam diameters of 0.4–0.8 mm or more are the norm for color CRTs.

Scan Rates

Many different raster scan rates are in common use today, beginning with the standard NTSC television frequencies of 15,750 Hz horizontal and 30 Hz (interlaced) vertical. Higher scanning frequencies are used in medical and computer displays to produce more horizontal scanning lines and, hence, more vertical addressability by breaking the vertical scan into more individual elements. Horizontal frequencies of up to 160 kHz are becoming utilized as the demand for ever-increasing resolution continues. Higher vertical refresh frequencies reduce perceptible flicker of the image, which is most visible with brighter displays and at the periphery of the field of vision. This is particularly a problem where two or more displays are located in close proximity, as in medical imaging systems. Generally, vertical refresh, or "frame," (one complete picture) rates above about 72–75 Hz will not have apparent flicker to most of the population, although some individuals are more sensitive to it than others. As the vertical refresh rate is increased to reduce flicker, the horizontal scan frequency, video amplifier bandwidth, and pixel rates must be increased proportionately in order to maintain the same resolution. See the formulas in Appendix C for the relationships between scan rates, video bandwidth, and resolution.

Interlace

Interlacing is a method used first for television raster displays to reduce apparent flicker without requiring higher horizontal scan rates and greater video bandwidth. It usually consists of two "fields" to make one frame, referred to as 2:1 interlace. Every other scan line is displayed in field 1 while the remaining lines are filled in for field 2 (Figure 3-10). Persistence of vision makes it appear to be a continuous picture, and since a partial picture is displayed twice as often, the effect of flicker is reduced by one-half. There is a trade-off, however, in that small movements of the eye or of the head cause the image to momentarily appear to break up into horizontal bands. Another drawback is that minor electrical and magnetic disturbances cause "line pairing," where some scanning lines are closer together than others, which can be objectionable at close viewing distances. For these reasons, interlacing is best suited for lower cost displays that will be viewed from a distance.

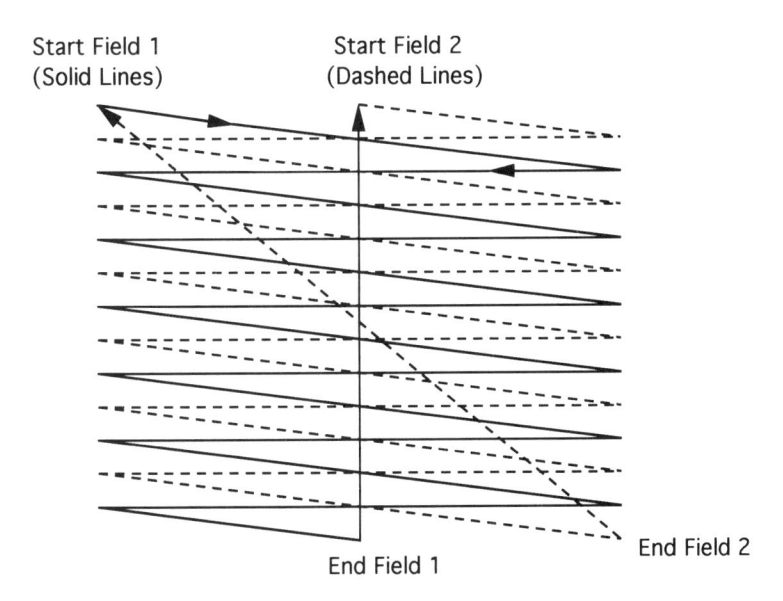

Figure 3.10. Simplified drawing showing 2:1 interlaced raster formation.

Video Bandwidth

Video bandwidth is directly related to the horizontal addressability of a display since it determines how many individual points may be controlled along each horizontal scan line. It also must be "flat" or have equal output at all frequencies throughout the bandwidth range in order to produce the correct shading for all sized objects on the screen and must not "ring" or have excessive "overshoot," which can cause problems in displaying sharply defined objects. Greater video bandwidths do not come easily, and 200 MHz is about the current practical limit. The difficulty of driving the CRT gun, which is both capacitive and inductive, to the 40–60-V levels necessary to produce bright displays with good contrast makes greater bandwidths very difficult. Wider bandwidth can be obtained using "one-bit" video where only two brightness levels exist, either on or off. The obvious trade-off is lack of gray scale.

Signal Inputs

Many types of signal inputs are available to interface CRT displays to various systems. The most common are as follows:

Composite Video. The video information is combined with the vertical and horizontal synchronizing signals to provide a single-input connector to the display.

Block Sync. The vertical and horizontal synchronizing signals are combined on one input connector while the video information is on a second input connector. This is also called "composite sync."

Separate Sync. Separate input connectors are used for the video, vertical sync, and horizontal sync.

RGB. Color displays usually use three separate video signal inputs for each of the three color primaries (red, green, and blue). The vertical and horizontal sync signals are combined with the green video input similar to the composite video described above. This is referred to as "sync on green."

Digital Input. Some CRT displays contain digital circuitry to interface to a host computer via a digital format such as the industry RS-422 standard. Internal memory is used to store one complete frame without the need to continually feed information to the display from the computer. Many image enhancement and presentation operations may be done by the computer, including contrast enhancement, image subtraction, windowing, negative images, scrolling to view selected areas of a large image, and simultaneous display of several smaller images.

Update speed for new images is an important consideration for digital displays and is determined by both the host computer and the display. Greater resolution and faster update rates both require faster data transfer rates in order to produce a complete image on the screen in the same amount of time. Doubling the resolution requires four times the interface speed to produce a complete image. Increasing the refresh rate to reduce flicker requires a proportional increase in interface speed.

Phosphors

Several phosphor screens are most commonly used for image and data displays. The most popular color screens are designated XXA–XXG for the various combinations of blue, green, and red phosphors that were previously designated as P22 sulfide/silicate/phosphate, P22 all-sulfide, P22 sulfide/vanidate, P22 sulfide/oxysulfide, P22 sulfide/oxide, and so on. Common monochrome phosphor types include WWA (previously P4), which was widely used in black-and-white television receivers; WBA (previously P45), used often for medical imaging displays; and "paper-white," which is similar to WWA but with a more natural color balance that is similar to a piece of white paper viewed in an office environment.

Good brightness and a "crisp" bluish-white color, which made WWA phosphor the best choice for black-and-white television are the reasons for its suitability for image displays. The only drawback is the fact that it is actually a blend of two different phosphors having yellow and blue colors respectively to produce white. This makes it a little more difficult to produce consistent screen color, and each manufacturer's WWA phosphor will be slightly different. Also, close examination of the screen will show a slight yellow and blue mottling. The fact that it is a blend of two phosphors also has a positive side since it is possible to tailor the blend to a customer's specification to produce a particular shade of white. This blending is exactly how paper-white screens are achieved. Paper-white has more yellow, which is sensed more easily by the eye, and is correspondingly brighter.

WBA phosphor screens are made of a single-component phosphor and will have better color consistency. WBA phosphor is used for a number of medical imaging applications to present a color similar to that of X-ray film. The trade-off, which is fairly severe, is a loss of 30–40% in luminance.

Approximately 200 other phosphor screens are available commercially, but most are not suitable for general-purpose displays because of low efficiency, colors that are objectionable with prolonged viewing, and more rapid aging.

Geometric Distortion

Geometric distortion is unique to CRT displays: Flat panels have pixel geometry defined by the physical location of each pixel. Several forms of geometric distortion may be present to some degree simultaneously in a CRT display. The ability to tolerate distortion of an image is highly dependent on the application, and as expected, higher precision displays have prices to match. Displays used for photogrammetry or medical diagnostics must be as geometrically accurate as possible while displays used for word processing or text editing do not require such a degree of refinement.

The following are the more common distortions, some of which are illustrated in Figure 3.11:

Pincushion distortion is common in larger screen displays, especially those with wide CRT deflection angles used to shorten tube length. Electronic pincushion correction circuits are included in higher performance displays. These allow adjustment for nearly straight edges on the raster. Low-cost displays use permanent magnets mounted on the deflection yoke to correct for pincushion distortion but at the expense of spot size and shape.

Barrel distortion is less common and can be the result of overcorrecting for pincushion distortion.

Trapezoidal, or "keystone," distortion may be the result of differences between the two pairs of opposing deflection yoke windings in one deflection axis.

102   DISPLAYS

PINCUSHION   BARREL

TRAPEZOIDAL OR KEYSTONE   ORTHOGONAL

TILT   HOOK OR FLAGGING

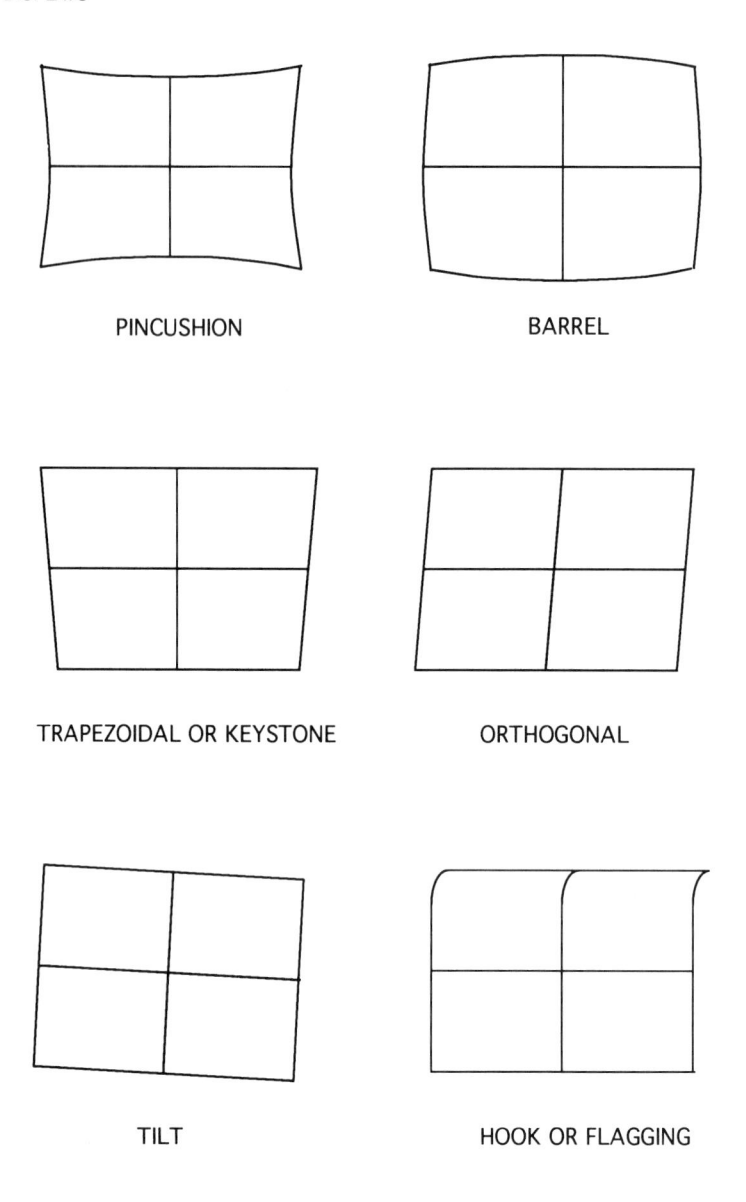

CRT DISTORTIONS

Figure 3.11. Common CRT display distortions.

Orthogonal distortion is the result of misalignment of the vertical and horizontal deflection yoke windings relative to one another. In electrostatically deflected CRTs, it is caused by similar misalignment of the two pairs of deflection plates.

Tilt is probably the easiest fault to remedy. It is caused by rotation of the deflection yoke on the CRT neck.

Hook, or flagging, is a bending of vertical lines, usually exhibited in the upper left corner of the screen. This is caused by parts of the horizontal scanning circuitry not being quite stabilized at the beginning of the vertical scan.

Nonlinearity is particularly noticeable when a large circle or crosshatch pattern is displayed. Nonlinearity will cause flattening or stretching of portions of the circle or variations in the crosshatch line spacing. Adjustment may be provided for linearity, at least in the vertical direction.

Line pairing, particularly common with interlaced displays and those with more scan lines, is a bunching of adjacent horizontal scan lines that show up as brighter and darker areas at normal viewing distances. The cause may be either electrical or magnetic disturbances within the display or nearby.

With *ringing,* a deflection yoke that is improperly matched to the horizontal deflection circuits or is of poor quality will cause a series of dark and light shaded bands at the left side of the screen that die out after a few percent or so of horizontal scan.

CRT Projection Displays

The CRT color projection displays are now used extensively for flight simulators, commercial sports video installations, and "home theater" entertainment systems. These displays employ three separate CRTs and optical systems, one for each of the three additive color primaries, red, green, and blue (Figure 3.12). Either front-projection or rear-projection screens may be used depending on system location requirements and available space. Figure 3.13 illustrates the basic Schmidt optical system that has enjoyed considerable success in various forms, including an optical system that is built into the CRT itself.

Screens for projection displays are usually directional rather than Lambertian (cosine) surfaces. Directional screens concentrate their light in one direction, that of the intended viewing location, instead of dispersing it over a wide viewing angle. Measurement of luminance, contrast, and chromaticity must be made from that direction in order to accurately characterize display performance.

Front- and rear-projection systems each require somewhat different measurement techniques. Front projection requires that the measurement instrumentation and its operator not shadow the area being measured. Rear projection may be measured similar to conventional CRT and flat-panel displays but with adequate consideration for the directional properties of the screen.

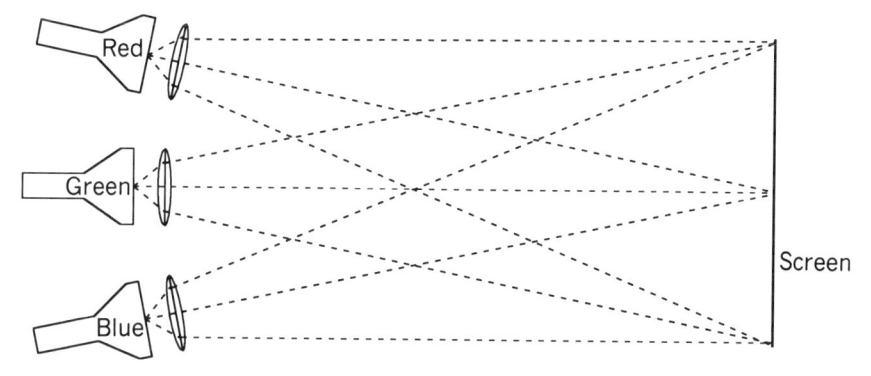

Figure 3.12. Color projection display system.

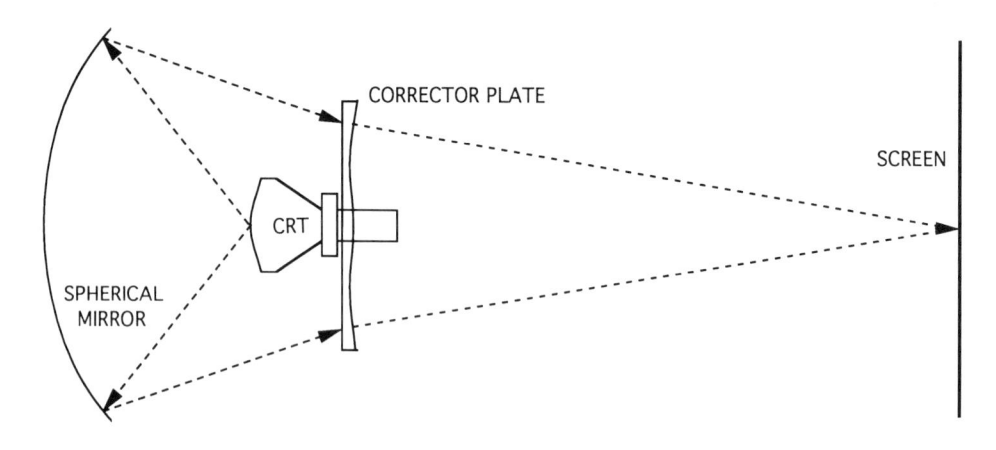

SCHMIDT PROJECTION SYSTEM

Figure 3.13. Basic Schmidt projection system.

Flat CRTS

Flat CRTs have been under development since 1951 when work was begun on the "Thin Tube" by W. Ross Aiken at the University of California Radiation Laboratory.[2,3] The long-standing goal of "picture-on-the-wall" television has still not been achieved commercially but is still "just around the corner," as it has been ever since the 1950s. Fortunately, there is enough intense development on several fronts aimed at this goal, and laptop computer, military, and workstation needs are beginning to generate enough volume to make it appear possible in the near future. Whether it is some form of flat CRT or one of the many competitive flat-panel technologies remains to be seen.

Several flat CRT concepts from the Aiken tube until the 1980s used beam addressing with various methods of bending the electron beam from a gun mounted parallel to the screen. During the 1980s, several small-screen, monochrome, flat CRTs were announced by Futaba,[4] Matsushita,[5,6] Philips,[7] Sanyo,[8] Sinclair,[9-11] and Sony.[12-16] Some were commercially produced for hand-held pocket monochrome television receivers beginning in 1983. Most used a more-or-less conventional electron gun mounted below the screen and electrostatic bending of the beam to cause it to land somewhat perpendicularly to the screen (Figure 3.14). None has had a major impact on the market.

Matrix-addressed flat CRTs appeared in about 1962 beginning with the Northrop Digisplay.[17,18] This device used a large-area cathode and switching grids to control the areas of the phosphor screen to be illuminated. It operated at high accelerating voltages similar to those of a conventional CRT. From the Digisplay evolved the low-voltage vacuum fluorescent display (VFD) by ISE in Japan around 1966. The VFD will be discussed in the next section of this chapter.

Combinations of beam-addressing and matrix-addressing have also been researched.[19] These have not had the emphasis or success that matrix-addressing has had. It appears that if any flat CRT technology is ultimately successful, it will be some form of matrix-addressed CRT.

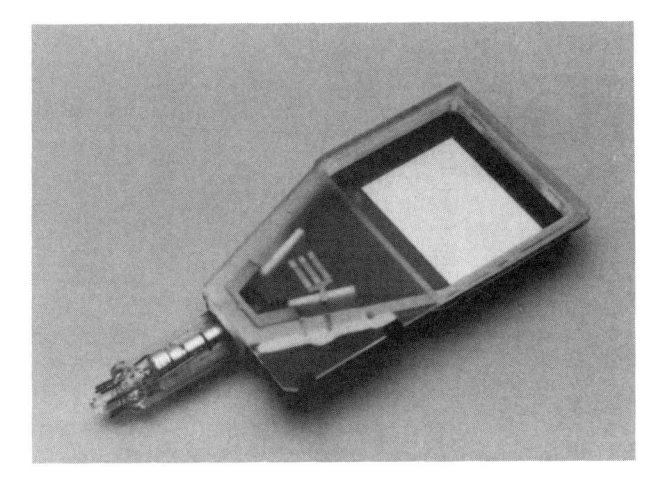

Figure 3.14. Flat CRT using beam bending techniques. (From the collection of James Richardson.)

In the late 1970s, the field emission display (FED) was developed by Stanford Research Institute (SRI) and is currently undergoing development by at least Candescent Technologies (formerly Silicon Video Corporation), Micron Display Technology, and PixTech, an alliance of Motorola, Texas Instruments, Raytheon, and Futuba.[20]

The FED uses a tiny field emitter cathode for each pixel (Figure 3.15). These are cold cathodes and do not require heaters with their consequent high-power consumption, as did the earlier matrix-addressed flat CRTs. The strong positive-voltage gradient produced by the nearby extraction grid to the sharply pointed tip of the field emitter causes electrons to be pulled from the tip toward the phosphor screen. The color subpixel to be addressed has a positive voltage applied to it to attract these electrons. The FED has many of the positive CRT attributes: high luminance, wide viewing angle, relatively low power consumption, full color, good resolution, and relatively simple processing using existing CRT fabrication techniques. It combines these with the precise addressability common to most other flat-panel devices.[21,22]

3.3 VACUUM FLUORESCENT DISPLAYS

The VFD is a hybrid display device intermediate between the conventional electron tube and the CRT. Simple triode structures with patterned, transparent anode substrates and low-voltage phosphors in an evacuated envelope comprise the basic structure of the VFD (Figure 3.16). Both alphanumeric character and matrix-addressed VFDs are available. Operation of VFDs is at very low anode voltages, typically on the order of 50–100 V. Phosphors such as zinc oxide (ZnO) have efficient low-voltage characteristics and good resistance to darkening with age. These are used in place of conventional CRT phosphors for VFDs. The VFDs usually are monochrome only. Low cost combined with high luminance and pleasing color make it well suited for consumer electronics displays, although it is now being largely displaced by backlit LCDs in this application. Because of the physical constraints of the evacuated envelope, it is limited in the size to which it may be scaled.[23–25]

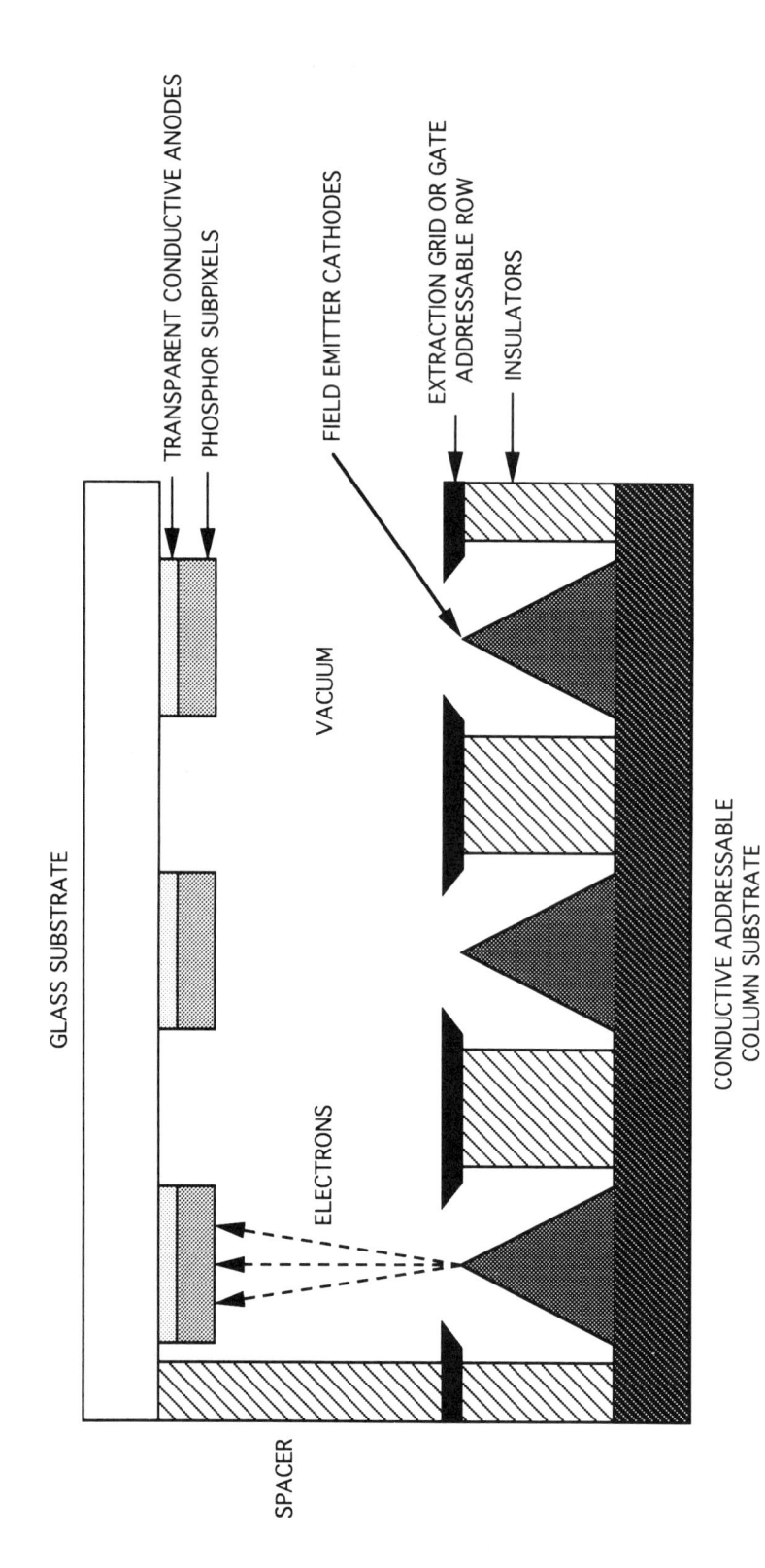

Figure 3.15. Field emission display (FED).

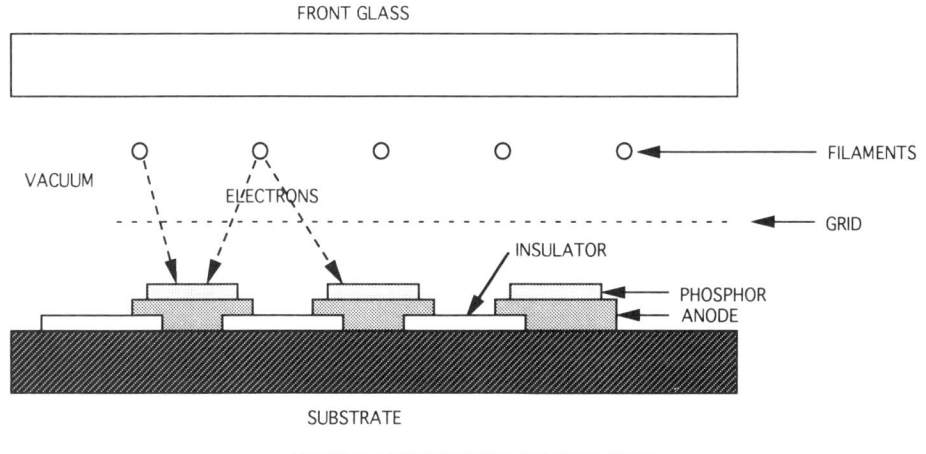

Figure 3.16. Vacuum fluorescent display (VFD).

3.4 LIQUID CRYSTAL DISPLAYS

Liquid crystal displays are second only to CRTs for information displays. If small LCDs for digital watches, calculators, and electronic instruments such as digital multimeters (DMMs) are included, they probably take first place in quantity although not yet in revenue.[26] High-quality LCDs for data displays and eventually television have received their greatest impetus from the lap-top computer industry. The LCD's greatest assets—thinness, light weight, low power consumption, and ease of interface to low-voltage digital electrical signals—have eminently qualified it for the lap-top computer.

Many liquid crystal materials exist. Without venturing into the chemistry of liquid crystal materials, which is beyond the scope of this book, liquid crystals consist of elongated organic molecules that may have their alignment altered under the influence of an electric field. At rest, the molecules tend to align themselves parallel with respect to each other. Applying an electric field to the molecules causes them to change their alignment orientation. Light passing through the liquid crystal material is thus polarized, and the polarization angle is controlled by the applied electric field. Optical polarizers with fixed orientation are used in conjunction with the liquid crystals to selectively pass or block light depending on orientation of the molecules.

For all of their qualities, LCDs are not without some faults. Bright, full-color LCDs are becoming available but cost is still very high compared to the shadow-mask color CRT. The LCD cost will decrease as production quantities increase. Display size is another limitation. Lap-top computer displays of about 250 mm are very practical and LCDs of 500 mm or so should follow in the near future but will probably not be inexpensive. The 1000-mm displays required for HDTV are another matter, although *tiling* is a possible solution for very large displays if the borders of each individual display tile can be reduced. Multiple small displays are butted together to form one large display in tiled displays. An important shortcoming is the somewhat restricted viewing angle of LCDs. Luminance, contrast, and color fidelity degrade at angles other than on the optimum

Direct Addressing

Many types of LCDs exist depending on cost and viewing requirements as well as the LCD processing required to meet these needs. The simplest monochrome LCDs such as are used in watches, calculators, and DMMs use *direct addressing*. These have electrodes in the shape of each character, character group, or symbol to be displayed at the desired location, as shown in Figure 3.17. Numbers are usually formed by selectively applying voltage between the appropriate segments of the seven segments comprising one digit on the display and the backplane electrode. Separate electrical connections are made for each individual character, group of characters, symbol, and numeric segments. These are driven by a microprocessor and LCD driver.

Multiplexed Addressing

Often, multiple backplanes are used to reduce the number of electrical connections to the LCD as well as the number of drivers. Going from one backplane to two cuts the number of segment connections and their drivers in half while adding only one additional contact and driver for the backplane itself. This is called *multiplexed addressing*. Multiplexing involves some performance compromises that become more severe as the number of multiplex levels increases. These compromises include a narrower operating temperature range, degraded contrast, and reduced viewing angles. Increased drive coding complexity is another factor.

Figure 3.17. Direct-addressed LCD.

Matrix Addressing

Rows and columns of dots may be used for the most versatility to present complex alphanumeric and graphic data. *Matrix addressing* is used to control the excitation of the rows and columns to cause the intersecting pixel to be turned on. Dot-matrix LCDs range from monochrome single-line text displays to full-color XGA displays.

Twisted Nematic LCD

The twisted nematic field-effect (TN-FE) LCD of the early 1970s became the first liquid crystal device to produce displays with sufficient contrast for commercial products. The advent of the TN LCD combined with large-scale integration (LSI) of the driving electronics made possible the tremendous growth in the use of digital watches and hand-held calculators, and these were the dominant display for low-to-medium information density products by the mid-1970s.[27]

Electrically, the LCD may be thought of as an array of parallel-plate capacitors with a liquid dielectric.[28] Figures 3.18 and 3.19 illustrate the construction of a portion of an LCD. The TN LCD consists of a thin sandwich of two parallel pieces of glass substrate with a transparent conductive pattern of indium–tin oxide (ITO) for the characters on one interior glass surface and a backplane of the same material covering the other. The space between the glass substrates is filled with the liquid crystal fluid. Polarizers aligned with their planes of polarization 90° to each other are applied to the front and rear outer surfaces of the sandwich.[29,30]

The LCDs may be operated in the reflective, transmissive, or transflective modes. In the reflective mode, ambient light from the viewing side of the display is absorbed by the display or reflected by a reflecting material behind the LCD to produce the displayed information. Because the reflective LCD does not generate light but merely utilizes ambient light, it is easily viewed in most environments, even in direct sunlight, where most LED or CRT displays become "washed out." Reflective LCDs are poor in low ambient illumination levels or in total darkness, where they become unreadable. This is overcome

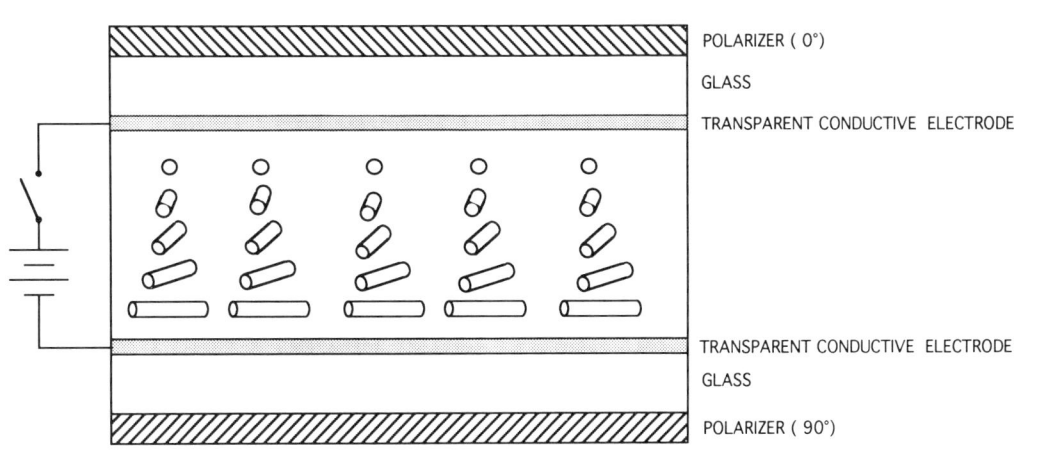

Figure 3.18. Portion of a twisted nematic LCD illustrating the "off" condition.

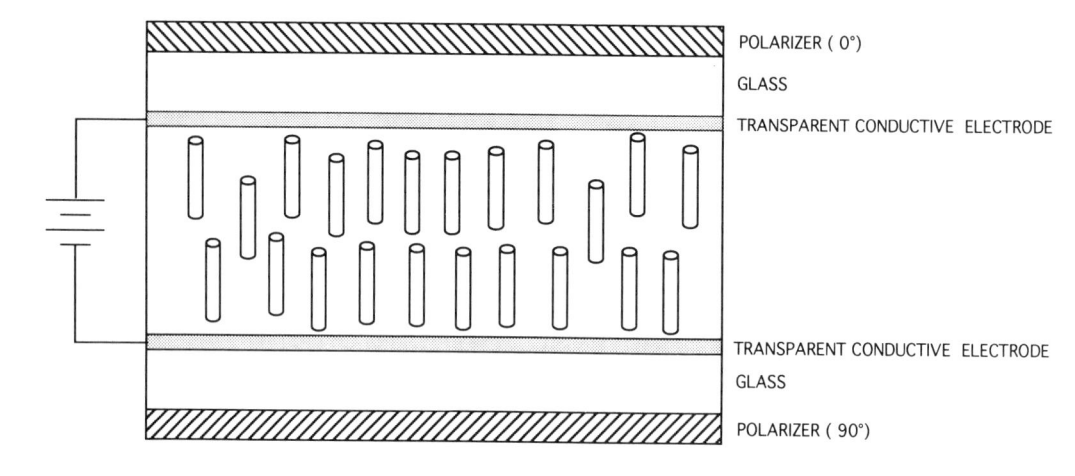

TN LCD (ON STATE)

Figure 3.19. Portion of a twisted nematic LCD illustrating the "on" condition.

by the use of an LCD designed for operation in the transmissive mode and the addition of a backlight. The backlight may be one or more incandescent lamps, LEDs, EL lamps, or cold-cathode fluorescent lamp (CCFL) depending on efficiency, color, space, and cost constraints of the product. Naturally, the addition of a backlight defeats some of the low power consumption advantage of the LCD. This shortcoming may be partially resolved by adding a switch so that the backlight is normally off but may be turned on by the user for those few times when use in subdued ambient illumination levels is required. Often, even this minimal use is further restricted by timing functions controlled by the microprocessor that automatically turn off the backlight after a predetermined time interval, usually on the order of 15–60 s or so. Transflective LCDs combine the features of both reflective and transmissive devices to allow use in any ambient illumination environment.

Supertwisted Nematic LCD

The supertwisted nematic (STN) and the similar supertwisted birefringent effect (SBE) LCDs were developed to provide display of high information density at high contrast with wider viewing angles. They also help overcome to a considerable degree the crosstalk that occurs when multiplexing large arrays by the sharper knee of the polarization and transmittance versus drive voltage curve for the STN LCD (Figure 3.20) compared to the TN LCD (Figure 3.21).[31–36]

The STN LCD uses materials, fabrication techniques, and driving electronics similar to those used for TN LCDs. Instead of the 90° twist of the TN devices, the STN LCD uses a twist angle of up to 270°, as shown in Figure 3.22.[37] The STN LCDs operate in either of two modes that usually exhibit a blue or yellow coloration. Several improvements have helped reduce the coloration of STN devices. The first is a double-layer STN (DSTN) employing a second LCD as a "compensator cell" cascaded with the active display cell. Later, a simpler stretched polymer compensating film (FSTN) was devised for the same purpose. Reduced cost, thickness, and weight as well as improved image appearance were obtained, although with a penalty of lack of temperature compensation. Further improvement has been obtained by the use of two film compensators.[38]

3.4 LIQUID CRYSTAL DISPLAYS   111

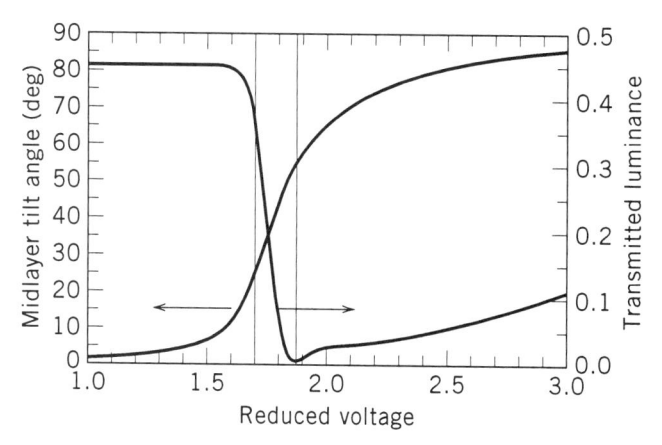

Figure 3.20. Polarization and transmittance versus drive voltage for a STN LCD. (Courtesy of Terry Scheffer.)

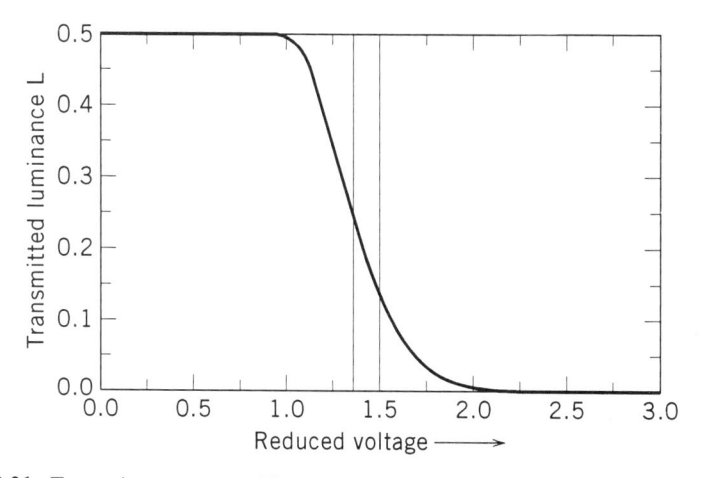

Figure 3.21. Transmittance versus drive voltage for a TN LCD. (Courtesy of Terry Scheffer.)

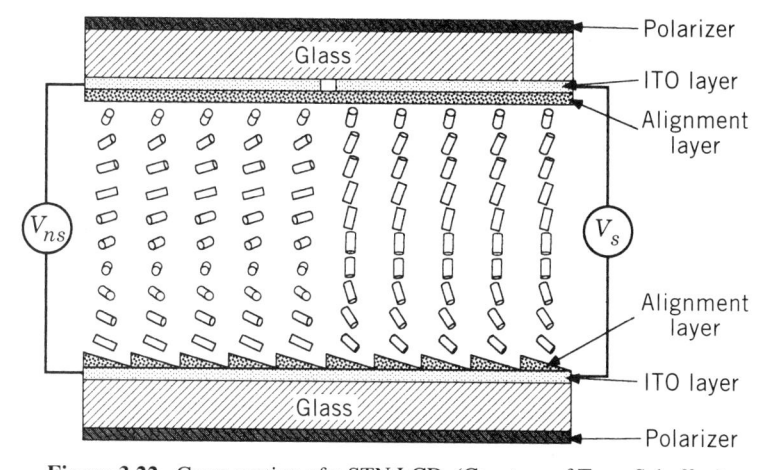

Figure 3.22. Cross section of a STN LCD. (Courtesy of Terry Scheffer.)

The passive matrix STN LCD has found wide application in lap-top computer displays but lacks the speed necessary for video displays. In lap-top applications, the higher display cost and reduced viewability of the STN LCD was more than offset by the size, weight, and power savings when compared to CRT displays. Indeed, the lap-top computer owes its existence to the LCD.

Passive Matrix-Addressed LCD

The next step up in LCD complexity is the passive matrix-addressed LCD. These are similar in construction to the previously described directly addressed LCDs and consist of conductive patterns in vertical columns and horizontal rows. The columns and rows may be patterned to provide, for example, a number of 7×9 dot arrays, the elements of which may be selectively driven by addressing along the vertical and horizontal axes. Each of these arrays can display any single alphanumeric character by addressing the appropriate combination of elements. Furthermore, a number of these 7×9 dot arrays are arranged in rows and columns themselves to allow lines of alphanumerics to be displayed. Alternately, many uniformly spaced columns and rows may also be used in a similar manner to display graphics and/or text. Passive matrix-addressed displays require a large number of external drivers for each axis.

Two problems are associated with passive matrix-addressed displays. One problem is the relatively long response times, typically 100–250 ms, for updating displayed information. This precludes its usefulness for video displays. The other shortcoming is the loss of contrast as the number of multiplexed lines increases. This is because the voltage on one element of the LCD cannot be changed without affecting all of the other elements connected to it.

Active Matrix-Addressed LCD

As the number of multiplexed elements increases, the time available for addressing each one decreases until there is not enough time for the pixel to respond to the driving voltage. The high information densities and fast refresh rates for full-color video and information displays preclude the use of any LCD except the active matrix-addressed LCD (AMLCD) for these applications. Active matrix-addressed LCDs contain on-board diode or transistor circuit components for each pixel that form a switch (Figure 3.23).[39] Thus, a brief pulse on the order of microseconds in duration can turn the pixel on and the switch will maintain it in the "on" state until reset.

The thin-film transistor (TFT) has been the most successful approach to AMLCDs so far, especially for full-color displays. Three TFT elements, each with a red, green, and blue filter covering it, respectively, make up one complete pixel.[40,41]

The processes and materials used to produce transistors directly on the LCD substrate have matured, and image quality for TFT AMLCDs is approaching that of the venerable CRT. Screen size, resolution, and cost are still the principal limitations for TFT AMLCDs, but considerable effort is still being devoted to overcoming these obstacles. At this time, 400 mm diagonal, full-color TFT AMLCDs with up to 1280×1024 pixel addressability, 100 cd/m^2 (nits), and approximately 25 W power consumption are commercially available from several manufacturers. Larger devices are just now becoming commercially available and higher addressability has been demonstrated.[42]

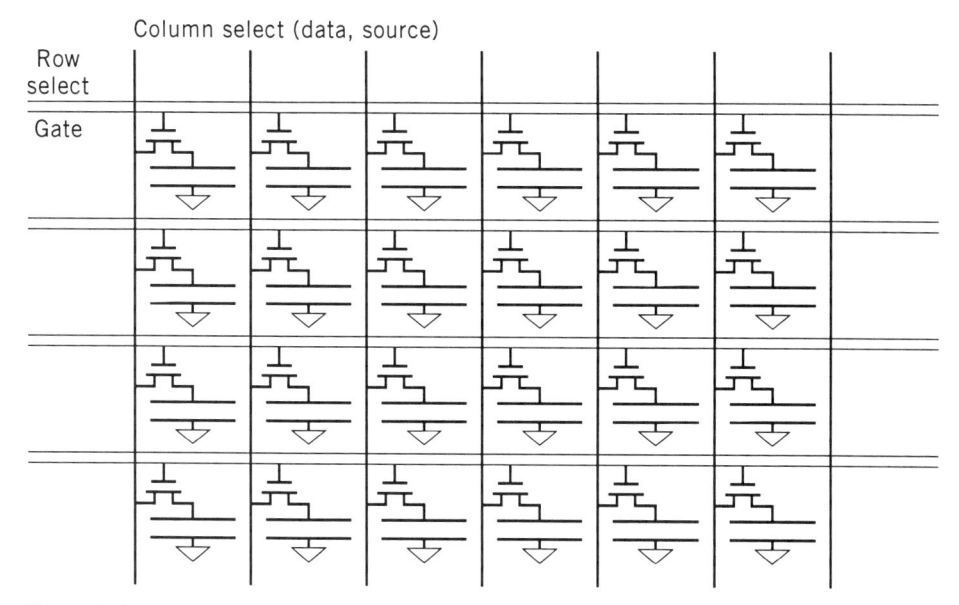

Figure 3-23. Functional diagram of several pixels in an active matrix LCD. (From R.G. Stewart, "Active Matrix LCDs," Soc. for Info. Disp. Seminar Lecture Notes, San Diego, CA, May 13, 1996, pp. M5/11. Reprinted by permission.)

LCD Projection Displays

Matrix-addressed LCDs are well suited to projection displays since they depend on an external light source for visibility. The triple-layer STN (TSTN) has been extensively used for overhead projector presentation systems so that computer displays may be viewed in conference rooms, in classrooms, and for other large gatherings. The LCD is placed on the projector and connected to a computer that generates the displayed information. Three stacked LCDs and a combination of neutral and colored polarizers produce subtractive color, as shown in Figure 3.24.[43] LCDs are being replaced by transmissive and reflective light valves using active matrix addressing.

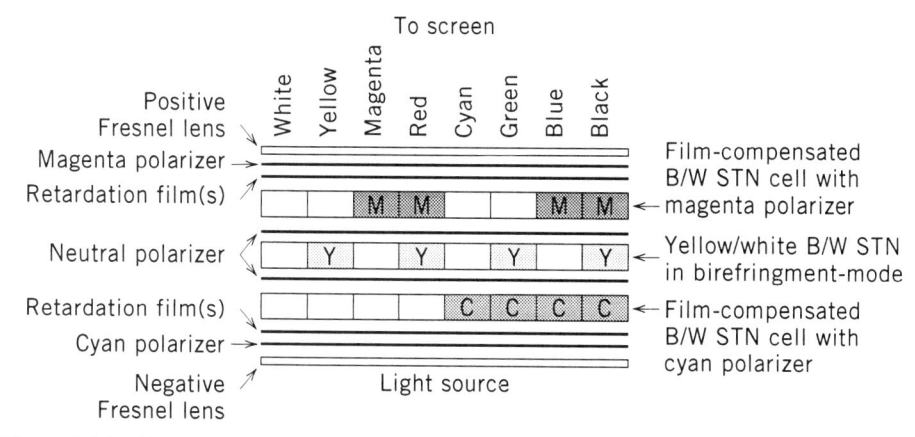

Figure 3.24. TSTN LCD as used for color displays using an overhead projector. (Courtesy of Terry Scheffer.)

Liquid Crystal Color Shutter

The liquid crystal color shutter (LCCS), or more correctly a *color switch,* combines a CRT and a liquid crystal panel to provide high-resolution limited or full-color displays from a simple monochrome CRT. Unfortunately, this is not a flat-panel display but does provide an alternative to the shadow-mask CRT. The LCCS approach to a color display combines the unique features of the liquid crystal and the CRT to produce a high-resolution color display using a conventional single-gun monochrome CRT. The color shutter is based on the field sequential system whereby one complete raster field is displayed for each of two color primaries, usually red and green. The two fields make up one complete frame.

The color shutter in its basic form consists of a liquid crystal "π cell" half-wave retarder sandwiched between two color polarizers and a linear polarizer (Figure 3.25). The color polarizers permit white light to pass through in one axis and only permit the particular color to pass for which it is designed in the axis oriented 90°. One filter passes red light and the other passes green light in the latter orientation. These filters are oriented 90° apart so that when one is passing its color, the other is allowing white light to pass; thus each filter permits light of the color selected by the other to pass. Light from the CRT screen is polarized by the linear polarizer and is rotated by 90° when the π cell is energized. This causes the light from the screen to be polarized such that only light of the color selected by the color polarizer oriented in that plane passes. When the π cell is not energized, the light is polarized 90° from that plane, which causes the other color polarizer to determine the color of light transmitted to the viewer. By energizing the π cell during alternate raster frames, one frame will present red information and the next green.[44-46]

The resolution of the CRT is essentially unimpaired and color misregistration is nonexistent. These advantages are not without some performance compromises. The brightness is about 15–25% of that of the CRT itself because of absorption and polarization by the various optical elements, and of course, the color gamut is limited to colors that can be produced by mixtures of red and green. The substantial loss of light through the LCCS and polarizers has its "bright side" so to speak; ambient light must make two passes through the assembly and very high contrast between the displayed information

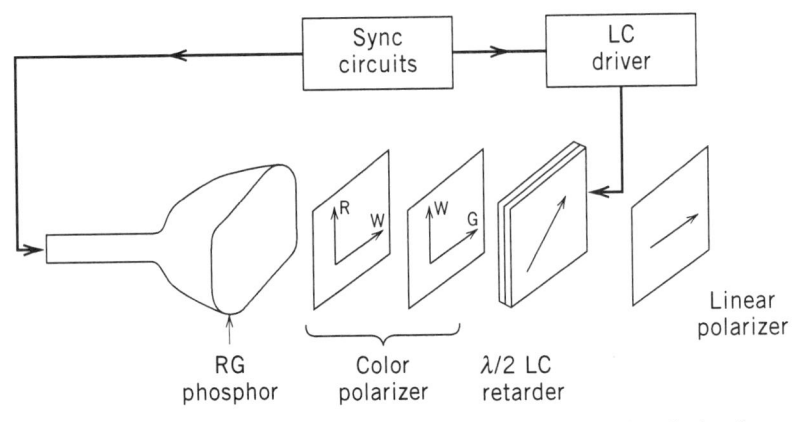

Figure 3-25. Diagram showing liquid crystal color shutter for a two-color display from a monochrome CRT display. (From R. Vatne, P. A. Johnson, Jr., and P. J. Bos, "A LC/CRT Field-Sequential Color Display," SID Intl. Symp. Digest of Tech. Papers, 1983, p. 29. Reprinted by permission.)

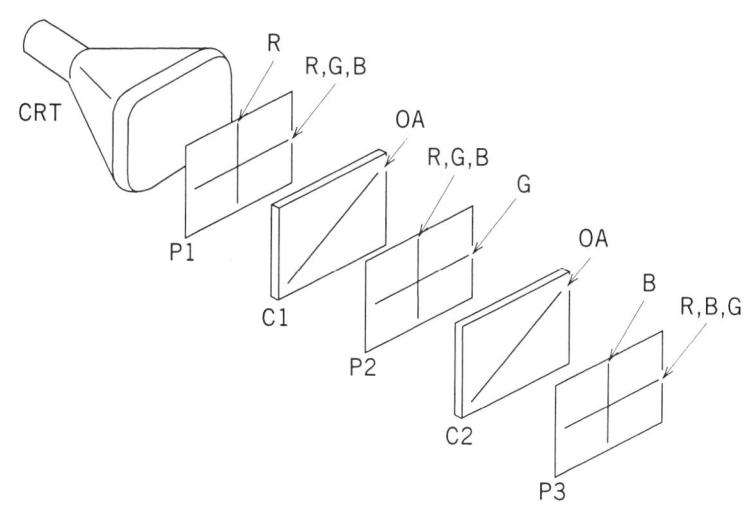

Figure 3.26. Diagram showing liquid crystal color shutter for a three-color display from a monochrome CRT display. (From P. Bos, T. Buzak, and R. Vatne, "A Full-Color Field Sequential Color Display," Eurodisplay 84 Proc., Paris, Sept. 18–20, 1984, p. 7. Reprinted by permission.)

and the background is the result. The LCCS does impose a resolution limit, although it is not spatially limited, as is the color shadow-mask CRT. The requirement to divide the video signal into the two color components for alternate frames reduces the amount of picture detail that may be displayed by a factor of 2 for a given video amplifier bandwidth and choice of scan rates.

Full-color displays using two LCCS (Figure 3.26) are a further refinement.[47,48] Improvements in shutter and CRT efficiency permit full-color pictures similar in color gamut to shadow-mask CRTs. Resolution is significantly greater for the LCCS method, but similar viewing angle problems to those of LCDs exist. A niche market for test instruments, military simulation, and virtual reality has developed using LCCS displays for screen sizes less than 150 mm or so.[49] Below this size, it becomes difficult to produce shadow-mask CRTs having acceptable resolution. Above this size, LCCS panels become expensive, although the monochrome CRTs used with it are inexpensive.

Similar LCCS techniques have been employed for stereoscopic displays using liquid crystal shutters, polarizers, and polarized viewing glasses.[50–52] This approach has the advantage of using passive viewing glasses that eliminate electrical connections required by some other methods.

3.5 PLASMA DISPLAYS

Plasma displays are gas discharge devices that are similar in operation to the familiar neon lamp, sometimes called a "glow" lamp (Figure 3.27). The basic neon lamp uses two electrodes in a tubular glass envelope filled with neon gas. An applied voltage of 65 V or more, either AC or DC, causes ionization of the gas and the region surrounding the positive electrode (both if AC is applied) to glow with a series of narrow-band orange emission lines from 580 to 640 nm that are characteristic of ionized neon gas at low pressure.

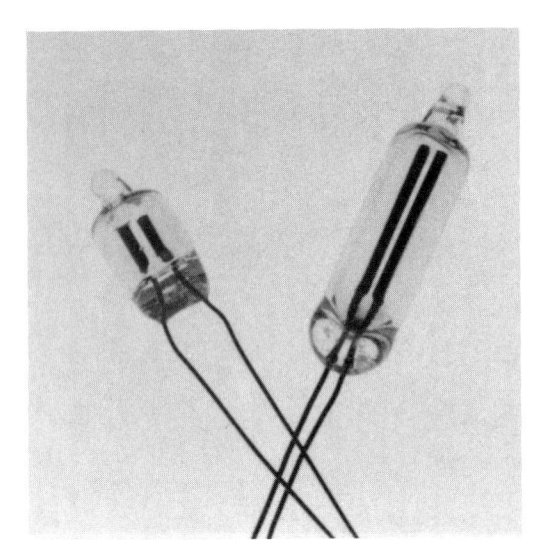

Figure 3.27. Common neon "glow lamps" that form the basis of plasma displays.

Discrete neon lamps were used for some of the earliest information displays in columnar form with each row of neon lamps representing a single digit from 0 to 9. Multiple columns allowed the display of numeric values several digits in length. The Beckman EPUT (events per unit time) meter/frequency counter of the 1950s is perhaps the best remembered example of the columnar neon lamp display. This was followed in the early 1960s by the Burroughs Nixie tube, having the digits 0 to 9 formed from wire electrodes superimposed one behind the other in a small tubular vacuum tube envelope filled with neon gas (Figure 3.28). Around this time, it was found that a

Figure 3.28. Burroughs "Nixie" tube employing a neon gas discharge to produce a digital display.

small amount of mercury vapor added to the tube increased life by reducing sputtering of the cathode, which left a metallic film on the glass envelope and reduced the light output of the display.[53] The Nixie tube evolved to a seven-segment flat display in the late 1960s with a construction similar to today's plasma display panel (PDP). Each of the seven segments could be addressed individually to produce the numbers 0–9.

The modern PDP employs two glass substrates with transparent conductive electrodes on the inside surface in the shape of the characters or matrix elements to be displayed (Figure 3.29). The panel edges are sealed, and the device is backfilled with a low-pressure mixture of gases such as neon, argon, xenon, krypton, helium, nitrogen, and/or mercury vapor after removal of air.[54] Some devices have used the visible light emission of the fill gas, especially neon, while others use ultraviolet-emitting fill gases such as argon or mercury vapor to excite phosphors applied to the interior walls of the device. The latter has the advantage of being able to produce light of any color depending on the phosphor used. Both AC and DC voltages have been used to drive plasma displays. Direct-current PDPs are simpler in fabrication, but AC PDPs have so far shown higher luminance.

Alternating current PDPs have the additional benefit of "memory" by adding capacitors in series with each pixel and driving them with high-frequency AC. In practice, the capacitors are constructed by adding dielectric material between the conductive electrodes and the gas fill. Bidirectional, repetitive "sustain" pulses of less than the voltage required to initiate a discharge are applied across each cell. A higher voltage pulse is used to "write" the selected pixels. This ionizes the gas and deposits charges on the cell walls, which reduces the voltage applied across the cell until the discharge is extinguished. About this time, along comes a sustaining pulse of opposite polarity. The voltage from the wall charges adds to the sustaining pulse to provide enough voltage to ionize the gas again. This process is repeated at a rapid rate until other pulses are applied that cause the discharge to be extinguished. Pulsed techniques have also been used for memory in DC PDPs.

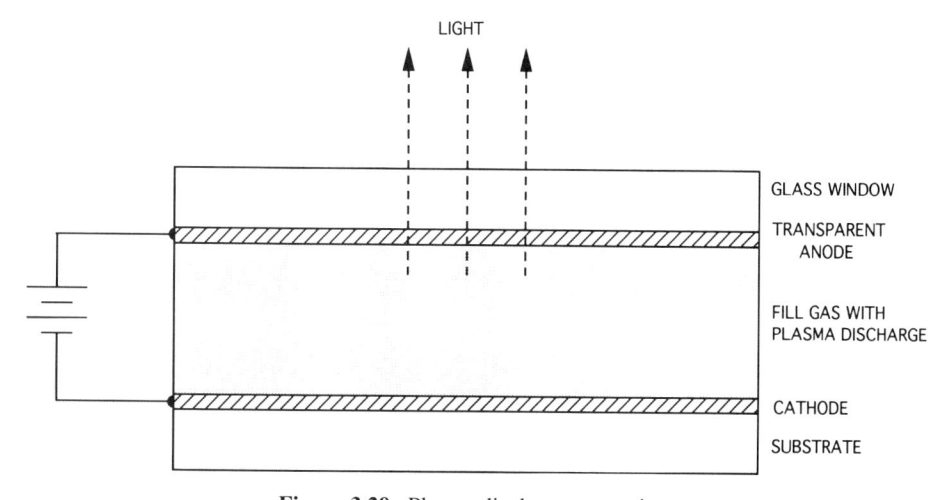

Figure 3.29. Plasma display construction.

Full-color 1000-mm (40-in.) diagonal DC PDPs for HDTV have been demonstrated experimentally by Nippon Hoso Kyokai (NHK).[55–57] The PDP has a niche market in military applications where high resistance to shock, wide operating temperature range, immunity to electromagnetic interference (EMI), and long life are important. Plasma display panels are also used for large displays for computer workstations, business, and publishing.[58] In the future, it is possible that PDP could find another niche in large-screen HDTV, where their high luminance and contrast are an advantage and their higher power consumption is not of great concern.[59,60]

3.6 PLASMA ADDRESSED LIQUID CRYSTAL DISPLAYS

The plasma-addressed liquid crystal (PALC) display combines features of both plasma and liquid crystal displays to produce an alternative method of active-matrix (AM) addressing for full color at video refresh rates.[61–63] The plasma produces no visible light itself but instead acts as a switch to address individual LCD pixels in a matrix of rows and columns (Figure 3.30). Processing is simpler than equivalent TFT AM-LCDs, and the efficiency is good, approximately 1 lm/W, because of the high cell transmittance. On the other hand, higher driving voltage is required for the plasma portion of the panel, although the drivers are low cost. It is anticipated that PALC displays could be a lower cost approach to flat-panel HDTV since the process scales better to large screen sizes than TFT AM-LCDs.[64] Sony has demonstrated large-screen PALC devices under the name Plasmatron.

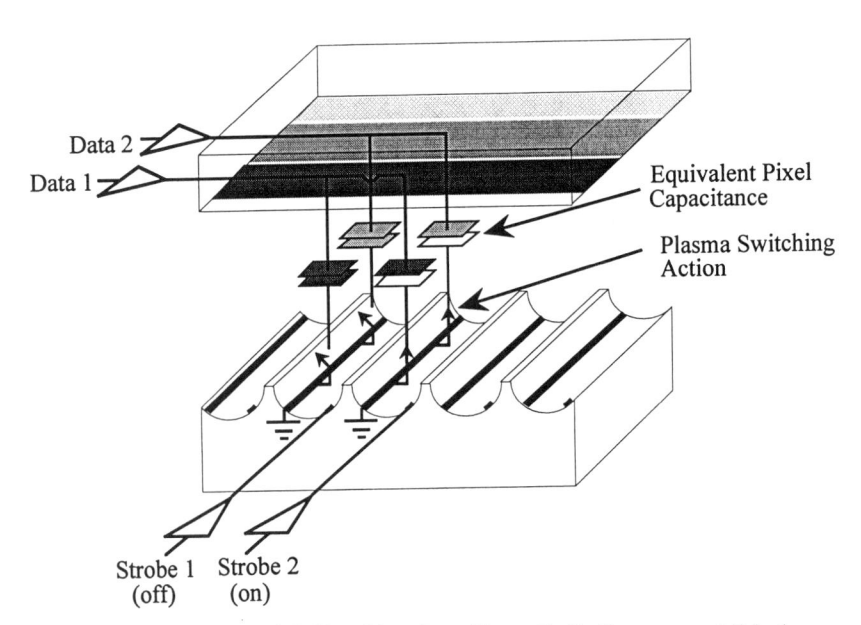

Figure 3.30. Plasma-addressed LCD addressing. (From T. S. Buzac, unpublished manuscript, 1995. Reprinted by permission.)

3.7 ELECTROLUMINESCENT DISPLAYS

Electroluminescent displays have been most successful in monochrome so far. Progress in EL devices has not been as rapid as with LCDs, and they have found their greatest usage in niche markets that require the features in which they excel. Subclassifications of EL displays include both powder and thin-film phosphors combined with either DC or AC operation. Further distinctions include monochrome, multicolor, and full-color EL displays. Full-color EL devices have not yet enjoyed much success in commercial applications due to their lower efficiency blue phosphors. Power requirements for AC EL and resolution limitations have prevented its use for notebook computers.

Electroluminescent devices fundamentally consist of two electrodes, at least one of which is transparent, with a layer of EL phosphor between them. An outer encapsulation, often plastic, protects the phosphor from absorption of moisture and insulates the assembly. Light is produced when voltage is applied to the device, which may be thought of as a "light-emitting capacitor." The most efficient EL phosphor is yellow-emitting manganese doped zinc sulfide (ZnS:Mn), which is well matched to the wavelength of maximum sensitivity of the human eye.

DC EL Displays

Direct-current operated phosphor powder techniques are the oldest and have been essentially replaced by AC EL devices. The operating voltage for DC EL displays is in the range of 120–180 V. The DC EL devices exhibit shorter lifetimes than provided by AC EL devices but are not prone to failure of individual pixels because of their resistive nature.

AC EL Displays

The AC thin-film electroluminescent (TFEL) device has high luminance, high contrast, wide viewing angle, high resolution, low power, long life, and ruggedness.[65] The simple construction of AC EL devices is illustrated in Figure 3.31. Currently, EL displays of 640 × 480 pixels are commercially available. Military, medical, industrial control, and

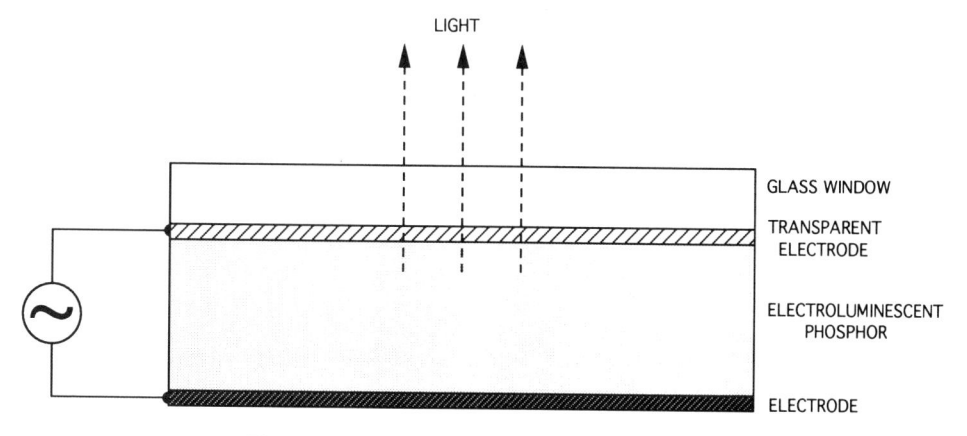

Figure 3.31. AC electroluminescent construction.

transportation applications, where performance, viewability, and ruggedness are important and cost is a less important consideration, have predominated.[66] Drivers for matrix addressing of EL devices are more costly than for LCDs because of the need to switch voltages up to 250 V. The power consumption of TFEL displays is somewhat high to be used for portable applications, although progress is being made in providing those with lower voltage and power. Gray scale is limited by the steepness of the luminance/voltage characteristics of the phosphor, but 16-bit gray scale has been shown. Also, AC EL devices are more prone to catastrophic failure of individual pixels because of dielectric breakdown caused by the high driving voltages.

Multicolor EL Displays

Limited-color TFEL panels have been commercially produced in 250 mm (10 in.) size with 640 × 480 pixels for VGA displays. By using just red and green (RG multicolor) pixels, both of which are relatively bright, eight colors may be displayed with good luminance and contrast at a power consumption of only 10 W. In this example, an array of red and green filters is used with the efficient yellow ZnS:Mn EL phosphor.[67]

Full-Color EL Displays

Two approaches to full-color EL devices have been used. One uses three phosphors, red, green, and blue. The other method uses a single broadband white phosphor and a pattern of red, green, and blue filters.[68] Full-color EL displays have been demonstrated, but an efficient blue-emitting phosphor, either as a separate phosphor or in a white phosphor with enough blue emission to allow filtered operation, has been elusive. The problem is aggravated by the fact that as the wavelength of the phosphor approaches those of deep blue, which is required for the greatest color gamut, the sensitivity of the eye is greatly diminished. At the wavelength of 450 nm commonly used for blue television phosphors, the eye has only 4% of the sensitivity that it has at its peak of 555 nm in the yellow-green region. New pure blue phosphors such as $SrS:Ce$, $SrGa_2S_4:Ce$, and $CaGa_2S_4:Ce$ have reportedly improved EL performance to some degree recently.[69–71]

3.8 LIGHT-EMITTING DIODE DISPLAYS

Light-emitting diodes are used in extremely high volume in the simplest display form, the discrete on–off indicators found in all categories of electronic products. They have almost completely replaced the venerable incandescent lamp for this function, even to the degree that LED assemblies are available with bases that directly fit incandescent indicator lamp sockets for older equipment. Light-emitting diodes offer many advantages. They are extremely rugged, have extremely long life, are inexpensively priced, produce high luminance, have relatively low power consumption, and generate very little heat. Arrays of series–parallel connected red LEDs are even replacing the traditional incandescent lamp for automobile brake lights and traffic signals. Also, LEDs are particularly well suited to driving fiber-optics for remote indicator applications because of the narrow emitted-light beam that some devices produce. They are also well matched to digital driving electronics since most require less than 5 V for operation.

Multiple LEDs were easily adapted to one of the earliest data displays, the seven-seg-

Figure 3.32. Seven-segment LED numeric display.

ment numeric display (Figure 3.32). The more complex dot-matrix, alphanumeric display (Figure 3.33) is less widely used. The first generation of electronic pocket calculators used red LED displays, although this was soon supplanted by the LCD with its infinitesimal power consumption. The appearance of the non-emitting LCD display was not as aesthetically pleasing as that of the LED display, but its longer battery life and viewability even in direct sunlight were deemed of far greater importance. The LCD did not completely displace the LED display for battery-powered devices: They are often used to backlight LCDs for viewing in darkness.

Larger matrix-addressed red, yellow, and/or green LED displays have found applications for aircraft cockpit displays, for military command centers, and in signs for displaying advertising or other messages. Matrices comprised of both multiple discrete LEDs and arrays of LEDs on a common substrate have been used depending on the application. The LEDs are most suitable for applications where commercial power or vehicle power is available rather than self-contained battery operation for portable products.

Now that blue LEDs are just becoming commercially viable and their efficiency increasing, they may also find application for large video displays in combination with the already inexpensive and efficient red and green LEDs available today. Cost is still high for blue LEDs, and their visual efficiency is low, primarily because of the insensitivity of the human eye to blue but also because they have not yet had the benefit of the years of evolutionary improvements of red, yellow, and green LEDs with their high-volume production.

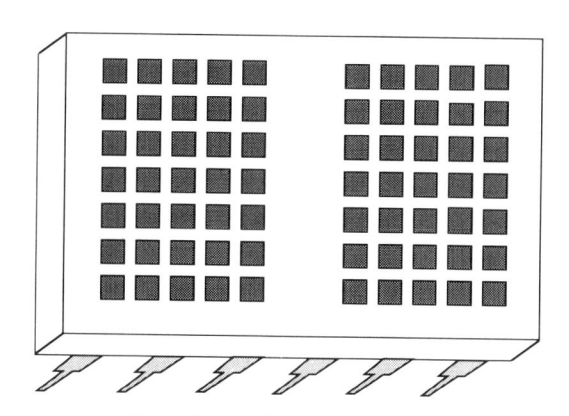

Figure 3.33. Matrix LED display.

3.9 OTHER DISPLAYS

The previously discussed displays account for an estimated 99% of the commercially available displays that are likely to be encountered. Several other display technologies exist that are obsolete, in limited use, or under development. Only those that are in current use or near commercialization will be discussed, and then only briefly for the sake of completeness, since they may be encountered in the literature or developmental situations.

Light Valve

Light valve displays have been with us since the introduction of the the Skiatron and Eidophor systems in 1939–1940.[72] Light valve displays are used to this day for large-screen projection systems with up to theater-size screens. They combine CRT technology with solid or liquid crystal cells and an external light source. Since the light is produced by a high-power external arc lamp rather than by conventional CRT phosphors that are limited by thermal damage that occurs at high input power levels, much higher screen luminance is possible.

The Skiatron used a single thin, transparent, flat crystal wafer of an alkali-halide. This crystal was coated on both sides with transparent conductive films. When scanned with a conventional electron beam raster, dark color centers formed near the surface of the crystal. A voltage applied across the cell would draw the charges forming the color centers through the insulating crystal until they reached the far side and dissipated. By selection of the proper crystal thickness and applied voltage, the transit time duration of the color centers could be matched to the frame rate of the raster. Light from a high-intensity lamp was imaged through the crystal and projected on the distant screen.

The Eidophor system was similar except that Schlieren optics and a fluid layer were used in place of the thin crystal. Deformation of the fluid surface by charges deposited by the electron beam formed images through a multiple slit arrangement.[73-75] Later developments by the General Electric Company added color to the same basic device, which they produced until recently under the trade name Talaria.[76-78] Other related approaches included the use of a deformable metal film by RCA[79] to achieve similar results to the Eidophor system; the Lumatron by CBS Laboratories, which instead used distortion of a thermoplastic film;[80] a version using tiny mirrors by Westinghouse;[81] and one that evolved to electron beam addressed liquid crystal light valves by Hughes Research Laboratories[82] and Tektronix (Figure 3.34).[83-85]

Finally, the LCD is becoming widely used for presentation projectors in both passive and active matrix-addressed versions with full color. The latter give better results for displaying video. These are still considered light valves, just the method of addressing is different. InFocus Systems and Sharp are among the manufacturers of such systems, which either may consist of a separate LCD panel for use with conventional overhead projectors or may be packaged as a self-contained projection system.

A few of the theater-size light valve projectors have achieved a degree of market success. The large-screen light valve is a good example of a niche market. The market is somewhat limited, but no other device has the ability to produce theater-size displays with adequate luminance. It appears that the market for intermediate screen sizes between that of direct-view displays and theater screens will ultimately be the most successful light valve application. Presentations of graphics are an everyday need at most companies and conferences. And the ability to interface to a computer to display anything that can be displayed on the computer itself makes it a powerful tool for such presentations.

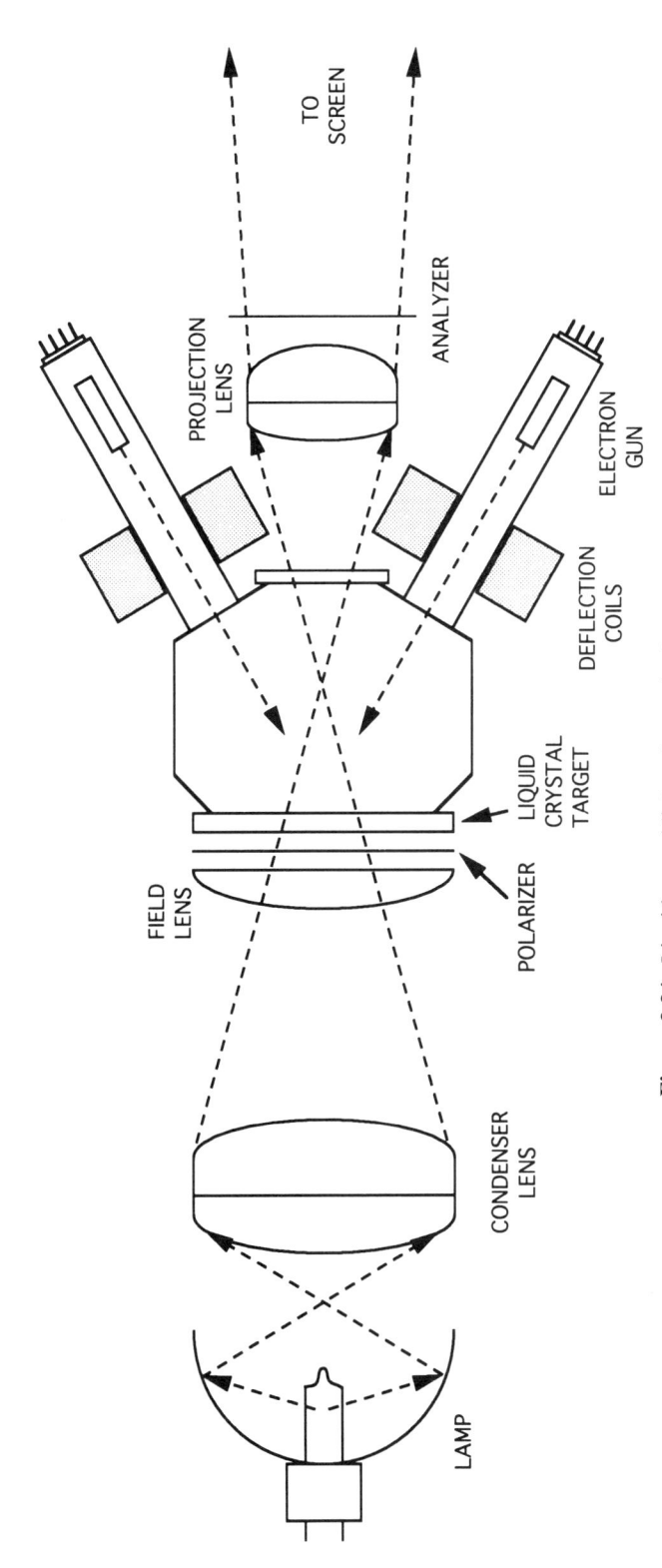

Figure 3.34. Liquid crystal light valve projection system.

Digital Micromirror Displays

The Texas Instruments digital micromirror display (DMD) is a recent approach to bright, high-resolution, high-contrast-projection displays.[86–88] The DMD uses a matrix array of small aluminum mirror elements, each one representing one pixel, suspended above SRAM cells by thin metal torsion hinges (Figure 3.35). Each mirror element is about 17 μm in size. Electrostatic forces developed between the mirrors and addressing electrodes twist the mirrors one way or the other depending on whether the digital voltage is in the on or off state. The movement of the mirror in both directions is limited by stops to allow a precise deflection angle to be produced in either state. Alignment of the optics is such that in one state light is projected to the screen and in the other state it is reflected away from the aperture of the projection lens and prevented from reaching the screen.

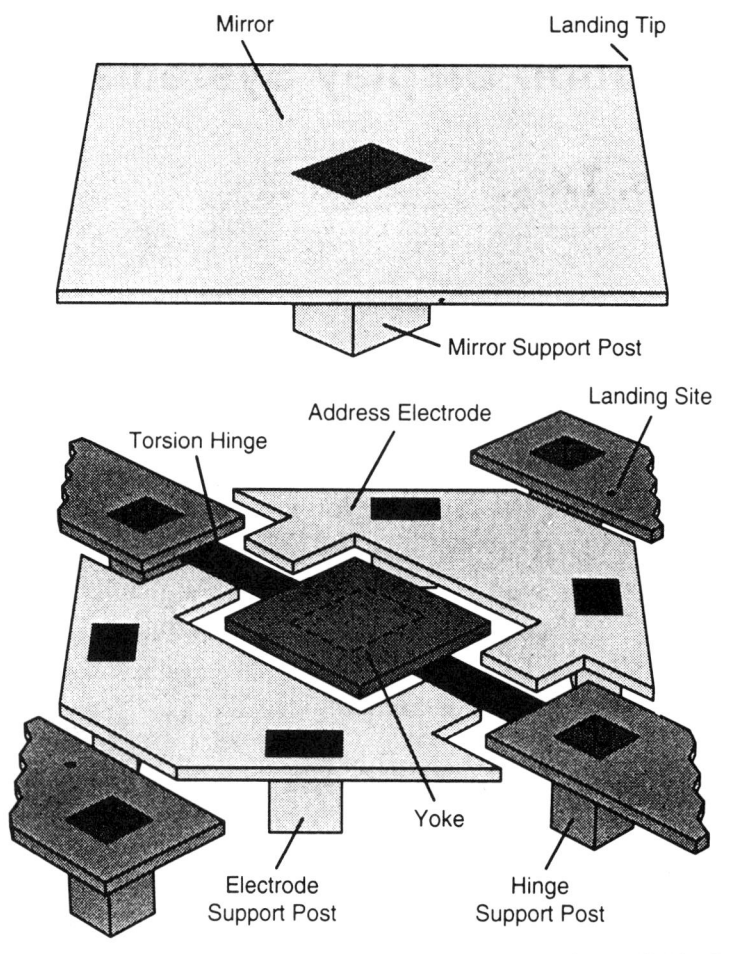

Figure 3.35. Digital micromirror display (DMD) element. (From J. B. Sampsell, "An Overview of the Performance Envelope of Digital-Micromirror-Device-Based Projection Display Systems," SID Intl. Symp. Digest of Tech. Papers, 1994, p. 670. Reprinted by permission.)

Gray scale is obtained by pulse width modulation of the driving voltage. Devices and a system for HDTV use have been fabricated jointly by Texas Instrument and Sony, although commercial production units are not yet available at the time of this writing. Demonstration units have been made in the UK by Rank-Brimar.

Jumbo Displays

Very large matrix-addressed video displays suitable for sports stadiums, arenas, and other entertainment applications have come into their own in recent years. These may be made up of a large quantity of individual incandescent lamps or light-emitting tubes similar in principle to a vacuum fluorescent display or deflectionless CRT. In some cases, modules containing several pixels may be used. Probably the most successful jumbo display has been the Sony Jumbotron.[89-91] These have been produced up to 25 × 40 m but are easily scalable for other sizes. Matsushita initially developed a system using triads of individual red, green, and blue light-emitting tubes called Diamond Vision.[92] This was superseded by their Diamond Vision Mark II, which uses modular pixel assemblies.[93] An example of a large matrix-addressed incandescent lamp display is the Panasonic Astrovision.[94]

Jumbo displays require high luminance for outdoor viewing. The relatively high cost is not an obstacle since the income from the events where they are used is considerable. This is another example of a niche display market.

REFERENCES

1. P. A. Keller, *The Cathode-Ray Tube,* Palisades, New York, 1991.
2. W. R. Aiken, "Cathode Ray Tube," U.S. Patent No. 2,795,731, filed Dec. 4, 1953, issued June 11, 1957.
3. W. R. Aiken, "A Thin Cathode-Ray Tube," *Proc. IRE*, Vol. 45, No. 12, pp. 1599–1604, 1957.
4. S. Itoh, M. Yokoyama, T. Tonegawa, K. Tsuburaya, T. Niiyama, and T. Kishino, "A Cathodoluminescent Flat Display with a Beam-Guide Deflection System," SID Intl. Symp. Digest of Tech. Papers, 1989, pp. 274–277.
5. "Monochrome Flat Tubes with 1,000-Line Resolution," *Japan Electronic Engineering,* p. 20, June 1987.
6. D. Lammers, "Matsushita Flattens CRT," *Elec. Eng. Times,* p. 29, May 29, 1989.
7. K. Smith, "Electrons Make U-Turn in Flat CRT," *Electronics,* pp. 81–82, Oct. 20, 1982.
8. M. Yamano, K. Hinotani, H. Hayama, S. Kishimoto, S. Sugishita, and M. Matsudaira, "A Color Flat Cathode-Ray Tube," *IEEE Trans. Consumer Elec.,* Vol. CE-31, No. 3, pp. 163–173, 1985.
9. K. Smith, "CRT Slims Down for Pocket and Projection TVs," *Electronics,* pp. 67–68, July 19, 1979.
10. C. Sinclair, "Small Flat Cathode-Ray Tube," *SID Dig.,* Vol. XII, pp. 138–139, 1981.
11. "Sinclair Releases Flat CRT TV as Sony Rethinks Its Watchman," *Displays,* p. 40, Jan. 1984.
12. C. Cohen, "Sony's Pocket TV Slims Down CRT Technology," *Electronics,* pp. 81–82, Feb. 10, 1982.
13. A. Ohkoshi, H. Sato, T. Nakano, T. Natori, and M. Hatanaka, "A Compact Flat Cathode-Ray Tube," *IEEE Trans. Consumer Elec.,* Vol. CE-28, No. 3, pp. 431–435, 1982.
14. "Sony Engineers Develop High-Resolution, Flat CRT," *Elec. Eng. Times,* p. 51, July 18, 1983.
15. A. Matsuzaki, H. Ikegami, M. Hatanaka, H. Kobayashi, and S. Wakita, "A Color Flat CRT and Its Application," IEEE Intl. Conf. on Consumer Elec., June 1986, pp. 76–77.

16. A. Matsuzaki, H. Ikegami, M. Hatanka, H. Kobayashi, and S. Wakita, "A Color Flat CRT and Its Application," *IEEE Trans. Consumer Elec.,* pp. 194–201, Aug. 1986.
17. L. E. Tannas, *Flat Panel Displays and CRTs,* Van Nostrand Reinhold, New York, 1985, p. 180.
18. L. A. Jeffries, "Digitally-Addressed High Brightness Flat Panel Display," SID Intl. Symp. Digest of Tech. Papers, 1973, pp. 96–97.
19. F. Noda and M. Fukushima, "Fundamental Study of a Flat CRT Using Multiple Vertical Deflection Plates," *Proc. SID,* Vol. 31, No. 3, pp. 209–213, 1990.
20. "The Grand Alliance in Flat Panels," *Business Week,* Aug. 28, 1995.
21. H. F. Gray, "The Field-Emitter Display," *Information Display,* pp. 9–14, Mar. 1993.
22. D. A. Cathey, "Field-Emission Displays," *Information Display,* pp. 16–20, Oct. 1995.
23. L. E. Tannas, *Flat Panel Displays and CRTs,* Van Nostrand Reinhold, New York, 1985, pp. 228–229.
24. S. Matsomoto, *Electronic Display Devices,* Wiley, New York, 1990, pp. 221–265.
25. S. Sherr, *Electronic Displays,* 2nd ed., Wiley, New York, 1993, pp. 204–205, 240–242, 296–297.
26. S. Zorb, "The Next Five Years," *Information Display,* pp. 18–20, Sept. 1992.
27. L. E. Tannas, *Flat-Panel Displays and CRTs,* Van Nostrand Reinhold, New York, 1985, p. 418.
28. A. R. Kmetz, "Liquid Crystal Displays: Device Characteristics and Applications," Soc. for Info. Disp. Seminar Lecture Notes, San Francisco, CA, June 8, 1984, p. 7.2-2.
29. S. Matsomoto, *Electronic Display Devices,* Wiley, New York, 1990, pp. 29–84.
30. S. Sherr, *Electronic Displays,* 2nd ed., Wiley, New York, 1993, pp. 242–259.
31. T. J. Scheffer and J. Nehring, "A New, Highly Multiplexable Liquid Crystal Display," *Appl. Phys. Lett.,* Vol. 45, pp. 1021–1023, 1984.
32. T. J. Scheffer and J. Nehring, "Investigation of the Electro-Optical Properties of 270° Chiral Nematic Layers in the Birefringence Mode," *J. Appl. Phys.,* Vol. 58, No. 8, pp. 3022–3031, Oct. 1985.
33. T. J. Scheffer and J. Nehring, "Supertwisted Nematic (STN) LCDs," Soc. for Info. Disp. Seminar Lecture Notes, Seattle, WA, May 17, 1993, pp. M7/1–M7/63.
34. T. J. Scheffer and J. Nehring, "Supertwisted Nematic (STN) LCDs," Soc. for Info. Disp. Seminar Lecture Notes, San Jose, CA, June 13, 1994, pp. M1/1–M7/84.
35. T. J. Scheffer and J. Nehring, "Supertwisted Nematic (STN) LCDs," Soc. for Info. Disp. Seminar Lecture Notes, Orlando, FL, May 22, 1995, pp. M2/1–M2/87.
36. T. J. Scheffer and J. Nehring, "Supertwisted Nematic (STN) LCDs," Soc. for Info. Disp. Seminar Lecture Notes, San Diego, CA, May 13, 1996, pp. M2/1–M2/69.
37. J. A. Castellano, *Handbook of Display Technology,* Academic, New York, 1992, pp. 196–199.
38. J. A. Castellano, *Handbook of Display Technology,* Academic, New York, 1992, pp. 199–204.
39. R. G. Stewart, "Active Matrix LCDs," Soc. for Info. Disp. Seminar Lecture Notes, San Diego, CA, May 13, 1996, pp. M5/1–M5/35.
40. S. Morozumi, "Active-Matrix LCDs," Soc. for Info. Disp. Seminar Lecture Notes, Seattle, WA, May 17, 1993, pp. M8/1–M8/31.
41. R. L. Wisnief, "Active-Matrix LCDs," Soc. for Info. Disp. Seminar Lecture Notes, Seattle, WA, May 17, 1994, pp. M8/1–M8/31.
42. F. Cuomo, J. Atherton, A. Ipri, D. Jose, R. Stewart, G. Taylor, M. Batty, M. Spitzer, D. P. Vu, B. Ellis, H. Franklin, B. Rhoades, M. Tilton, and B. Y. Tsaur, "A High Density 1280 x 1024 Transferred-Silicon AMLCD," *SID Intl. Symp. Dig.,* pp. 77–80, May 1995.
43. J. A. Castellano, *Handbook of Display Technology,* Academic, New York, 1992, pp. 204–205.

REFERENCES

44. R. Vatne, P. A. Johnson, Jr., and P. J. Bos, "A LC/CRT Field-Sequential Color Display," SID Intl. Symp. Digest of Tech. Papers, 1983, pp. 28–29.
45. P. J. Bos, P. A. Johnson, Jr., and K. R. Koehler-Beran, "A Liquid-Crystal Optical-Switching Device (π Cell)," SID Intl. Symp. Digest of Tech. Papers, 1983, pp. 30–31.
46. P. J. Bos and P. A. Johnson, "Field-Sequential Color Display System Using Optical Retardation," U.S. Patent No. 4,582,396, filed May 9, 1983, issued Apr. 15, 1986.
47. P. Bos, T. Buzak, and R. Vatne, "A Full-Color Field Sequential Color Display," Eurodisplay 84 Proc., Paris, Sept. 18–20, 1984, pp. 7–9.
48. T. J. Haven, "Reinventing the Color Wheel," *Information Display,* pp. 11–15, Jan. 1991.
49. C. Sherman, "Field Sequential Color Takes Another Step," *Information Display,* pp. 12–15, Mar. 1995.
50. P. J. Bos, "Stereoscopic Imaging System with Passive Viewing Apparatus," U.S. Patent No. 4,719,507, filed Apr. 28, 1985, issued Jan. 12, 1988.
51. P. Bos, T. Haven, and L. Virgin, "High-Performance 3D Viewing Systems Using Passive Glasses," SID Intl. Symp. Digest of Tech. Papers, 1988, pp. 450–453.
52. P. Bos and T. Haven, "Field-Sequential Stereoscopic Viewing Systems Using Passive Glasses," *Proc. SID,* Vol. 30, No. 1, pp. 39–43, 1989.
53. J. H. McCauley, "Indicator Tube," U.S. Patent No. 2,991,387, 1961.
54. J. A. Castellano, *Handbook of Display Technology,* Academic, New York, 1992, pp. 111–141.
55. K. Werner, "Emissive Flat-Panel Displays," *Information Display,* pp. 10–12, Dec. 1993.
56. J. D. Birk, "Emissives Get Brighter—with Colors," *Information Display,* pp. 20–22, Dec. 1994.
57. J. Koike, "Long-Life, High Luminance 40-in. Color DC PDP for HDTV," Asia Display '95, 1995, pp. 943–944.
58. S. Zorb, "The Next Five Years," *Information Display,* pp. 18–20, Sept. 1992.
59. S. Berry and R. Sherman, "Other Flat-Panel Displays," *Information Display,* pp. 14–16, Feb. 1995.
60. P. S. Friedman, "Are Plasma Display Panels a Low-Cost Technology?" *Information Display,* pp. 22–28, Oct. 1995.
61. T. S. Buzak, "A New Active-Matrix Technique Using Plasma Addressing," *SID 90 Digest,* pp. 420–423, 1990.
62. T. S. Buzak, P. J. Green, S. J. Guthrie, S. C. Harley, K. Hillen, G. R. Lamer, P. C. Martin, D. L. Nishida, T. L. O'Neal, W. W. Stein, K. R. Stinger, and M. D. Wagner, "16-in. Full-Color Plasma-Addressed Active-Matrix LCD," *SID 93 Digest,* 1993, pp. 883–886.
63. T. Buzak, unpublished manuscript, 1995.
64. T. S. Buzak, K. J. Ilcisin, and P. C. Martin, "The Suitability of Plasma Addressed Liquid Crystal Displays for HDTV," 2nd Liquid Crystal Seminar, Chiba, Japan, Sept. 1994.
65. R. T. Tuenge, "Recent Progress in Color Thin Film EL Displays," ITE/SID Asia Display '95, 1995, pp. 279–282.
66. J. Vieira, "More Contrast and Luminance for EL," *Information Display,* pp. 18–20, Feb. 1995.
67. R. T. Tuenge, "Recent Progress in Color Thin Film EL Displays," ITE/SID Asia Display '95, 1995, pp. 279–282.
68. J. Haaranen et al., "512($\times$ 3) $\times$ 256 RGB Multicolor TFEL Display Based on 'Color by White,'" *SID 95 Digest,* 1995, pp. 883–886.
69. C. N. King, "Electroluminescent Displays," Soc. for Info. Disp. Seminar Lecture Notes, San Jose, CA, June 13, 1994, pp. M9-1-38.
70. R. T. Tuenge, "Recent Progress in Color Thin Film EL Displays," ITE/SID Asia Display '95, 1995, pp. 279–282.

71. A. Kato, M. Katayama, K. Sugiura, K. Inoguchi, H. Ishihara, N. Ito, and T. Hattori, "An RGB 8-Color EL Display in a Stacked Panel Configuration Using Unfiltered SrS:Ce," ITE/SID Asia Display '95, 1995, pp. 287–290.
72. A. H. Rosenthal, "A System of Large-Screen Television Reception Based on Certain Electron Phenomena in Crystals," *Proc. IRE*, pp. 203–212, May 1940.
73. E. Labin, "The Eidophor Method for Theater Television," *J. SMPTE*, Vol. 54, pp. 393–406, 1950.
74. E. Baumann, "The Fischer Large-Screen Projection System," *J. SMPTE*, Vol. 60, pp. 344–356, 1953.
75. G. J. Chafaris, "Display Technology—Today and Tomorrow," *SID Proc.*, Vol. 10, No. 1, pp. 3–8, 1969.
76. T. T. True, "Color Television Light Valve Projector System," IEEE Intl. Conv., Session 26, 1973.
77. W. E. Good, "Recent Advances in the Single-Gun Color Television Light Valve System," SID Intl. Symp. Digest of Tech. Papers, 1975, pp. 24–25.
78. T. T. True, "Recent Advances in High-Brightness and High-Resolution Color Light-Valve Projectors," SID Intl. Symp. Digest of Tech. Papers, 1979, pp. 20–21.
79. J. A. van Raalte, "A New Light Valve for Television Projection," *J. SMPTE*, Vol. 80, pp. 474–476, 1971.
80. R. J. Doyle and W. E. Glenn, "Lumatron: A High-Resolution Storage and Projection Display Device," *IEEE Trans. Elec. Dev.*, Vol. ED-18, No. 9, pp. 739–747, 1971.
81. A. S. Jensen, "The Mirror Matrix Tube: A Novel Light Valve for Projection Displays," SID Intl. Symp. Digest of Tech. Papers, 1975, pp. 28–29.
82. A. D. Jacobson, J. Grinberg, W. P. Bleha, L. S. Miller, L. M. Graas, and D. D. Boswell, "A New Television Projection Light Valve," SID Intl. Symp. Digest of Tech. Papers, 1975, pp. 26–27.
83. D. A. Haven, "Electron-Beam Addressed Liquid-Crystal Light Valve," *IEEE Trans. Elec. Dev.*, Vol. ED-30, No. 5, pp. 489–492, 1983.
84. D. Haven, D. Whitlow, and T. Jones, "Full-Color Electron-Beam-Addressed Light-Valve Projector," SID Intl. Symp. Digest of Tech. Papers, 1986, pp. 372–374.
85. T. S. Buzak, R. S. Vatne, G. A. Nelson, and D. E. Whitlow, "Electron-Beam-Addressed Light Valve Operating Modes and Performance," SID Intl. Symp. Digest of Tech. Papers, 1987, pp. 64–67.
86. J. B. Sampsell, "An Overview of the Digital Micromirror Device (DMD) and its Application to Projection Displays," SID Intl. Symp. Digest of Tech. Papers, 1993, p. 1012.
87. J. B. Sampsell, "An Overview of the Performance Envelope of Digital-Micromirror-Device-Based Projection Display Systems," SID Intl. Symp. Digest of Tech. Papers, 1994, pp. 669–672.
88. G. Sextro, T. Ballew, and J. Iwai, "High-Definition Projection System Using DMD Display Technology," SID Intl. Symp. Digest of Tech. Papers, 1995, pp. 70–73.
89. J. A. Castellano, *Handbook of Display Technology*, Academic, New York, 1992, pp. 165–168.
90. H. Nakagawa and A. Ohkoshi, "A New High-Resolution Jumbotron," SID Intl. Symp. Digest of Tech. Papers, 1986, pp. 246–249.
91. M. Hayashi, T. Muchi, M. Ozeki, T. Masaloki, and T. Arae, "A 15-mm Trio-Pitch Jumbotron Device," SID Intl. Symp. Digest of Tech. Papers, 1989, pp. 98–101.
92. N. Fukushima and N. Terazaki, "A Light-Emitting Tube Array for Giant Colour Display," *Displays*, pp. 207–211, Oct. 1983.
93. N. Shiramatsu, M. Nakano, S. Iwata, M. Sanou, N. Terazaki, and S. Futatsuishi, "A High-Resolution High-Brightness Color Video Display for Outdoor Use," *Proc. SID*, Vol. 30, No. 4, pp. 309–312, 1989.
94. W. E. Glenn, "Television Displays for Stadiums," *Information Display*, pp. 22–25, Jan. 1987.

CHAPTER 4

MEASUREMENT INSTRUMENTATION

Electronic displays present information that may be characterized by a number of photometric, radiometric, colorimetric, and spectroradiometric measurements. No single measurement can describe how the human eye, a very complex device, sees a display and perceives its quality. Luminance, contrast, chromaticity, resolution, uniformity, registration, and viewing angle all play a part. To fully evaluate a display, several instruments and a test generator may be required, although the most important parameters may be evaluated with less instrumentation depending on individual requirements. Display developers usually require the most accurate (translation expensive) instrumentation for research and design while manufacturing lines do not usually require extreme absolute accuracy. Ruggedness, repeatability, and cost will be the highest ranking factors for the latter situation. Finally, the end user may only require an easy-to-use instrument capable of maintaining consistency of display appearance from day to day for a single display.

Available instruments range from the basic $50–100 incident-light meter to $50,000 spectroradiometers. Some photometers use separate heads to measure different parameters, thus avoiding expensive duplication of the electronics portion where more than one function must be measured. They also allow expansion of a basic photometer at a later date as needs grow. We will now look at the different classes of light measurement instruments and the performance/cost trade-offs involved in selecting suitable instruments for specific display measurements.

There are a number of modern photometric instruments available today. These cover a wide range of measurement needs at an equally wide range of prices. The application will determine just how much instrument capability is required today, but keep in mind the need for flexibility for tomorrow's requirements. The state of the art in displays and instruments is rapidly advancing, and photometric and colorimetric testing will play an increasingly important role in the future for both the manufacturer and the customer.

4.1 PHOTOMETERS

Photometers are most commonly designed to measure either illuminance (light incident on a surface) or luminance (light emitted or reflected from a diffuse surface). A third measurement, luminous intensity, is often used to quantify LED performance. Two characteristics of a photometer are of greatest importance: spectral and spatial. Photometers for each of the three measurement types must have spectral (color) response similar to that of the human eye. Just how close a spectral match is required depends on the application. The three measurements differ primarily in the geometric relationships between the light source and the photometer. Note that almost all of the performance parameters, except the geometric characteristics, that will be discussed under Illuminance Photometers apply equally well to luminance and LED photometers.

Common Characteristics

Spectral As previously mentioned, the spectral sensitivity should be similar to that of the human eye rather than that of the sensor alone, most often a silicon photodiode. Only in cases where the photometer is calibrated with a light source with an identical spectrum to that being measured may a spectrally uncorrected sensor be used.[1] When only incandescent lamps were used for both calibration and illumination, it was possible to use simple lightmeters. This is rarely possible today with the plethora of modern light sources and displays encountered and their widely differing spectra. These include CRTs, plasma displays, liquid crystals, EL devices, LEDs, and conventional fluorescent, trichromatic fluorescent, tungsten halogen, mercury, metal halide, and both low- and high-pressure sodium lamps. Even the age-old illumination source, daylight, is anything but stable spectrally. It varies greatly with time of day, degree of overcast, altitude, angle of incidence, pollution levels, and ratio of skylight to direct sunlight. To avoid inaccuracies from sources of different spectra, colored filters are placed in front of the sensor to alter its spectral sensitivity to simulate that of the internationally agreed upon 1931 CIE Standard Observer or *photopic* curve. This is sometimes known as *color correction* or more frequently and more correctly as spectral or photopic correction. Early photometers used the human eye itself to compare known light sources to those being measured, thus avoiding some but not all of the spectral pitfalls. Visual comparison of different colored sources is difficult without special techniques, and different individuals have different spectral sensitivities, especially with different ages. The 1931 CIE Standard Observer was a means to provide for standardized photometric measurement techniques. See Chapter 1 for more information on the CIE Standard Observer.

Spectral correction filters for early phototube and selenium photovoltaic cells began with cells containing colored chemical solutions that could be spectrally controlled by ratios of the solutions and by their concentrations.[1] This progressed to the use of gelatin Viscor filters by Weston and Wratten-type gelatin filters by Kodak. The Wratten 102 filter was a reasonable photopic simulation for selenium cells while the 106 was useful for vacuum phototubes with cesium–antimony photocathodes. Lack of stability with humidity and time and the variations of spectral response from one detector to the next were serious limitations of gelatin-type filters for precision measurements. Gelatin filters were standard for many years until the advent of more sophisticated light sensors and sources spawned by a combination of technology advances and energy conservation pressures.

The silicon photovoltaic solar cell of the 1960s formed the basis of the silicon photodiode for measurement purposes in the 1970s. It and the closely related CCD have grown to be the most important of all detector types today for most photometric, radiometric, colorimetric, and spectroradiometric applications. The silicon photodiode offers many advantages over earlier light sensors. These include excellent long-term stability, a wide linear dynamic range of 8–10 decades, lack of hysterisis or "memory" effects, and a wide spectral sensitivity range from the ultraviolet to the infrared. Actually, the latter is sometimes considered a fault for photometric applications. The silicon photodiode's high sensitivity in the infrared can sometimes result in anomalous response from incandescent lamps, which are rich in infrared energy if the sensor is not adequately filtered by proper photopic filter design. Silicon photodiodes have turned out to be so stable that they have become standards in themselves, replacing the standard calibration lamps previously used in many cases. See Chapter 2 for further information about the characteristics and operation of silicon photodiodes.

With the introduction of the silicon photodiode, the need for accurate photopic correction for photometry became apparent. Multielement colored glass filters proved to be the answer. These are found in two forms, the homogeneous (stacked) filter and the mosaic filter. Their physical construction is illustrated in Chapter 2. Their various glass filter elements selectively reduce or remove portions of the spectrum to produce a sensor having photopic response. The stacked filter has advantages for portable photometers of lower cost and spectral correction that is spatially symmetrical about the central axis. The mosaic filter is best suited to laboratory use because of its higher transmission and greater spectral accuracy.

Accuracy of the sensor's match to the CIE Standard Observer curve has been historically specified as integrated accuracy, a meaningless specification since almost any sensor may be made to have an integrated photopic inaccuracy of 0% merely by choosing the optimum calibration wavelength to which to normalize the curve. Deviations above the photopic curve are averaged with those below. A curve with deviations of +20% for half of the curve and −20% for the remainder would have 0% integrated inaccuracy and would thus appear better than one that was +1% through the entire spectrum. An actual spectral curve of the sensor compared to the CIE Standard Observer (Figure 4.1) or the percent deviation from the Standard Observer (Figure 4.2) is a better approach. Deviations from the ideal response may easily be seen, and the wavelength and magnitude of such deviation may be used to estimate their effects on a particular measurement if the spectrum of the source is known. A still better method of specifying photopic accuracy is described in a recent CIE standard.[2,3] The deviation of relative spectral responsivity from the photopic curve is specified by a value, f_1' in percent, which essentially is a sum of the errors *regardless of sign* for each wavelength. This is an excellent way to specify photopic

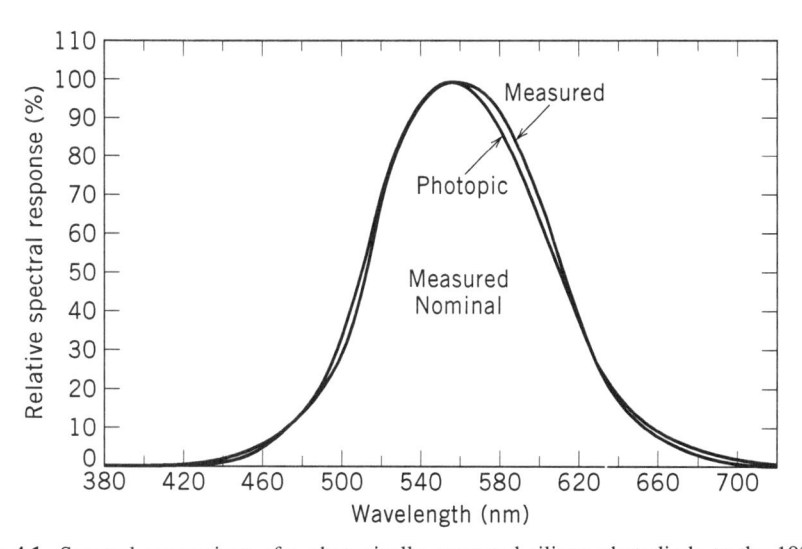

Figure 4.1. Spectral comparison of a photopically corrected silicon photodiode to the 1931 CIE Standard Observer curve.

132 MEASUREMENT INSTRUMENTATION

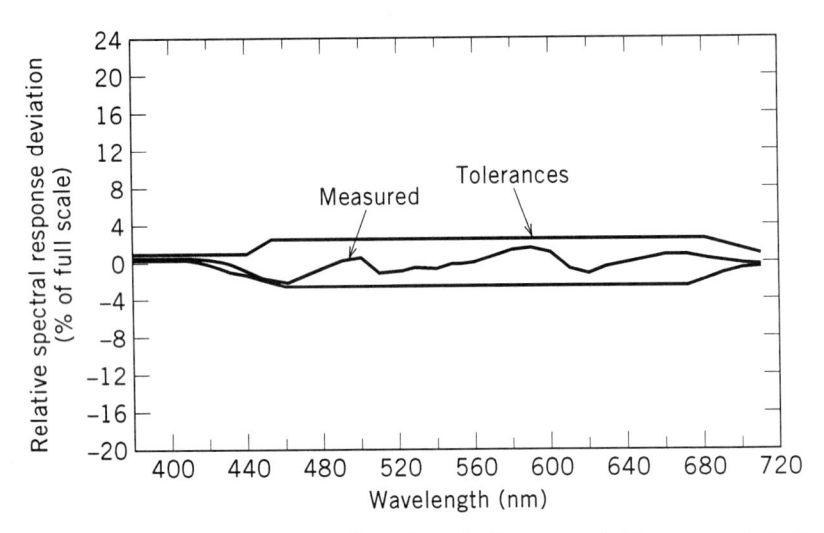

Figure 4.2. Spectral deviation (percent) of the photopically corrected silicon photodiode illustrated in Figure 4.1 from the 1931 CIE Standard Observer curve. The manufacturer's limit specifications are shown for the visible spectrum.

accuracy, but it is just beginning to be found in photometer specification sheets. If you do not see this value in a manufacturer's specification sheet, ask for it. Probably because the numbers are larger than the 1 or 2% integrated values historically used, they have been a little slow to be accepted. The f_1' value might be more like 4–5% and would be at an unfair competitive disadvantage when compared to an instrument specification of integrated accuracy by a user that does not understand the difference between the two specifications. Laboratory instruments using mosaic filters are available with f_1' values of under 1.5% for illuminance photometers and 2% for luminance photometers (DIN class L) while quality portable instruments that use stacked glass filters should be under 3% (DIN class A).[4] Low-cost lightmeters may be up to 6% (DIN class B) or worse (DIN class C, 9%).

Compared to electrical measurements, photometry is a very inexact science. Even such long-established institutions as the internationally agreed upon CIE photopic response curve originally adopted in 1924 is still subject to much debate as to its actual shape. And beware of accuracy specifications that are much better than 5% for a general-purpose photometer. The electrical accuracy of the instrument (typically 1–2%) is sometimes emphasized, but photometric accuracy is limited by a number of spectral, spatial, and temporal variables that make 5% a more realistic accuracy specification.

Two additional tests are specified in CIE Publication No. 69 on the subject of spectral response. These are the amount of unwanted response to infrared (r) and ultraviolet (u) since, by definition, a sensor matched to the CIE Standard Observer response curve should not respond to either. The most common problem is with a poorly made photopic filter used with a silicon photodiode. Some of these filters may exhibit considerable infrared transmission. Silicon photodiodes have their highest sensitivity in the near infrared. Combine these two factors with the very high infrared output from the tungsten or tungsten–halogen lamps commonly used for calibration of photometers and you have the potential for considerable error. If only incandescent light sources were to be measured, the errors would be minor unless the color temperature differed greatly from that of the calibration source (usually 2850–3200 K). But displays and fluorescent lighting usually have little if any infrared present and would read lower than the true value if such a sensor was used. The same sensor could prove entirely adequate *if* calibrated with a light source spectrally identical to that to be measured. The values of *r* and *u* are not often available from illuminance meter manufacturers. The best way to avoid these and other potential problems is to use good-quality instruments from reputable manufacturers.

There is a relatively simple test for infrared response that will approximate the CIE *r* value, although the actual values obtained will be somewhat different because of the additional mathematical analysis used in the CIE derivation. A Schott RG830 glass filter of 3 mm thickness is inserted between a source of illuminant A and the sensor in a way that precludes any unfiltered light from the source reaching the sensor. Illuminant A is an incandescent lamp operating at 2854 K and is conveniently simulated using nothing more complicated than a standard 60-W light bulb at its rated operating voltage. A source-to-sensor distance of 1 m is usually convenient. Measurements are made with and without the filter in place in a manner that does not alter the lamp-to-sensor geometry. With the filter in place, the reading should be less than about 2% of the reading obtained without the filter.

Other important considerations for photometers addressed by CIE Publication No. 69 include directional response (f_2), linearity (f_3), display error (f_4), fatigue (f_5), temperature dependence (f_6), response to modulated light (f_7), polarization error (f_8), response to nonuniform illumination (f_9), peak-overload capability (f_{10}), settling time, range change errors (f_{11}), and focus error (f_{12}). Not included in CIE 69 are warmup, influence of supply voltage, aging, zero drift, environmental effects, ruggedness, portability, safety, traceability, interfaces, and convenience of use. You can expect to see increasing use of the CIE specifications in the future once vendors and customers get over the initial shock of specifications that, on the surface, appear worse than older products.

Range and Linearity Photometers usually have several ranges. The best silicon photodiodes may span light levels of eight decades or more,[5] although this is far more than the average user requires. The photometer selected should ideally have "headroom" of approximately 5–10 times the maximum anticipated light level to ensure operation in the region of best linearity and to allow for the ability to measure evolutionary improvements in future displays. At the other extreme, a noise floor and resolution of at least one-tenth the minimum expected light level will maximize accuracy. This may not always be possible, especially for luminance photometers, which trade off sensitivity for narrower acceptance angles.

Nonlinearity, often a handicap associated with the older selenium cell, photoconductive cell, phototransistor, and phototube sensors, is largely an artifact of the past for moderate- and premium-priced photometers. Today's silicon photodiode is capable of less than 1% nonlinearity over several decades of sensitivity.[6] The only factor of concern is the nonlinearity which may occur near the maximum short-circuit current (I_{sc}) of the photodiode. The I_{sc} is principally limited by series resistance of the contacts to the silicon wafer and, to a lesser degree, by heating of the silicon by light sources bright enough to cause saturation to occur. Because of the rounding of the current-versus-light-intensity curve that occurs near saturation (Figure 4.3), manufacturers of photometers try to operate at a maximum of one-tenth of the saturation current. Also, some additional headroom is desirable since I_{sc} is slightly different for every photodiode, even from the same lot. The lower usable limit is determined by the photodiode shunt resistance combined with the input bias current of the transimpedance amplifier that it feeds.

Silicon Photodiode Linearity

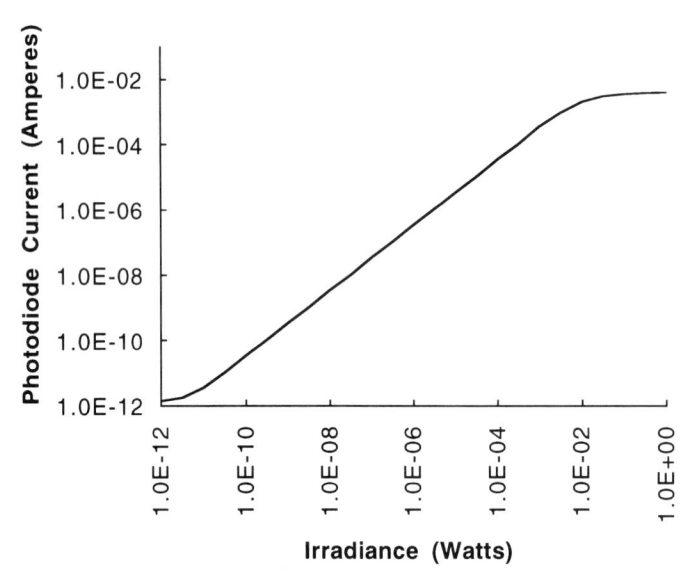

Figure 4.3. Typical silicon photodiode linearity. Note deviation from linear response above approximately 1 mA caused by series resistance losses. Deviation also is noticeable below about 10 pA due to bias current offsets in the operational amplifier combined with the shunt resistance of the photodiode. This will depend to a large degree on the amplifier itself.

Range change errors (CIE f_{11}) are present to at least a small degree in most photometers. The cause is mostly from slight differences in amplifier gain from the ideal value for each range. For instance, a reading of 90.0 lx on the range displaying the maximum resolution without overranging might read 89 lx on the next (less sensitive) range. According to CIE Publication No. 69, the f_{11} value would be $[(89/90.0) - 1] \times 100\%$, or 1.1%. Gain determination resistors of 1% accuracy or better are inexpensive and commonly used. Since overall photometer accuracies are about 5%, using less commonly available 0.1% resistors will improve accuracy only marginally. Beware of photometers claiming 1% accuracy. This is electrical accuracy rather than photometric accuracy and is misleading.

Many instruments today employ "autoranging" as a convenience feature. Autoranging is a useful feature, but the ability to override it can be equally useful at times. It may be defeated to speed up the time to take a measurement if the readings are always of the same magnitude. Each time that the instrument automatically switches ranges on its way to displaying a final value, a certain amount of settling time will be added before stabilizing and displaying a reading. This is particularly noticeable when a sensor that has been in darkness is suddenly exposed to a much higher level.

Fatigue *Fatigue* is a temporary change in sensor sensitivity with recent history of light exposure. The selenium cell, photoconductive cell, and phototube usually suffer a loss of sensitivity or inability to return to a reading of zero after exposure to bright light. Carried to an extreme, the degradation may be permanent. Fatigue disappeared with the advent of the silicon photodiode, which is capable of operation even at light levels that cause some heating without permanent damage or even temporary alteration of its characteristics.

Usually silicon sensors can be exposed to any light level that one is willing to expose one's hand to at the same distance. It may be way above the measurement range of the instrument but will not result in sensor damage.

One other sensor, the photomultiplier tube (PMT) should be mentioned at this point. The PMT, although fairly expensive, less stable, and requiring a high voltage power supply, is a good choice for extremely low light level measurements or for "spot photometers" measuring very small areas (minutes of arc) at more moderate light levels. The high current amplification of the PMT, as high as ten million times, makes it unforgiving of overexposure to light. At the very least, temporary changes in dark current (noise level), sensitivity loss, and/or spectral response changes usually result. Often these changes will be permanent. Never expose a PMT directly to ambient light with voltage applied to it and use care to always operate it at the lowest possible light level consistent with the task at hand. Even exposure to fluorescent light with no power applied to it will often require a recovery period in darkness before its full performance capability returns. This is especially true for the popular S20 photocathodes.

Response Time Cathode-ray tubes and some other displays as well as most sources of ambient light, with the exception of daylight, are of a pulsed nature. Display refresh rates are usually in the range of 50–80 Hz while fluorescent and other lighting exhibit either 100 or 120 Hz modulation since maximum light is emitted at both the positive and negative peaks of the line voltage waveform. The human eye becomes sensitive to flicker at refresh rates below about 75 Hz. Therefore, higher refresh rates where the eye responds strictly to the average illumination level are desirable to minimize flicker. Photometers for illumination and displays should have time constants similar to or a little longer than that of the human eye in order to integrate the light intensity adequately for an average reading. This is especially important for photometers having digital displays. If the time constant of the photometer is not long enough, the readings will tend to vary rapidly as it samples the fluctuating light level at different points in time. Too long a time constant will cause slow response each time that a different light level is measured since the photometer will have to take a number of readings before settling to the correct average reading. A time constant of about 100 ms approximates that of the human eye.

Other factors to consider when measuring the average intensity of light sources with very high peak intensity and relatively low average intensity are the overload characteristics of the sensor and amplifier. This is often not a problem with modern photometers and radiometers for most common light measurements since there is usually considerable "headroom" because of the wide dynamic range of silicon photodiodes. If a problem is suspected, it is easy to check for using a calibrated thin-film neutral density filter. Placing the filter in front of the sensor while making the measurement in question should result in the reading decreasing almost exactly by the transmission of the filter. CIE Publication No. 69 prescribes methods for evaluating and specifying the response to modulated light (f_7) and the peak-overload characteristics (f_{10}) of photometers, although these are not often specified in photometer catalogs.

This is not to say that all photometric measurements dictate the need for slow response times. There are times when fast response times are required for photometry or radiometry. These include measurement of peak intensity and rise- and falltime characteristics of phosphors, strobe lamps, and pulsed lasers. These measurements require special techniques that will be elaborated on in Chapter 9.

Power Supply Variations Power supply voltage variations are not usually a concern in good-quality instrumentation. Precision voltage regulator integrated circuits are usually used for all critical circuits and add very little manufacturing cost to photometric instruments. Battery-powered instruments may be a bit more sensitive to supply voltage since the range of useful battery voltage may change by 20–30% from a fresh battery to the point at which the low battery indicator comes on. The instrument should meet all specifications within that range; however, best accuracy and minimum drift will be obtained in the "plateau" region of the discharge curve (Figure 4.4). Never trust measurements taken after the low battery warning comes on even for those instruments that appear to function normally beyond that point. If stable line power is readily available, it is advisable to use AC power adaptors for battery-powered instruments that have provision for them. The savings in battery cost and avoidance of the inconvenience of having the batteries expire right in the middle of a series of important measurements make this an attractive alternative, and the additional cost of wall-mount power adaptors is minimal.

Aging Aging is the irreversible change in responsivity of a photometer with time. This may be due to changes in electrical gain of amplifiers or spectral response of the sensors. The electrical gain is easily corrected during periodic recalibration. Always follow the manufacturer's recommendation for the interval between calibrations. This is particularly important with the advent of International Standardization Organization (ISO) 9000 requirements but has been required in the past for instrumentation used on U.S. government contracts as well. Not so obvious is the need for periodic verification of spectral responsivity of photometric sensors. Humidity, temperature, atmospheric contaminants, and physical abuse can cause changes in photopic correction filter spectral response over time, especially for lower cost photometers where gelatin filters are used rather than glass. Even

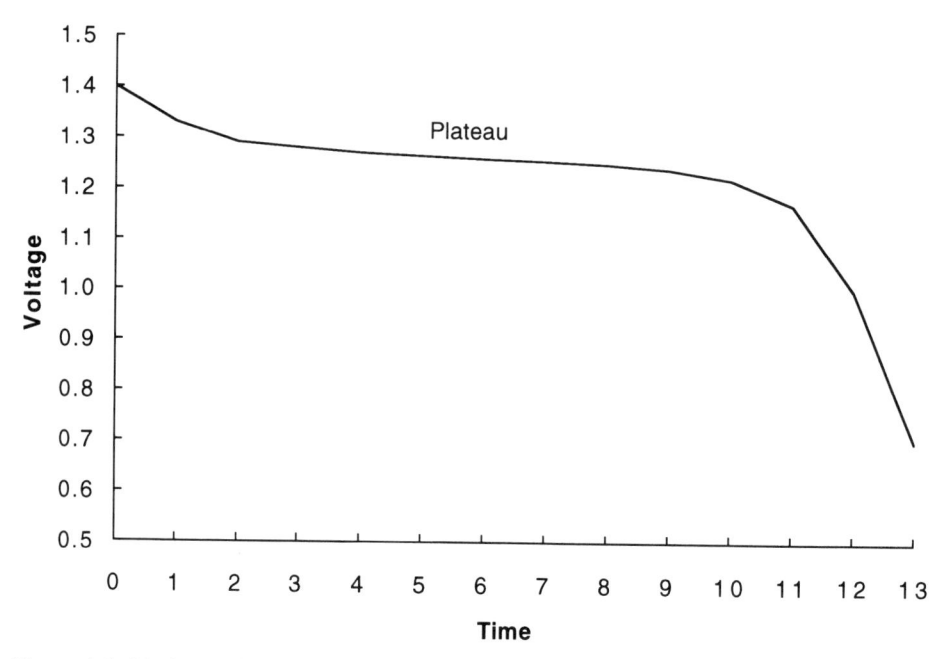

Figure 4.4. Discharge characteristics of a typical NiCd battery with a constant load. Note the relatively level "plateau" where most stable instrument operation is found.

glass filters are not totally immune to aging. Severe operating conditions can cause delamination of the filter stack or, in extreme cases, portions of the glass may turn frosty on outer surfaces or to powder for certain hygroscopic glasses. At the time photometers are recalibrated for sensitivity, sensor assemblies should be visually inspected and spectral response curves taken if possible to verify meeting the original specifications. A substantial change in instrument calibration since the previous calibration is indicative of a possible deterioration of the sensor assembly and should be investigated further.

Zeroing Zero drift is usually related to temperature changes, either internally during warmup or in the ambient operating environment. Autozeroing is designed into some instruments; others require manual adjustment. Zero drift is most troublesome when measuring very low light levels and can cause large offsets in some cases. Always follow the manufacturer's procedure for zeroing photometers. Verify the zero by blocking all light to the sensor and making sure the reading is near zero. Some fluctuations of the zero reading may be observed on the most sensitive range and are to be expected at the picoampere sensor output levels encountered at less than about 0.1 lx, or 3 cd/m^2. Experience with a particular photometer will allow you to judge whether the amount of zero drift or noise fluctuations with the sensor covered are normal. The average reading obtained with the sensor covered may be nulled out using the "zero" adjustment if one is present or subtracted from any readings taken. Be sure to note the polarity of the reading to be subtracted. Even though we do not have negative light, we certainly can have negative electrical offsets. Also, the zero reading may be different for each range. Use the zero offset for the particular range being used when subtracting it out. Ideally, the zero should be rechecked before and after each reading taken on the most sensitive range. This will compensate for a drifting zero. No specification for zero error or drift has been recommended by CIE yet.

"Autozeroing" is found in a number of photometers and related instruments. Autozeroing may occur either at turn-on and/or periodically during operation. Usually instruments that autozero upon turn-on require the sensor to be covered so that no light reaches them while that function is performed.

Environmental Environmental effects on photometers cover a multitude of considerations. These include temperature, humidity, altitude, shock, vibration, transit, drop, electromagnetic emissions, electromagnetic susceptibility, and electrostatic discharge (ESD). The environmental specifications are stated by the manufacturer in their catalogs or manuals. Generally speaking, the wider the range of environmental specifications that an instrument will meet, the better it will perform under normal operating conditions since to meet environmental requirements additional attention must be devoted to the design. Of course, this is also manifested in a higher price.

The temperature range of photometers is often limited by the LCD, which may have slow response time at low temperatures.

The temperature dependence of photometers is of less concern for the sensor than for the high-sensitivity, transimpedance operational amplifiers used to convert the low current levels (sometimes picoamperes or even femtoamperes) from the sensor to a voltage for the digital voltmeter circuitry used to display the reading. Silicon photodiodes do have a temperature coefficient, but the resultant sensitivity changes occur mostly in the ultraviolet and infrared regions. Figure 4.5 illustrates the changes in spectral sensitivity with temperature for a typical photodiode.[7] Note that very little change is evident in the visible

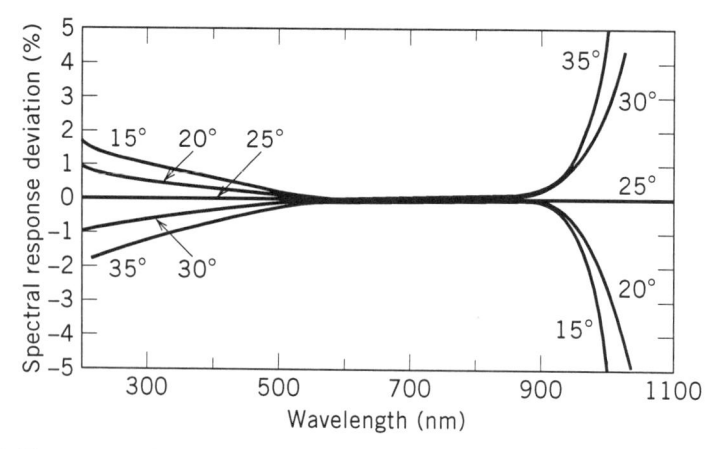

Figure 4.5. Temperature dependence of a silicon photodiode versus wavelength. (From EG&G Electro-Optics, "Silicon Photovoltaic Detectors and Detector/Amplifier Combinations," Application Note D3011C-8, Jan. 1984, Fig. 3.)

spectrum of interest for display measurement. Because most photodiodes for photometric measurements are operated in the "photoamperic" mode, which is free of dark-current effects, the temperature coefficients of silicon are not much of a consideration either. One temperature effect of the silicon photodiode is of concern for photometry. The bulk resistivity of silicon decreases inversely with temperature. Typically, a 10°C temperature increase will cause the silicon photodiode shunt resistance to decrease by 50%. By itself this is not significant, but since this is connected directly to the input transimpedance amplifier, it causes a doubling of offset due to the effect of the amplifier input bias current. This causes the zero offset of the amplifier to change as well as the sensitivity of the zero adjustment to become "touchier." The same change in bulk resistivity versus temperature is the cause of dark-current effects in biased silicon photodiodes. Here, a greater current will flow from the bias power supply through the photodiode and its load resistor, which feed the input amplifier. Biased operation is only a concern for measurements requiring fast response times. This is usually in the sub-microsecond realm.

"Warm-up" of photometric instrumentation is no longer a major concern in these days of digital, low-power, solid-state electronics. One or 2 min usually is sufficient for the active devices to stabilize their temperature and any leakage currents. A few instruments use cooled detectors to minimize dark current and noise. Adequate time must be allowed for them to reach equilibrium in this case. In any event, make sure to follow the manufacturers' recommendations if they specify a minimum warm-up time. Of greater concern is the need to allow time for the instrument to stabilize its temperature when moved from a cold ambient environment to a warm one or vice-versa. The instrument should not be operated during this period of time. Condensation may also be a problem with change of ambient temperature in a humid environment. The input amplifiers, often capable of measuring mere picoamperes at their highest sensitivity in instruments having a wide dynamic range, may develop zeroing or gain errors if condensation is present. Allow at least $\frac{1}{2}$ to 1 h for stabilization of temperature and drying of condensation.

The maximum humidity that the instrument should be exposed to will usually be limited by electrical leakage when measuring low light levels and/or photopic filter degradation. Altitude is usually of little concern since no high voltage is present that might cause arcing at high voltage except for those photometers utilizing photomultiplier tubes. Shock, vibration, drop, and transit testing are usually based on MIL-T28800 and are mostly a concern for portable instruments where physical ruggedness is a necessity.

EMC and ESD Requirements for European Community (EC) certification and marking require electromagnetic emissions and susceptibility to be tested to European Norm EN-50081-1 and EN-50082-1, respectively, before they may be sold into the European Community. Electrostatic discharge testing to International Electrotechnical Commission (IEC) standard 1000-2 at a minimum of 8 kV and preferably 15 kV ensures that instrument damage will not result for normal static-induced discharge such as generated from carpets and clothing on cold dry days. The IEC standard 1000-2 is expected to be replaced by another European Norm, EN-61000, in the near future.

Safety Safety certification to Underwriters Laboratories (UL) and the Canadian Standards Association (CSA) have been common. Many other countries have had their own safety standards, but these have often not been mandatory or have been simply ignored due to low sales volume into some of these countries. The European Union is changing all of that with harmonized standards and strict adherence to them before they are allowed to be imported. Another European Norm, EN-61010-1, will replace the current IEC standard 1010-1, which in turn has replaced UL 3111-1 and CSA C22.2 #231. The newer international standards are achieving worldwide status, which is helping to avoid the testing to a multitude of standards that was becoming necessary in a global marketplace.

Traceability Traceability to National Institute of Standards and Technology (NIST), formerly the National Bureau of Standards (NBS), standards is often required in the United States. Surprisingly, some well-known, imported products do not meet this requirement. That does not imply that they are necessarily inaccurate, just that documentation is not available from the manufacturer showing such traceability. In such cases, independent testing laboratories can usually provide the certification and the recalibration if necessary to meet traceability to the instrument's specified accuracy. The fewer the steps in the chain of traceability from NIST or other national standards laboratories, the less will be the uncertainty of measurement accuracy. Most users cannot afford to have calibrations done at national standards laboratories, but secondary calibration laboratories are relatively plentiful and provide calibrations only one step removed from those of the national laboratories.

Other Features Analog and digital interfaces are often featured even in some portable photometers. These range from the simplest analog output to RS-232, GPIB (IEEE-488), and some nonstandard digital interfaces. An analog output can be useful for data logging with a chart recorder. A common misapplication of the analog output is to connect it to an oscilloscope for measuring rise- or falltime of pulsed light sources. The response time of the analog output is usually limited to the response time of the human eye, about 100 ms, by the input amplifier that must emulate the eye to avoid erratic readings from rastered displays and AC-operated ambient light sources. Pulsed light sources require special measurement techniques. See Chapter 9 for information on these techniques. High-speed photometers or pulse-integrating photometers are also available for this purpose.[8] Proper use of the analog output requires measurement of the no-light condition to set the chart recorder to zero and the measurement of the anticipated light level to set the recorder sensitivity to match the reading on the photometer. If the recorder has no provision for adjustment of sensitivity, note the recorder deflection at a known light intensity to determine the relationship. For example, one division deflection might be obtained at a luminance of 23 cd/m^2. Three divisions would thus be 69 cd/m^2. Photometers with autoranging should be used in the manual ranging mode to avoid the complication of range changes causing ambiguous recordings due to not knowing what range was being used and when range changes occurred during unattended operation.

Digital outputs vary considerably in format. Some permit only readings to be transferred to the computer at a fixed rate while others may allow full control of all photometer functions and to transfer readings on command. The variants are too many to be discussed in detail here. Follow manufacturers' recommendations and keep their customer support phone numbers close by.

Portable photometers are mandatory for work in the field and offer much in the way of convenience in even the manufacturing or laboratory environment. Some compromise in performance and features is usually dictated to achieve portability, but the miniaturization made possible by modern electronics makes possible hand-held instruments with performance previously unobtainable in all but the most costly instruments.

Convenience features of photometers take many forms. Digital readouts are almost universal, with conventional electromechanical meters relegated to the "lightmeter" category where low cost is the keyword. Liquid crystal displays are most often used for quality instruments because of their low power consumption and ease of interfacing to the microprocessors used to provide versatility. Backlighting of the LCD is sometimes provided and is an important feature if the photometer is to be used in subdued ambient illumination such as television control rooms. Subdued lighting is also often required in display photometry to avoid errors from the ambient light falling on the display screen under test.

The widespread use of microprocessors has allowed the ease of inclusion of a number of convenience features in many photometers. One common function is a "hold" button. This allows a reading to be stored and displayed indefinitely and can be useful when trying to read light levels in an awkward position. Both metric and English photometric units are widely encountered in display specifications. While easily converted using the conversions in Appendix D, it is often a convenient feature if the photometer displays either unit at the touch of a button.

Interchangeable sensors for illuminance, luminance, luminous intensity, and other units can present a distinct cost savings if more than one quantity must be measured or if measurement needs are expected to change in the future. The electronics portion of photometers for the different units is essentially the same; only the sensors and their calibration are different.

Illuminance Photometers

The need for illuminance measurements for displays lies almost solely in the need to define the operating environment for a display for certain applications to ensure adequate viewing contrast. Most displays are operated in "normal" office environments. The ambient light level may be expected to be within the range of about 300–700 lx (30–70 fc) in the vicinity of the display. Other users may prefer operation in subdued light to avoid glare, reflections, and the loss of color gamut that occur when ambient light falls on the display screen. Television control rooms are an example of the need for critical viewing with a minimum of ambient light. Generally, ambient light measurements are unnecessary in either of these viewing situations. It is in daylight or other high-light-level viewing applications such as airport control towers and aircraft cockpits that measurement of ambient light levels becomes important. Here, the very viewability of the display hinges on the type of display and its contrast enhancement technology.

As with all photometers, the spectral sensitivity of an illuminance meter is important because of the different spectral outputs of today's light sources that may affect display viewability. Because of this, the previously discussed CIE f_1' specification is of paramount importance.

Next in importance is the *spatial response*, or sensitivity to incoming light according to angle. A perfectly diffusing surface (Lambertian surface) receives light in proportion to the cosine of the angle of incidence of light on it. Only 70.7% is received at a 45° angle and 50% at a 60° angle. To visualize this, think of a piece of matte white paper lying on the surface of a desk. Light is received not only from directly overhead but also from many other directions by reflection and scattering from walls and light entering windows. All light from the 180° hemisphere contributes to some degree to the illumination of the paper. To measure the useful light falling on such a surface, an illuminance meter should accept light in a similar manner. This is called cosine correction. In a typical office environment approximately 20–25% of the useful light is from shallow angles to the surface so errors of that magnitude may be expected from a sensor that is not cosine corrected. Cosine correction is the second most important parameter for an illuminance photometer in all applications except where a point light source at a distance from and perpendicular to the sensor is to be measured.

Cosine correction is accomplished by the use of a milky white plastic or glass diffuser over the sensor that allows light to be received from a full 180°. The surface texture and the shape are chosen to achieve a cosine angular response. The height of the diffuser edge above the surface of its housing and use of raised concentric annular rings are additional means of optimizing the cosine correction. Figure 4.6 is an example of maximum allowable deviation limits for a commercial illuminance photometer. The f_2 specification determined according to CIE Publication No. 69 is recommended for specifying cosine correction. Values of f_2 from 3 to 5% are acceptable for most display applications.

Illuminance photometers may be called upon to measure ambient light levels from less than 1 lx for subdued lighting environments up to 100,000 lx or more for direct sunlight. If the high end of a particular photometer is not high enough for the expected light levels, it may be increased through use of a calibrated thin-film neutral-density (ND) filter. NDs of 1.0 (10% transmission) and 2.0 (1.0% transmission) are usually sufficient for all practical applications. To determine the exact correction for a particular filter, measure the same light source or one of similar geometry at a lower intensity and note the readings both with and without the filter. Divide the reading without the filter by the reading with it. That factor may be used to multiply all readings taken with the filter in place to obtain the actual illuminance. This works best for "point" light sources such as a single lamp at some distance from the illuminance photometer as the angular response of the filter will alter the cosine correction accuracy. It may yield acceptable results for nonpoint sources if the correction factor as determined above is done with the actual measurement geometry anticipated.

Illuminance meters span the gamut from the under $100 lightmeter (Figure 4.7) to the laboratory photometer costing several thousand dollars. Midpriced instruments (Figure 4.8) are suitable for all but the most demanding display measurement requirements.

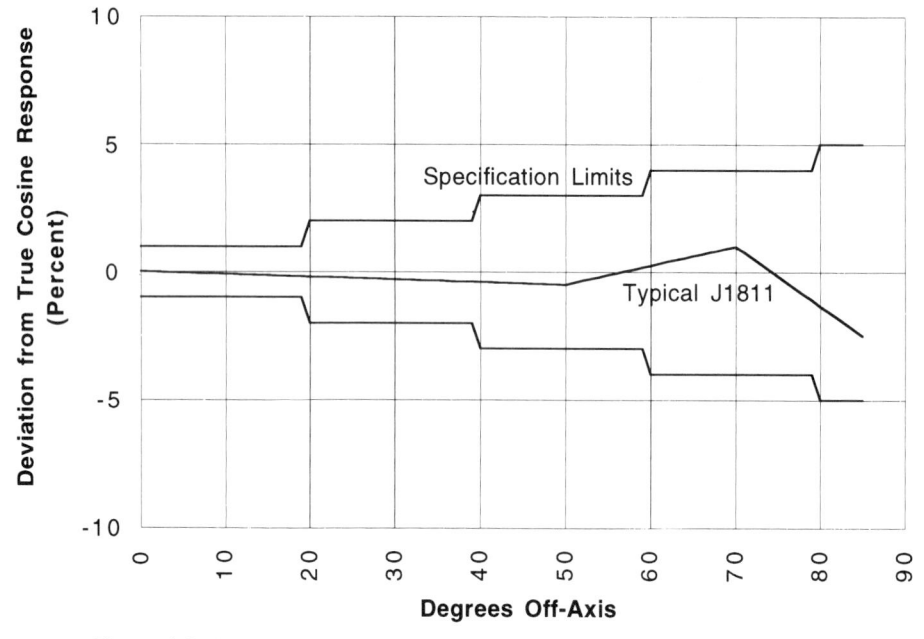

Figure 4.6. Cosine response of an accurately corrected illuminance photometer.

Figure 4.7. Inexpensive basic illuminance meter in the "lightmeter" category.

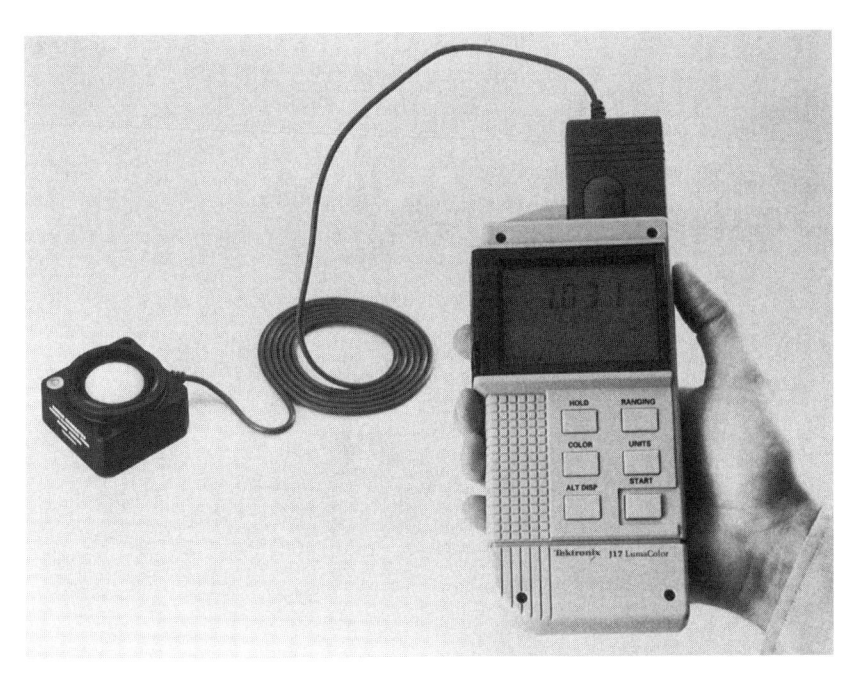

Figure 4.8. Moderate cost illuminance photometer suitable for display measurements. Interchangeable sensor heads allow other photometric, radiometric, and color measurements to be made. (Tektronix, Inc. All rights reserved. Reproduced by permission.)

An often-asked question is how to convert footcandles (usually measured with an inexpensive lightmeter) to foot lamberts for display measurements. While there is a convenient relationship between the two units, in real life it only applies in the measurement of surface diffuse reflectance under ambient illumination. Incident light (illuminance) upon a surface in footcandles multiplied by the diffuse reflectance (1.0 represents a perfectly reflecting diffuse surface) equals the surface luminance in footlamberts. *If* the display screen were flat and infinitely large *and* closely approximated a Lambertian surface with a 180° angle of emission according to the cosine of the angle, *then* a good illuminance meter could be placed in close proximity to the screen for a measurement. Unfortunately the index of refraction within a glass display panel causes light at extreme angles to become trapped, and this considerably reduces the total light output. Using such a technique is recommended when only an approximation is required. Even then the accuracy is very limited.

Luminance Photometers

Probably the most common display measurement, luminance defines how bright a surface, in this case the display screen, appears to the average human eye. As described previously, the sensor must have the same spectral response as the human eye. This is especially important for displays in order to accurately measure different phosphor colors.

These may include TV white at 9300 K, studio monitor white at 6500 K, or paper-white computer displays. Even more extreme are the individual red, green, and blue phosphors used to produce white for emissive color displays. Errors of 50% or more are possible when using an inexpensive photometer to measure the individual phosphors.

The lowest cost instruments (under $500) may use selenium sensors (largely made obsolete by silicon sensors), gelatin filters, a hood to define the acceptance angle or measurement field, and simple electronics such as an analog current meter with a very limited sensitivity range. Low-cost meters are usable only for the most basic of measurements where relative readings will suffice. They should never be used to compare different display or phosphor types and colors.

Most display luminance measurements can be performed with a moderate-cost instrument ranging from $1000 to $5000. These usually employ a silicon sensor for good long-term stability, ruggedness, freedom from fatigue effects, and wide dynamic range, typically 6–8 decades. Multielement glass filter assemblies are used to closely match the CIE photopic response curve for the average human eye in higher performance products. Glass filters have much better long term stability than the much less expensive gelatin or Wratten-type filters. Calibration certificates and actual spectral sensitivity curves of the sensors are sometimes supplied or available as options. The calibration certificates will satisfy requirements for NIST traceability but provide little in the way of useful information on accuracy limitations. The actual spectral curves of the sensor, however, depict deviations from true photopic response and permit estimation of expected errors for the individual CRT phosphors. The electronics of instruments in this price range will also be much more sophisticated with back-lit digital readouts, microprocessor control, automatic ranging, automatic zero, push-button selection of metric or English measurement units, interchangeable sensor heads, RS232 output, and battery operation, often included for improved accuracy and convenience.

After the photopic spectral correction accuracy, the next most important characteristic of a photometer for luminance is its acceptance angle. The least expensive approach is the "apertured" system where a field defining aperture is located some distance in front of the sensor (Figure 4.9). This technique is usually used for acceptance angles of 8°–20°. Sensitivity is traded off inversely with the square of the acceptance angle and limits the minimum practical field size that may be measured. There is also usually no optical sighting with the aperture method, which precludes it from being used to measure small areas.

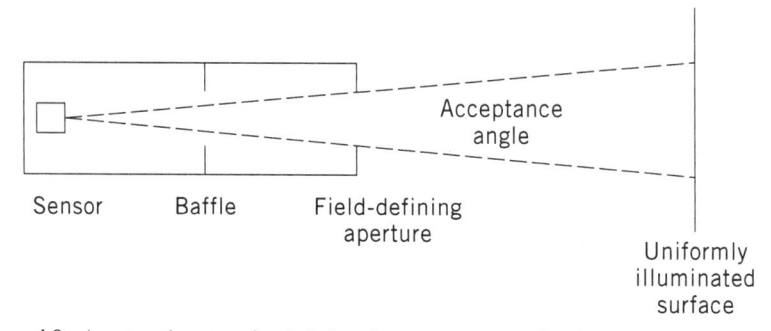

Figure 4.9. Apertured system for defining the acceptance angle of a luminance photometer.

There are, however, some clear advantages to the apertured photometer other than just cost. If a uniformly illuminated measurement area of 1 or 2 in. in diameter is available, it is easy to place the head against the surface to be measured. No focusing or sighting is necessary. Also, the light from the relatively large measurement area is spatially averaged, thus reducing the possibility that one might be measuring a localized bright or dark spot.

On the other hand, one may actually want to measure the small area variations that are present across an area of the display or the luminance of a single character stroke or individual pixel. Spot photometers, also called telephotometers, with fields of measurement of $1°$ and $\frac{1}{3}°$ are available for this purpose. These contain more sophisticated optical systems that gather more light than a simple aperture while simultaneously allowing the operator to view the actual area being measured within the larger viewing field, which is typically 8–10°. Because the total light flux intercepted by the detector varies with the square of the acceptance angle, the $\frac{1}{3}°$ instruments have 10 times less sensitivity than one designed for $1°$ although the optical gain of a $1°$ model can make it similar in sensitivity to an $8°$ apertured photometer. The focusing range is usually from 1 m or less to infinity. For a $1°$ acceptance angle this represents about 17.5 mm diameter per meter of distance. At 10 m it would measure a circle of 175 mm diameter. Commonly available photographic close-up lens sets may be used to allow measurement of smaller areas; a +10 diopter lens or combination of lenses will measure areas as small as 0.38 mm diameter. When using close-up lenses, readings should be made of a uniformly illuminated surface with and without the lens in order to determine the lens loss. Readings then taken with the lens may be corrected for this loss. Tripod mountings are usually provided for steady mounting and ease of use with such high magnification. An *x–y* positioner may be used for even more rigidity and precise positioning.

A factor not often considered with luminance photometers of both the apertured and optical types is *flare*, also referred to as *veiling glare* or *surrounding field effect*. Flare is due to unwanted reflections occurring within the region between the entrance aperture or lens and the sensor. The result is that a photometer with, for example, a $1°$ acceptance angle will have some response outside of that angle and can cause significant errors if a bright source is located outside the field being measured but at an angle that produces reflections into the sensor. This is a particular problem for display contrast measurements if a small dark area surrounded by a bright area is measured. The effect of flare problems is to degrade the measured contrast from what it actually is. Proper baffling, antireflection coated lenses, and use of ultraflat black paint on the internal parts of the optical system of the photometer are employed by the manufacturer to reduce flare. Flare may be tested by measuring the small central area of a large uniformly emitting area light source and reducing the larger area with progressively smaller apertured masks. See the excellent article by Boynton and Kelley[9] for further information on this frequently encountered but often unrecognized problem. Publication 69 of the CIE also describes tests to characterize photometers for flare.

Luminance photometers at the high-performance and high-price end are used in many display applications for military or other critical applications. Instruments such as the Photo Research Pritchard photometer (Plate 6) can measure the luminance of extremely small areas and light levels due to their use of a PMT as the sensor. The high gain of the PMT, 1×10^6 or more, is the key to their performance, although not without trade-off of cost, ruggedness, and stability. For many situations, they are worth their weight in trade-offs as they are virtually the only way to accomplish the task at hand.

Video Photometers

The video photometer is essentially a CCD television camera that has been optimized to make photometric measurements. In this way, it is possible to automatically take photometric data from many areas of a display screen simultaneously. This can be particularly important for calibration and performance verification of displays for high-volume manufacturing operations.

Video photometers are usually capable of measuring several display characteristics. These may include luminance, line width, misconvergence, linearity, and pincushion/barrel distortion. Video photometers provide considerable features and performance but are commensurately costly, being in the range of $50,000 or more for sophisticated units.

4.2 RADIOMETERS

Radiometers are not widely used for display measurement but are occasionally utilized peripherally in display manufacturing and processing applications such as measurement of laser power and infrared LED output. The ideal radiometer would have equal response to all wavelengths throughout the range of interest. Readouts are in power in watts or related units such as watts per square meter or watts per square meter per steradian. Two basic radiometer types are commonly available: one uses thermal detectors; the other uses semiconductor photodiodes often with multielement glass filters to flatten the spectral response. Each has its own distinct advantages and, of course, drawbacks.

Spectrally, the thermal detector based radiometer has nearly ideal flat spectral response that may extend from ultraviolet through the visible to infrared. The trade-off is sensitivity. Thermal detectors are well suited to light levels in the milliwatt and watt range but are poor at lower levels. They also tend to be relatively slow in response time since the thermal sensors must reach thermal equilibrium before displaying a stable reading. Thermal radiometers are the exception to the widespread use of digital displays for readouts. Several radiometer models use large analog meters that are entirely adequate for the response times and accuracies involved.

Semiconductor photodiode radiometers, on the other hand, have excellent sensitivity. Microwatt and nanowatt levels are easily measured with fast response times. Higher light levels are easily measured by use of calibrated thin-film ND filters. For silicon photodiodes, the most common of semiconductor detectors, the trade-off is a more limited spectral range and greater deviations from true flat response within that range. Figure 4.10 illustrates the spectral response of a typical silicon sensor with spectral flattening filters. The range of flattened response is about 450–950 nm. Within that range, the maximum allowable deviation from flat response is ±8%. These are calibrated using a monochromatic light source at a near-centered wavelength where the deviation is 0%. The actual spectral response curve is usually supplied with the instrument or, at worst, is available as an option. With the spectral response curve, it is easy to determine correction factors for other wavelengths so that the spectral response errors may be minimized. Broadband light sources having considerable spectral content below 450 nm or above 950 nm such as incandescent lamps, xenon lamps, or sunlight require special consideration since these portions of their output will not be measured properly. Calculation of the percentage of their light that is within the passband of the radiometer or the same determination made by comparing readings of the silicon sensor radiometer with a thermal radiometer may be required. In other cases, where the need is to quantify the radiance or irradiance available to a video camera or photographic film, no correction may be necessary since their response is also generally confined to the 450–950 nm range.

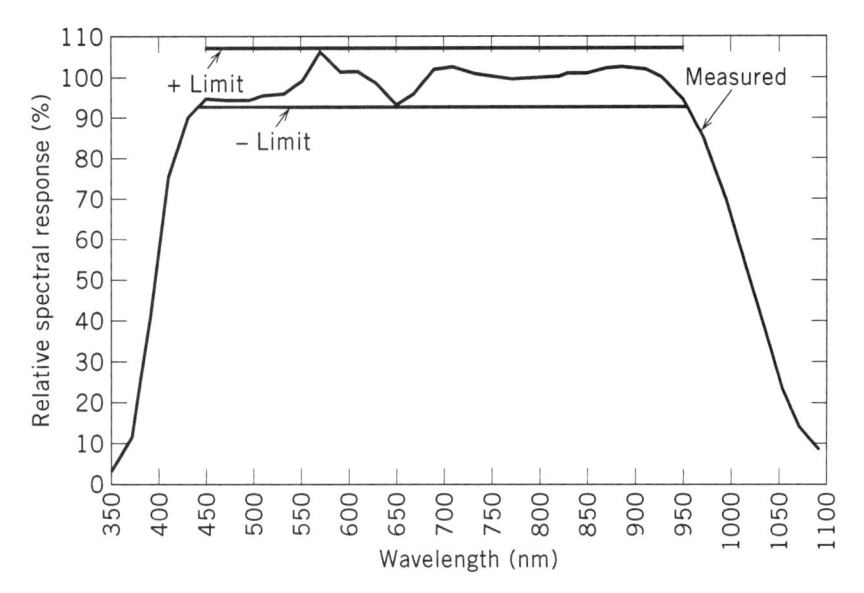

Figure 4.10. Spectral response of a filtered silicon radiometric sensor.

The semiconductor-photodiode-based radiometer will usually have similar functions and features to those of photometers. Fundamentally, the only differences are in the spectral response and units displayed.

4.3 FILTER COLORIMETERS

Chromaticity measurements often go hand in hand with luminance measurements. Most colorimeters also provide luminance measurements, although most luminance photometers do not conversely measure color. An exception is a recent instrument (Figure 4.11) that uses interchangeable heads to measure in a number of photometric, radiometric, and colorimetric units.

Colorimeters are divided into two basic classifications: filter colorimeters and scanning colorimeters. The latter are often called spectroradiometers since they also measure the spectrum of the light output from a display and compute the color from that data.

Moderate cost, from $2500 to $7500, and ease of use characterize the filter colorimeter. Absolute accuracy is the trade-off for cost when compared to scanning colorimeters, although repeatability and the ability to measure color difference can be quite good for the filter colorimeter.

Filter colorimeters use three or four silicon sensors with spectral correction filters (Figure 4.12) with construction similar to those described previously for illuminance and luminance photometers. Either gelatin or glass filters may be used for the same factors as for photometers—cost versus performance. The sensors simulate the CIE X, Y, and Z tristimulus functions (essentially sampling overlapping curves of red, green, and blue, respectively) as closely as possible within the constraint of sensitivity loss as the spectral match improves. Some of this sensitivity loss can be regained by use of high-quality input amplifiers (at higher cost, naturally). The inability to exactly match the tristimulus functions while retaining adequate sensitivity for low light levels is the reason for the absolute accuracy of filter colorimeters being inferior to that of scanning colorimeters.

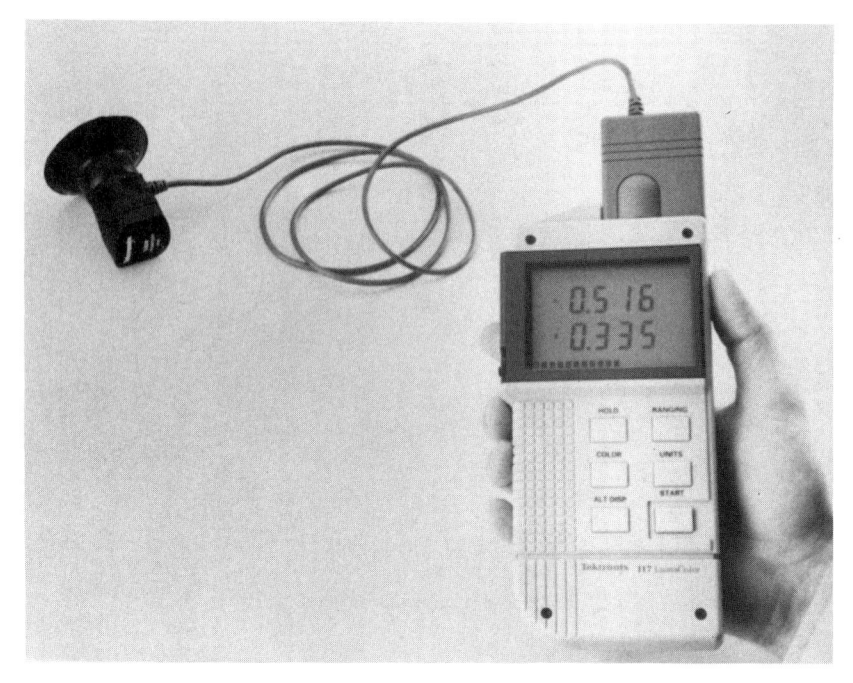

Figure 4.11. Filtered colorimeter for display chromaticity measurement. (Tektronix, Inc. All rights reserved. Reproduced by permission.)

Figure 4.12. Three-sensor assembly for a filter colorimeter. Each sensor mimics one of the CIE tristimulus functions, X, Y, and Z. The Y sensor also provides luminance information, and about 19% of the signal from the Z sensor is added to the X channel to match the secondary blue peak in the X curve. See Chapter 1 for the CIE tristimulus curves.

The Y tristimulus sensor has the same spectral sensitivity as the CIE photopic sensitivity curve and therefore does double duty to provide luminance data. The X, Y, and Z tristimulus data are used to compute the x and y or u' and v' chromaticity coordinates used to specify the color of displays.[10] See Chapter 1 and Appendix F for more detailed information on the tristimulus functions. Some filter colorimeters also display correlated color temperature (CCT) in Kelvin, although the accuracy may be suspect. This is due to the fact that there is no convenient algorithm to convert color coordinates to color temperature. A detailed look-up table in the microprocessor ROM is used, although the user is often better off by acting as the look-up table his or herself and plotting the color coordinates on an expanded CCT chart. Correlated color temperature measurement alone is not a valid means to use for adjusting or verifying a display. The concept of CCT is that the actual point on the chromaticity diagram lies *somewhere* along a line intersecting the Planckian locus. The term "somewhere" is the catch. If it is above the Planckian locus, it will appear greenish, and if below, it will appear purplish. The coordinates of the actual intersection of the CCT line and Planckian locus provide a unique point on the chromaticity diagram, and any monitor adjusted for that white point will appear the same as any other monitor adjusted to the same coordinates (ignoring luminance differences). For example, a color temperature of 6500 K is represented by, and only by, the coordinates $x = 0.313$ and $y = 0.329$. A *correlated* color temperature of 6500 K may be any point along a line between about $x = 0.301$ and $y = 0.403$ to $x = 0.325$ and $y = 0.245$. In almost all cases, the customer would be better off if the colorimeter manufacturer did not provide readout in CCT and simply included an expanded chromaticity chart depicting color temperatures along the Planckian locus.

As with photometers, and maybe more so, accuracy specifications of filter colorimeters must be used with a degree of caution. Chromaticity accuracy is often specified only at a specific color temperature such as 2854 K (Illuminant A) or 6500 K (D_{65}). Others specify the accuracy for a particular color temperature in the color difference mode. This looks much better than the absolute accuracy. Worst-case accuracy usually occurs near the outer edge of the chromaticity chart where near-monochromatic red and blue phosphors lie. A badly needed standard on colorimeter specifications similar to that previously published for photometers is in preparation by the CIE.[11,12]

Acceptance angle is an important consideration when choosing a colorimeter. It must be small enough to measure within the area available, such as a single color bar, but not so small as to be sensitive to localized small-area color variations. It should measure at least 30 or so complete pixels to minimize the effects of missing or nonuniform pixels by averaging the light intensity over that area.

Filter colorimeters may include battery operation, RS232 output, the ability to store one or more reference readings for measuring color difference, and suction cups for attaching the sensor head directly to the display screen. Other convenience features such as backlit LCD displays, autoranging, autozero, and battery power may also be found.

4.4 SPECTRORADIOMETERS

Scanning spectroradiometers have the edge where absolute accuracy and versatility are required. Prices start at about $10,000 and can approach $100,000 for specialized versions. Most employ a dedicated computer to do the considerable amount of number crunching required to determine chromaticity coordinates from the spectral scan. And since the scan is on a wavelength-by-wavelength basis, the computing ability of the system allows the data to be easily converted into many measurement units and displayed in several formats. Integration of the display's light output over selected wavelength or time intervals with appropriate computations can yield luminance (both metric and English), radiance, *XYZ* tristimulus values, *x–y* and $u'–v'$ color coordinates, color temperature, and spectral distribution curve, most of which may be displayed on the screen of the PC simultaneously. Data formats may include tabular or graphic data by wavelength, CIE chromaticity charts, and color difference presentations.

Two classes of spectroradiometers are common.[13] The first, the traditional scanning spectroradiometer (Figure 4.13), has been with us for some time. In this configuration, a single wideband detector such as a photomultiplier tube measures and records the amount of light at each wavelength through a monochromator. A more recent technique utilizes a diode array with a dedicated element for each wavelength to integrate all wavelengths simultaneously (Plate 7). In both types of spectroradiometers, corrections are made for the actual spectral responsivity of the sensor, and colorimetric data are computed from the corrected spectral data. Both types of spectroradiometers are usually considerably slower to measure color than filter colorimeters due to the need for the mechanical scanning of the entire spectrum or for diode array sensors to integrate enough photons at each wavelength for an adequate signal-to-noise ratio. The latter depends greatly on the light level being measured. As many as several minutes may be required for a complete spectrum scan. The traditional scanning spectroradiometer has one primary advantage: It has the inherent ability to record high-resolution spectral data, often with 1 nm resolution or less because of the high sensitivity of photomultiplier tube detectors. The light level at the sensor is lost in direct relationship to the spectral resolution. Diode array spectroradiome-

Figure 4.13. Laboratory scanning spectroradiometer using a PMT. (Courtesy of Photo Research.)

ters perform better with wider spectral resolution, such as 5–10 nm. Longer integration times and thermoelectrically cooled diode arrays permit better resolution but still cannot equal that made possible by photomultiplier tubes. The diode array spectrometer does, however, provide greater ease of use, ruggedness, portability, and long-term stability to offset the sensitivity disadvantage.

Obviously the skill levels required to operate either type of spectroradiometer are higher than those required for the simpler filter colorimeter. Some understanding of the basics of photometry and colorimetry are valuable, however, when any photometric, radiometric, and colorimetric measurements are to be made due to the many subtle factors that may affect the accuracy of measurements. The spectral output, geometry, and time characteristics of both the source and sensor as well as interposed filters and nearby objects all influence the accuracy of the measurement whether using a $50 lightmeter or a $50,000 spectroradiometer.

4.5 SPECTROPHOTOMETERS

The spectrophotometer is used to measure the transmittance or reflectance of filters and other materials rather than the light emitted by a light source. Spectrophotometers are very similar in design and construction to spectroradiometers. The primary difference is the inclusion of a stable, broadband light source. Light from this source passes through or is reflected from the material being measured. The remaining light is analyzed using a scanning monochromator and suitable detector such as a PMT or diode array. Use of a concentrated light beam of constant intensity reduces the problem of sensitivity and permits good spectral resolution. In the display industry, spectrophotometers are used primarily for materials and processing needs, although reflectance from hard-copy media also has become important for color matching to displays. Actually, the term *spectrophotometer* is somewhat of a misnomer since it is not in keeping with the use of the base word *photo-* to signify measurements based on visibility to the human eye as photometry implies.

4.6 GONIOPHOTOMETERS

The goniophotometer is most commonly used to measure light output versus angle of a light source but may also measure contrast. Goniophotometers have assumed greater importance with LCDs because of their directional properties. The goniophotometer consists of a swinging arm to which a photometer is attached and that pivots about the display to be measured. Conversely, the photometer may be in a fixed position and the display attached to the swinging arm. The actual configuration is usually individualized for the particular measurement need of the user and may range from a room-sized affair for measuring roadway luminaires down to desk top size or less for measuring LEDs.

The goniophotometer produces plots of light intensity versus angle.[14,15] These may be simple one-axis plots, two-axis plots, or full hemispherical plots of isointensity (equal intensity) or iso-contrast contours, as shown in Figure 4.14.[16] Isocolor plots are often used as well.

152 MEASUREMENT INSTRUMENTATION

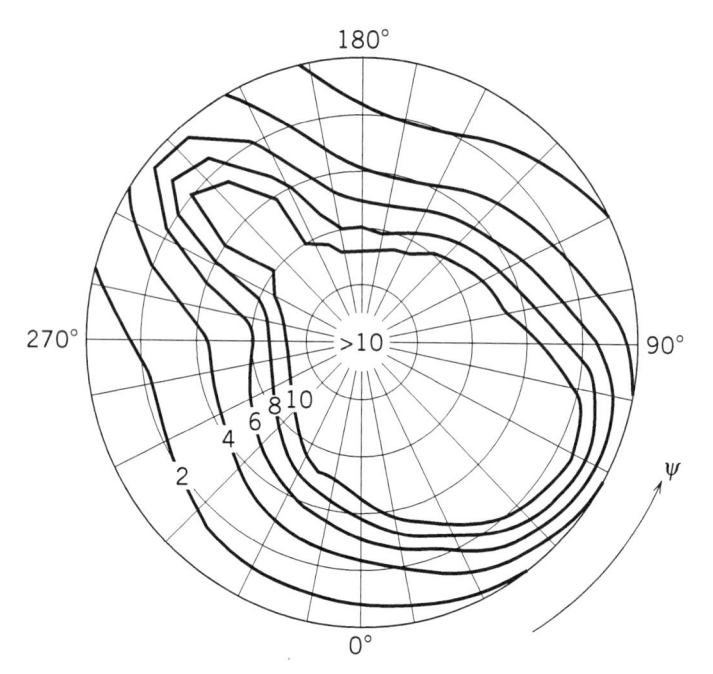

Figure 4.14. Isocontrast plot of a liquid crystal display. (From T. J. Scheffer et al., "24 × 80 Character LCD Panel Using the Supertwisted Birefringence Effect," SID Intl. Symp. Digest of Tech. Papers, 1985, p. 122. Reprinted by permission.)

Due to their specialized nature, few goniophotometers are commercially available and most are constructed by the user. LMT in Berlin, Germany, is one of the few commercial suppliers of such equipment. The photometer itself is usually an off-the-shelf unit with the mechanical fixturing customized to fit the requirements. Important considerations are proper photometer-to-display distance, automatic recording where a large number displays and/or data points must be measured, exclusion of all ambient light, and liberal use of ultraflat black paint on everything in the vicinity. This may even extend as far as avoidance of light-colored clothing by the operator if he or she must be close to the equipment during the test.

4.7 INTEGRATING SPHERES

The integrating sphere is an important tool for photometry, radiometry, and colorimetry, although for displays it is most likely to be confined to the standards or calibration laboratory. A brief description is included for familiarity in the event the term comes up in the course of light measurement discussions.[17-19]

The integrating sphere's primary application is to measure the total light output of a light source that may have directionally nonuniform light output. Lamp manufacturers rely on them for determining lumen output or mean spherical luminous intensity [previously mean spherical candlepower (MSCP)] from lamps. It consists of a sphere whose interior walls are uniformly coated with a highly reflective diffuse white coating. Light is uniformly diffused over the entire interior of the sphere, thus allowing the average light

Figure 4.15. 1.65-m-diameter integrating sphere. (Courtesy of Labsphere, Inc., North Sutton, NH.)

level to be sampled at any point on the interior. Knowing the diameter and hence the surface area allows the total light output to be computed. In practice, one or more sensor ports are provided in the sphere, and baffles prevent direct light from the lamp under test from reaching the sensor port. The sensor is usually aligned to view the area of the sphere walls opposite the port. Integrating spheres range in size from the walk-in size for testing 2.4-m (96-in.) fluorescent lamps to lemon size for laser or LED measurements. Figure 4.15 shows one of the larger integrating spheres, 1.6 m in diameter.

4.8 OPTICAL COMPARATORS

Small optical comparators or microscopes are used for simple visual measurement of resolution either by measuring line width or merging lines with the shrinking raster technique. A calibrated reticle optically superimposed on the area being measured provides a measurement scale. A comparator with $20\times$–$25\times$ magnification is probably the most useful. See Chapter 7 for further information on resolution measurement techniques.

4.9 CONVERGENCE MEASUREMENT INSTRUMENTS

Instruments for convergence measurement of color CRT displays, or more properly *misconvergence,* range from the simple optical comparator such as the Klein convergence gauge to elaborate video photometers. The Klein gauge (Figure 4.16) uses adjustable optics to optically align the misconverged lines of a displayed crosshatch or dot pattern on the screen (Figure 4.17). The amount of optical adjustment required for alignment may be read directly from the scales on the comparator's two adjustment knobs. This is the lowest cost approach (less than $1000) if the use of a basic comparator microscope is ignored. The latter requires considerable estimation and is especially prone errors between different operators.

Figure 4.16. Klein convergence gauge. (Courtesy of Klein Optical Instruments, Inc.)

Figure 4.17. Measurement of misconvergence with the Klein convergence gauge. (Courtesy of Klein Optical Instruments, Inc.)

More sophisticated misconvergence measurement instruments are based on CCD detector arrays to display magnified images of the crosshatch or dot convergence test pattern. These instruments often combine misconvergence with line width measurement. Expect to pay several thousand dollars for this capability. A number of these are designed to be used in conjunction with personal computers.

The most refined instruments for misconvengence measurement are the video photometers previously discussed, which may also include additional capability such as geometry, luminance, luminance uniformity, pincushion/barrel distortion, display size, and even a built-in test pattern generator. These are ideal for the production line and are fully computer controlled for high-speed testing. Provision may also be included to provide automated calibration adjustment of the display under test. Prices for such advanced systems may run $50,000 or more.

REFERENCES

1. J. W. T Walsh, *Photometry,* Constable and Co., London, 1958, p. 112.
2. Commission Internationale de l'Eclairage (CIE), "Methods of Characterizing Illuminance Meters and Luminance Meters," CIE Publication No. 69, 1987.
3. Lichtmesstechnik (LMT), "How to Select a Photometer," Application Note AN5-688, 1988.
4. Deutsches Institut fur Normung (DIN), Deutche Norm 5032 Teil 7, "Lichtmessung," Dec. 1985.
5. G. Eppeldauer and J. E. Hardis, "Fourteen-Decade Photocurrent Measurement with Large-Area Silicon Photodiodes at Room Temperature," *Appl. Opt.,* Vol. 30, No. 22, pp. 3091–3099.
6. W. Budde, "Definition of the Linearity Range of Si Photodiodes," *Appl. Opt.,* Vol. 22, No. 11, pp. 1780–1784, 1983.
7. EG&G Electro-Optics, "Silicon Photovoltaic Detectors and Detector/Amplifier Combinations," Application Note D3011C-8, Jan. 1984, Fig. 3.
8. K. Miller, "Matching Photometers to Applications," *Information Display,* pp. 7–9, 18, Sept. 1989.
9. P. A. Boynton and E. F. Kelley, "Measuring Contrast Ratio of Displays," *Information Display,* pp. 24–27, Nov. 1996.
10. P. A. Keller, *The Cathode-Ray Tube,* Palisades Press, New York, 1991.
11. H. Terstiege, "Characterizing the Quality of Colorimeters," *SID Dig.,* pp. 641–644, 1991, Vol. 22.
12. Commission Internationale de l'Eclairage (CIE). "Guide to Characterizing the Colorimetry of Computer-Controlled CRT Displays," CIE Technical Committee TC 2-26 Final Report, May 1994, p. 9.
13. R. Walker, K. Miller, and T. Bulpitt, "Spectroradiometric Instruments and Applications," *Test Measure. World,* Oct. 1983.
14. R. McCluney, *Introduction to Photometry and Radiometry,* Artech House, Boston, 1994, pp. 295–296.
15. M. E. Becker and J. Neumeier, "Measurement of Electro-Optic Properties of LCDs," *Proc. SID,* Vol. 31, No. 4, pp. 273–293, 1990.
16. T. J. Scheffer, J. Nehring, M. Kaufman, H. Amstutz, D. Heimgartner, and P. Eglin, "24 × 80 Character LCD Panel Using the Supertwisted Birefringence Effect," *SID Dig.,* Vol. 16, p. 122, 1985.
17. D. J. Lovell, "Theory and Applications of Integrating Sphere Technology," *Laser Focus/Electro-Optics,* pp. 86–96, May 1984.
18. H. D. Wolpert, "Unraveling the Mystery of Integrating Spheres," *Electro-Optics,* pp. 39–40, Aug. 1983.
19. Labsphere, Catalog II, 1996–97, pp. 100–116.

CHAPTER 5

Luminance and Contrast Measurement

5.1 GENERAL BACKGROUND

Light output, usually in the form of luminance, and contrast are primary display concerns for monochrome displays. They are almost equally as important for color displays, although *color contrast* is added to the contrast considerations in this case. The equipment required for luminance and contrast is a good-quality luminance photometer and sometimes an illuminance photometer where ambient light levels must be determined. The procedures for measuring luminance are generally straightforward and may be performed by relatively untrained personnel on the production line provided sufficient attention is devoted to instrument selection, test pattern generation, procedures, operator training, photometer calibration, and establishing correct ambient light levels. Contrast measurements are more difficult to determine accurately and require greater attention to the same details.

Luminance Photometer Selection

The most obvious considerations in selecting a luminance photometer for a given application are that it covers the range of luminance values expected and has a readout in the desired units. Most modern luminance photometers are intended for display measurement and have adequate range for all but the very lowest or highest luminances. The highest luminances are not much of a problem; if the photometer will not measure high enough, just add a calibrated thin-film neutral density filter. Provision is made on many luminance photometers to allow fitting external filters. Check with the manufacturer as to the recommended filters and mounting method. The low-light-level end is not as easily solved. Here, the specifications should be carefully examined, and if possible, a photometer should be tried to assure adequate stability and reading resolution for the requirements at hand.

A more subtle selection criteria is the acceptance angle. The luminance photometer selected should measure and average the luminance of an area of 30 or more complete pixels in order to minimize the effect of defective pixels. Usually it is easy to work with a measurement area of 25–50 mm diameter. Luminance photometers with acceptance angles of 8°–20° are suitable for such diameters. Spot photometers are indispensible for measurement of individual pixels or subpixels but are quite expensive, require more care in proper use, and may measure too small of an area to be useful for many routine display measurements.

Other important luminance photometer selection criteria, such as spectral correction accuracy, are discussed in more detail in Chapter 4.

5.1 GENERAL BACKGROUND   **157**

Test Pattern Generation

Several published and proposed test methods exist for luminance. These differ primarily in the test pattern used. The results using different methods should not be too dissimilar for a well-designed display. Either a full-screen display at full drive or a window pattern is quite useful for luminance measurement. Other patterns of different shapes are used but are more difficult to generate. Many other display measurements are not so forgiving of the techniques used as is luminance.

The full-screen, 100% white field is easiest to generate since many displays can produce this condition when connected to a computer used for word processing. Just put up a full-screen white page. This method has been used for television receivers, computer displays, and in modified form in EIA TEPAC Publication No. 105[1] for industrial CRTs. The full white field is particularly useful for measuring luminance uniformity since any area of a screen may be easily measured without changing position of the white area, and presumably all areas may be considered to have the same excitation conditions.

There are cases where the full-white-field method should not be used. If the video drive level is high and the power supplies not well regulated, the increased anode current may cause a change in anode voltage and screen power density. This changes the deflection sensitivity as well, which further affects the screen power density per unit of area. Check that the raster size does not change as the drive level increases before attempting luminance measurement with a full white field.

The same full-screen white field may be used for backlit LCDs. One difference in technique should be observed. For LCDs, the photometer should be at the optimum intended viewing angle. For all other displays, the luminance photometer should usually be perpendicular to the screen.

The window pattern (Figure 5.1) is often used for television studio monitors. This pattern more closely simulates actual use of a monitor with a television image, but a video test pattern generator is required to produce it. Specialized test patterns similar to this may often be easily generated for computer displays by using drawing or paintbrush software.

Figure 5.1. Window pattern for luminance testing.

Procedures and Operator Training

Once the photometer is selected and the measurement procedure determined, the procedure should be thoroughly documented if it is to be used for production line measurements. Test conditions, ambient illumination levels, test equipment, techniques, photometer zeroing, calibration requirements, and test limits should all be included in the procedures as required for the particular measurement to be made.

Even in nonproduction applications test reports can be extremely important if the measurements may need to be repeated at a later date, as when testing for display device aging, troubleshooting changes in performance in subsequent vendor shipments, quality assurance such as ISO-9000, verification of conformance to internal, customer, or government specifications, and just plain repetition of measurements because "something didn't look right the first time." The test report should include at least such data as device identification, display drive conditions, measurement location within the screen area, ambient illumination conditions, photometer type and serial number, date, and operator's initials.

Operators should be properly trained for photometric measurements. Include considerations such as setup, measurement location and distance, photometer zeroing procedure, avoidance of ambient illumination, care of the photometer, and do not overlook an explanation of why the testing is important.

Photometer Calibration

The luminance photometer should have its calibration verified at intervals no longer than those recommended by the manufacturer or required by contract for government requirements. Generally speaking, one year is a realistic calibration interval in most cases. The calibration lab performing the calibration verification should be well acquainted with the special considerations involved in photometric calibrations and be able to provide documentation showing traceability to NIST or other government standards.

Ambient Illumination

Since the objective is to determine the amount of light produced by the display, it stands to reason that any contribution to light from the screen by ambient light should be avoided. This may involve measuring the display in a dark room or at least a gray room (subdued lighting), use of a light occluder around the sensor if it is in contact with the screen, or subtracting the ambient contribution from the measured luminance. The latter is possible because of the additivity of light as expressed in Grassman's law (Appendix C). In this case, the luminance is measured under the prescribed drive conditions and another measurement made without moving anything but with the display turned off or no drive present in order to determine the ambient contribution. Just subtract the ambient reading from the luminance reading to find the luminance produced by the display alone.

5.2 LUMINANCE

Probably the most common of display measurements, luminance, relates strongly with viewability of a display in its intended environment. Luminance is the measurement equivalent of the subjective term *brightness*.[2] While the word "brightness" will be found in this text, its use should be taken to mean only the visual appearance of the display. The other legitimate use of the term is for the control found on many displays, which is labeled BRIGHTNESS.

The preferred unit for luminance is the candela per square meter, also called the nit. The candela per square meter is the SI unit of luminance and is preferred according to ANSI Standard 268-1992,[3] ANSI/IES RP-16-1986,[4] CIE 17.4,[5] IEC 50(845),[6] and NIST Special Publication 811.[7] In the face of such prestigious support of the candela per square meter, it is difficult to justify continued use of the nit; however, the very same documents all designate the lux as the preferred unit of illuminance in place of the lumen per square meter. The candela per square meter and lumen per square meter have the advantage of showing the precise meaning of the unit but do tend to be more confusing to photometry neophytes than the simple words nit and lux. In the United States, the long obsolete term footlambert still hangs on, but the candela per square meter (or at least the nit for the nonconformist) should be used in all cases. Footlamberts can be easily converted to candelas per square meter by multiplying by 3.426.

Information displays are used in ambients ranging from total darkness to direct sunlight. Generally speaking, the brighter the display is, the wider the range of ambient light conditions it may be used in, and the brightness control can almost always be turned down if desired for lower ambient conditions. Luminance is not the only concern; contrast plays an important part as well and will be discussed later in this chapter. Higher luminance may often degrade other display characteristics, including resolution, power consumption, color purity, and display device life. Moderate display ambient environments are the typical office illumination levels for which most displays are designed. This is typically 300–700 lx (30–70 fc). Displays intended for such use usually have luminances in the range of 50–100 cd/m^2 (approximately 15–30 fl). For viewing displays in direct sunlight, the luminance must be considerably higher and combined with the use of high-performance contrast enhancement filters. Liquid crystal displays are the one exception to the "brighter-is-better" guideline. Since reflective and transflective LCDs are usually viewed by reflected ambient light, they will automatically compensate for the light level under which they are viewed. The only problem is in very low ambient light levels where virtually any self-luminous display (EL, PDP, CRT, projection, etc.) may be easily viewed. The LCD will become invisible unless provided with a backlight. Luminance measurement of the backlight/display combination, which may be from less than 3 cd/m^2 to up to 30 cd/m^2, is often important and is performed similarly to emissive displays. Just make sure that the photometer sensor is aligned at the intended viewing angle.

Full-Field/Flat-Field/Area Luminance

Full-field, flat-field, and area luminances denote the measurement of a uniform screen area as opposed to line or dot luminance. The SAE standard ARP 1782[8] describes the method well, and although formulated for avionics CRT displays, it is pertinent to other display types and applications. The full white field is especially applicable to computer displays since documents for word processing are displayed in this manner. Although the luminance may be only 60 to 70% of full white, paint or drawing software can provide 100% levels. Proposals for EIA standards from the National Information Display Laboratory (NIDL) for monochrome CRT[9] and color CRT[10] displays are similar in technique to ARP 1782, and flat-panel display standards may be expected to similarly evolve. It is to the benefit of the industry as a whole to develop flat-panel and CRT display standards that have as much commonality as possible to allow meaningful display comparisons between the various competing display technologies. They all have the same human viewer to satisfy as the ultimate judge of viewability.

Luminance of a full- or near-full-field 100% white screen is normally measured in the center of the screen (Figure 5.2) where luminance is most uniform, unless there is a requirement to measure luminance uniformity. In the case of color CRT displays, the raster size should be set to the prescribed size and the color balance adjusted to display a white field of the specified color coordinates. In some situations, it may also be desired to individually measure the luminance of the red, green, and blue fields that comprise white when combined. Accurate photopic correction of the photometer is especially important for the blue field. The blue spectral region of the CIE Standard Observer curve is the most difficult to match accurately.

Full-field area luminance measurement is quite straightforward as long as the drive conditions are established correctly. Just make sure that ambient light is excluded during the measurement or compensated for by subtracting the background light from the readings. Some photometers for displays have light occluders or suction cups (Figure 5.3) that help shadow the screen area being measured from room light, but do not count on them to exclude all ambient light at low display luminance levels. The light received by other areas of the screen may diffuse throughout the entire screen due to scattering by the phosphor crystals.

Either spot or conventional luminance photometers may be used. The measurement area viewed by the photometer is not critical as long as three conditions are met. First, the area measured by the photometer should be uniformly illuminated since the photometer will average over that area. Second, the photometer field of measurement must be small enough so as not to extend beyond the edges of the display in any direction. This would result in luminance values that are lower than the actual luminance. Lastly, the measurement field must be large enough to cover at least 30 complete pixels so that variations in

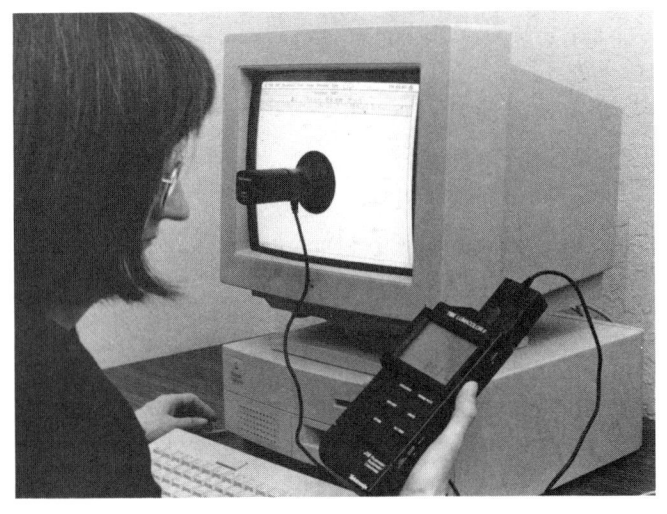

Figure 5.2. Display luminance measurement.

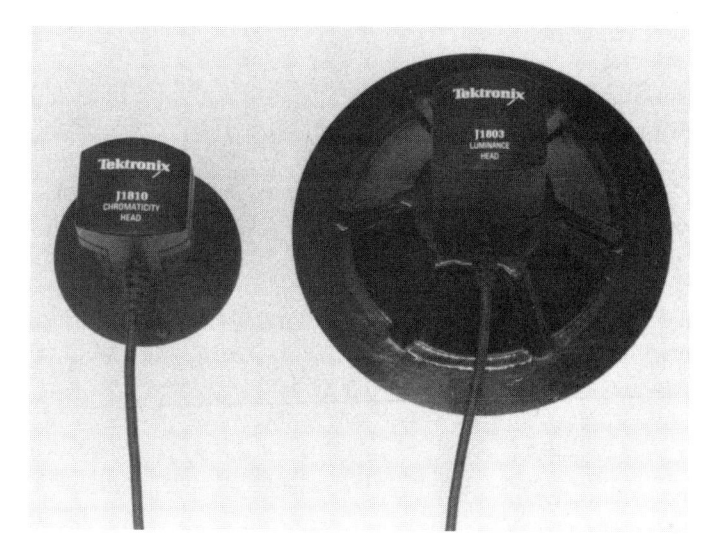

Figure 5.3. Suction cup and light occluder cone for photometer heads. Both tend to reduce the effect of ambient light on display measurements.

individual pixel luminance are averaged and defective or missing pixels do not unduly affect the reading. If spot or narrow-angle photometers are used, the measurement area may be increased by moving the photometer further from the screen, although this may make ambient illumination of the screen more troublesome if the photometer was shadowing the screen at the closer distance.

For all but liquid crystal and projection displays, the photometer should be perpendicular to the measured area. For LCDs and projection displays, the luminance should be measured at the specified optimum viewing angle for the display.

Some precautions are as follows:

Make sure the photometer is correctly zeroed according to the manufacturer's directions.

Check for ambient light effects by turning off the display to see if there is any residual reading. Alternatively, turn off the ambient lighting if possible to determine its effect on the luminance reading.

For CRT raster or other pulsed displays, possible photometer saturation at high light levels may be checked by interposing a calibrated thin-film neutral density filter between the photometer and the display. The luminance reading should change by the attenuation factor of the filter. Do not use photographic neutral-density filters for this purpose since they are not spectrally neutral or accurate enough for the purpose.

Be sure that the photopic correction error (f_1') is less than 3%. See photometer manufacturer's data or contact them for the information.

Make sure that the photometer has been verified or recalibrated within the recommended calibration interval.

Measure a known good display or luminance standard to verify proper readings.

5.3 LUMINANCE UNIFORMITY

Luminance uniformity is a closely related measurement to luminance itself. Backlit LCDs are particularly prone to luminance nonuniformity when incandescent, cold-cathode fluorescent (CCFL), or LED backlighting is used. Electroluminescent backlighting tends to be more uniform but may become less uniform at higher drive frequencies, where losses are greater, especially at a distance from the busbar edge connections. The CRT displays often have significant luminance drop-off in the corners where alumininizing is thinner and/or the electron beams strike the shadow mask at significant angles. Processing variables can introduce luminance nonuniformity in virtually any type of display. Measurement of these nonuniformities is generally just a matter of measuring the luminance at several locations on the screen rather than just the usual center screen location. Most common are five-point (Figure 5.4) and nine-point (Figure 5.5) uniformities. A full-screen white raster should be used for the uniformity test. Five points should suffice for most needs since they are located in the areas of greatest luminance nonuniformity.

5.4 LUMINANCE WARMUP AND AGING

The NIDL describes a procedure for measurement of luminance during the warmup period whereby luminance measurements are made for monochrome CRTs at 1-min intervals for the first 5 min, 5-min intervals during the first 30 min, and 10-min intervals thereafter.[11] The method is equally applicable to any backlit flat panel display as well as CRTs since backlights often have warmup characteristics of their own.

Luminance aging refers to long-term degradation. Luminance aging occurs to some degree in all displays except perhaps the reflective LCD. This aging is characterized by life testing of the display devices under prescribed operating conditions. Particular attention must be paid to maintaining those operating conditions as well as photometer calibration. It is desirable to use the same photometer during the entire life test or at least to overlap the measurements if two photometers are used to ensure similar readings or that correction factors may be applied to the data to make the data consistent. Consistency rather than absolute accuracy is of prime importance for aging tests.

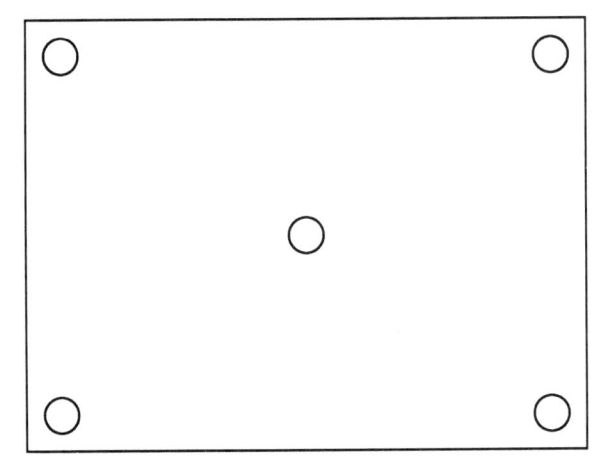

Figure 5.4. Locations at which luminance may be measured for a five-point luminance uniformity test.

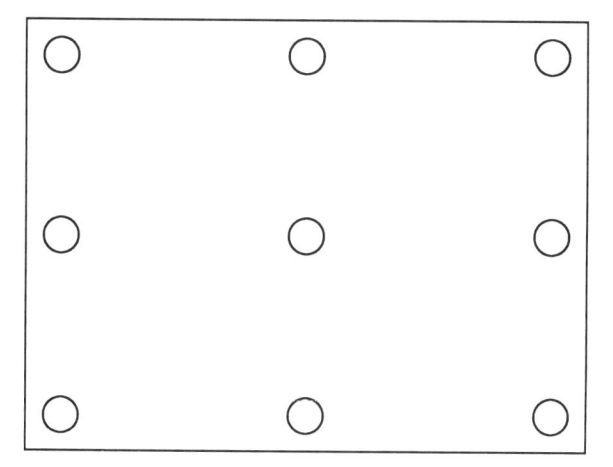

Figure 5.5. Locations at which luminance may be measured for a nine-point luminance uniformity test.

5.5 LUMINANCE STABILITY VERSUS FILL FACTOR

Luminance stability describes the variation of luminance that may occur as greater percentages of the screen area are written at the 100% white level. This is primarily a concern with low-cost CRT displays that may not have adequate high-voltage power supply regulation. The display may drop out of regulation with a 100% full white field and increase the scanned area ("blooming") due to decreased electron beam stiffness at lower accelerating voltage. At the same time, the luminance will change because the same amount of beam current is spread over a larger area and the accelerating voltage is decreased. The beam current may also change if other electron gun voltages are affected.

Luminance stability measurement is described in the NIDL proposal for monochrome display measurement.[12] The NIDL proposal uses a white square pattern (Figure 5.6) beginning with the size adjusted to 5% of the lesser of the active screen dimensions (vertical or horizontal depending on whether landscape or page format is used, respectively).

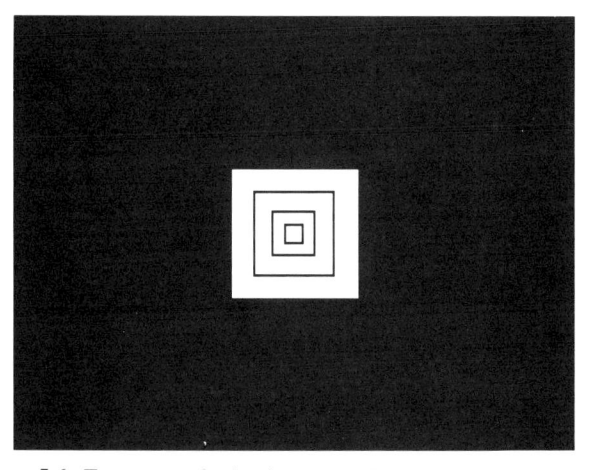

Figure 5.6. Test pattern for luminance stability. (After NIDL, ref. 12.)

164 LUMINANCE AND CONTRAST MEASUREMENT

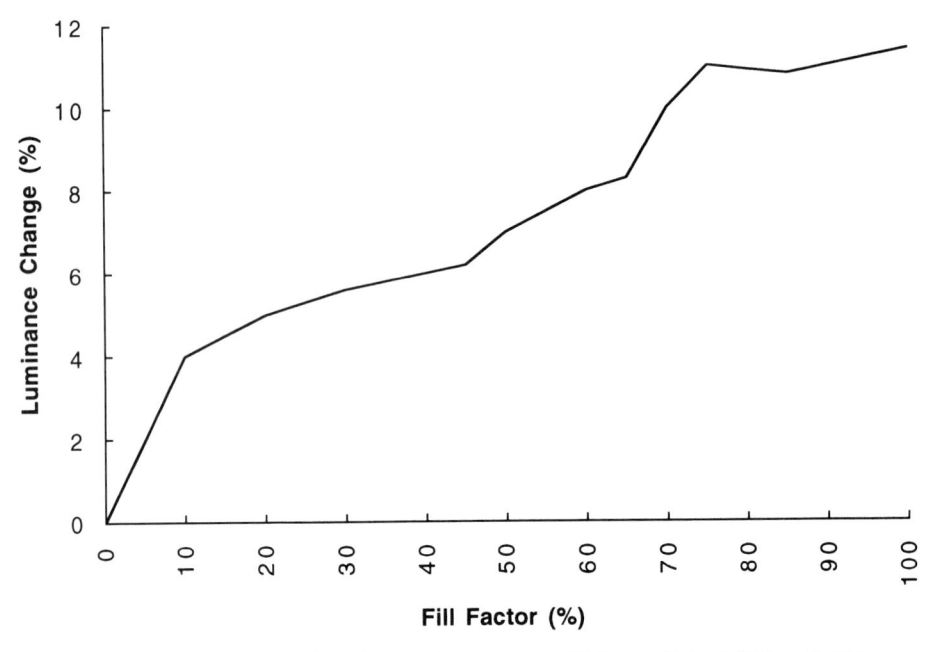

Figure 5.7. Example of luminance change versus fill factor. (After NIDL, ref. 12.)

The luminance is measured and the square adjusted in size by increments of 10% and luminance measured for each size up to and including full field. Luminance versus area is plotted as shown in Figure 5.7. The NIDL recommends that the luminance of the 5% square should be set at 5% of maximum luminance and the background adjusted to the minimum luminance (ideally 0% of maximum luminance). This author feels that a more realistic choice might be to set the 5% square to the maximum luminance attainable with the user-adjustable brightness control or in the event of a brightness control with a detented position, the detent setting should be used. As with all luminance measurements, any ambient illumination should be minimized.

5.6 LUMINANCE LINEARITY

Two classifications of display light output nonlinearity exist. The first is nonlinearity of light output from a phosphor screen as the current density is increased. This is usually caused by phosphor saturation where the efficiency of the phosphor degrades due to temperature increases that occur at higher current densities. This condition is reversible as opposed to permanent phosphor degradation or aging, which may occur if the threshold current for phosphor damage is exceeded. There is also a rarely encountered class of phosphors that is superlinear, that is, the light output increases at a steeper slope than that of the current density. The other nonlinearity is called gamma (see Section 5.7) and describes the change in light output versus applied drive or video signal level. Gamma is due to the superlinear characteristic of beam current relative to the grid or cathode drive voltage of the CRT electron gun.

The EIA publication TEP105-16 describes two methods for measuring CRT phosphor linearity versus current density using either an illuminance photometer or a luminance microphotometer.[13] The former is a relatively low cost method while the latter requires much more sophisticated instrumentation. When determining luminous efficacy values, the line width must be accurately known for the microphotometer method, whereas only the more easily measured raster size is needed for the illuminance method. The greatest accuracy will thus be realized with the lowest cost and simplest approach. Both methods as published require considerable range of adjustment of raster size in one axis or the other.

The illuminance meter method uses a raster of known size that is fully contained within the active area of the screen. The CRT beam current is measured and the beam current density computed using the raster scan area. Measurement of CRT beam current is often accomplished by inserting a floating microammeter in series with the CRT anode connection provided the CRT does not have any beam-limiting apertures connected to the anode. Extreme care must be exercised in making this measurement since the anode voltage, and hence the microammeter, may be elevated to as high as 30 kV or more. The horizontal raster lines should be fully merged so that no raster structure is visible with the largest size raster to be used, even when viewing with a hand magnifier. This may be done by either increasing the number of scanning lines or slightly defocusing the beam. The sensor for the illuminance photometer is placed perpendicularly to the center of the raster at a measured distance from it. This distance should be greater than 10 times the maximum dimension of the raster. Both the sweep rates and beam current are held constant during the test and the raster size is reduced to produce higher current/power densities. The illuminance is recorded versus each raster size chosen. The beam current density is then computed for each raster area. When plotted versus the beam current density, the luminance data will show the relative nonlinearity of the phosphor for those conditions. Although not specifically suggested in TEP105-16, the same test might be performed with a constant raster size while adjusting the beam current.

Since the distance to the illuminance meter is known, the luminous intensity in candelas may be obtained from the same data. The illuminance value in lux (lumens per square meter) multiplied by the square of the distance in meters equals the luminous intensity in candelas. This can be taken a step further if the overall accelerating voltage (cathode to anode) is known. The power density in watts per square meter may be computed by multiplying the current by the voltage and then dividing by the raster area in square meters. Dividing the computed candelas by the computed power density yields the luminous efficacy in candelas per watt. The efficacy may then be plotted against the power density to get the absolute phosphor characteristics for that particular phosphor screen. Differences in phosphor lots, aluminization, and screening techniques will cause differences in phosphor saturation characteristics from one manufacturer to another and to a lesser degree between CRTs made by the same manufacturer.

The luminance microphotometer method uses an expanded raster such that the individual horizontal scanning lines are completely resolved. Here, the length is adjusted to control the peak current/power density. The luminance microphotometer is focused on the center of one scanning line and the peak luminance is recorded for several raster widths. The peak input power density is computed by

$$W = \frac{VIfT}{Hd}$$

where

> W = peak power input density, W/m^2
> V = CRT overall accelerating voltage, V
> I = CRT anode current, A
> f = raster refresh rate, Hz
> T = time for one horizontal sweep, s
> H = horizontal length of the scan line, m
> d = beam spot diameter, m

Further computations of luminous efficacy are described in EIA publication TEP105-16.

5.7 GAMMA

Despite the linear characteristics of video amplifiers used to drive it, the CRT is a nonlinear device. Due to the emission characteristics versus drive voltage of the CRT cathode and control grid, a small increase in driving voltage will produce a larger change in beam current. The light output from the screen will thus increase in proportion to the beam current up to the beginning of phosphor saturation. *Gamma* for displays, as opposed to photographic gamma, is defined by the IEEE as "the exponent of that power law that is used to approximate the curve of output magnitude versus input magnitude over the region of interest."[14] It is expressed by the equation $B = KE^\gamma$, where B is the luminance (L would be more correct but B has been historically used), K is a constant, E is the CRT control grid drive voltage above cutoff, and γ is the exponent representing gamma. CRTs normally have gammas in the range of 2.0–3.0 with 2.5 being typical.[15] The gamma may be specified for a CRT alone or for the entire display system.

To measure gamma, conventional luminance measurements as previously described should be made at several drive voltages or current levels. The NIDL recommends a white square with an area of 1% of the total screen area and positioned in the screen center. For display devices that show good luminance stability versus fill factor, as described in Section 5.5, the area used should not be critical. The data obtained may be plotted on a log-log graph or used to compute the gamma using the equation

$$\frac{L_2}{L_1} = \left(\frac{v_2}{v_1}\right)^\gamma$$

where

> L_1 = luminance for the first drive voltage (v_1)
> L_2 = luminance at the second drive voltage (v_2)
> γ = gamma, dimensionless

The gamma may be different depending on the range of drive levels used for the computation. For instance, the gamma at high drive levels may differ from that at low drive levels. Additionally, individual points for digital drive signals may exhibit nonlinearities due to digital-to-analog converter errors. A sufficiently large range between the drive levels selected should minimize errors or variations from either cause. It is especially important to exclude ambient illumination when measuring luminance at the low drive levels as offsets caused by ambient light may result in substantial errors in computed gamma.

5.8 CONTRAST

Perceptual contrast is the psychophysical appearance while *physical contrast* is the measured value, much the same as brightness and luminance denote the appearance and measurement, respectively, of light output from a display. The contrast (either form) may be that of the display device alone or may include specified ambient illumination conditions. Scattered electrons in a CRT, diffusion of light from adjacent areas, ambient light, crosstalk in matrix-addressed flat panels, and inability to totally extinguish pixels that are not being addressed are causes of contrast degradation.

Display contrast computation may use any one of several equations, each with its own distinctions. There is also a lack of agreement among the many published definitions and standards that include the term "contrast" and its variations. It is important to ascertain that contrast has been computed the same way when making comparisons. While this author has attempted to use the most authoritative sources of definitions and equations below, there will be many exceptions encountered. Even the CIE definition leaves open to individual choice the equation used to compute contrast, listing only the simple contrast ratio equation shown below as a typical example.[16] The only way to be certain of valid comparisons is to ascertain the actual equation used for the computation of displays to be compared and the ambient illumination conditions used. While this text uses the terms L_{bright} and L_{dark}, other sources may use variations on this, such as L_1 and L_2, B_1 and B_2 (standing for the incorrect term "brightness"), object and background luminance, and so on. Contrast may also be specified for "detail" or "large-area" contrast. Detail contrast indicates closely spaced adjacent areas of the screen while large-area contrast is for separated areas. The former is the more critical since electron or optical scattering, crosstalk, wide low-level skirts on the beam profile, and so on, will usually have more effect under these conditions and will determine the ultimate legibility and viewability of the display.

In a totally dark viewing environment, the measured contrast may be expressed as *contrast ratio* and is simply the ratio of the luminance of the bright areas to that of the dark areas ($C_R = L_{bright}/L_{dark}$).[17] This is one of the more common ways of expressing display contrast and may also be referred to as the dynamic range. The range of possible values is one to infinity. Since it is always the bright area divided by the dark area (in this interpretation), the result can never be negative even for reverse video applications or LCDs. A contrast ratio of 1 would indicate invisibility of any difference between the driven and undriven areas, assuming there is no color contrast between the areas. Contrast ratio may also be used where ambient light is present, as described in EIA TEP105-10,[18] but the ambient illumination level must be specified as part of the contrast ratio specification and is only meaningful under that condition.

Luminance contrast,[19] or just "contrast," is perhaps the most widely used and is expressed as $C = (L_{bright} - L_{dark})/L_{dark}$. It is equal to the contrast ratio minus 1. The range of luminance contrast is zero to infinity with zero representing invisibility if no color contrast is present. This method is often recommended since it removes the background luminance, which is additive from the highlight measurement. A strong case can be made for the opposite since the eye sees the sum of the background and highlight luminance just as the photometer does. Also, a luminance contrast of 1 is more difficult to visualize than the equivalent contrast ratio of 2. The latter clearly indicates that the highlight appears double that of the background. Luminance contrast is specified in slightly modified form for avionics CRTs in SAE ARP 1782[8] and includes the effect of ambient illumination, which can be especially severe in the cockpit environment.

168 LUMINANCE AND CONTRAST MEASUREMENT

Another definition of contrast is *Michelson contrast,* also called *contrast modulation.* Here, the equation used is

$$C_m = \frac{(L_{bright} - L_{dark})}{(L_{bright} + L_{dark})}$$

The range of $C_m = 0-1$. With this method, the effect of measured contrast changes correlates better to the perceptual contrast of the display. A 25% improvement in C_m will represent a significant improvement in appearance regardless of where it is in the range of 0–1. Conventional contrast ratio and luminance contrast improvements of even 100% may only be important at the low end of their range.[20]

Shades of gray is the most complex of display contrast measurements. Here a series of luminance steps of proportional change are used to determine the number of discernable steps that may be displayed. The effect of poor contrast is to reduce the numbers of visible steps at the low end of the luminance scale. The steps are made to be in increments of $\sqrt{2}$, or about 1.414, which is clearly discernable by any normally sighted observer. Shades of gray is used more often as a visual test rather than as a photometric measurement. The observer needs only to count the number of discernable steps.

Recommended test patterns for contrast measurement vary as widely as do the definitions and equations. In their simplest form, measurement of the luminance of a full-screen display at full drive followed by a similar measurement at minimum drive may suffice. A 25-mm-wide black vertical bar (Figure 5.8) is recommended by EIA TEP105-10[21] for CRT displays with ambient illumination. Window patterns are suggested for measurement of contrast (Figure 5.9) and degradation of contrast by spot halation (Figure 5.10) for monochrome CRTs by NIDL and are used with a more complex equation because of the halation.[22] Finally, lowercase *e* and *m* are displayed and measured using microphotometer techniques to determine detail contrast in ISO 9241-3.[23]

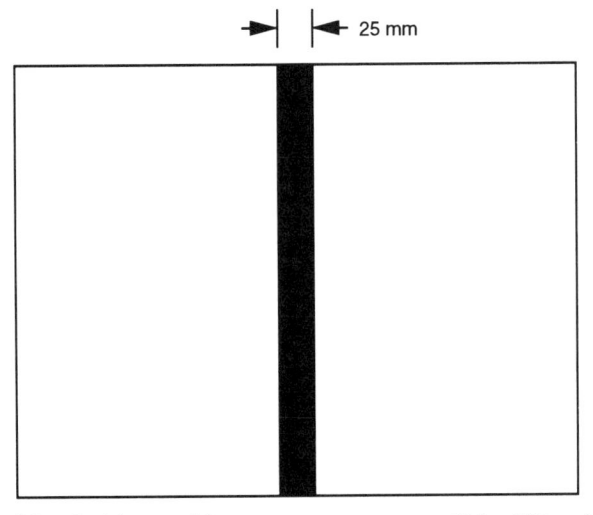

Figure 5.8. Black bar used for contrast measurement. (After EIA, ref. 21.)

Figure 5.9. Window pattern used for contrast measurement. (After NIDL, ref. 22.)

5.9 DIFFUSE REFLECTANCE

Diffuse reflectance is a measure of the light that is more or less uniformly scattered by the display screen (Figure 5.11) as opposed to the mirrorlike, or *specular*, reflections (Figure 5.12) that usually occur at the glossy glass surface. In the cases of the CRT and electroluminescent display, it is the phosphor and any surface roughness of the glass surface such as etching that diffusely scatter a portion of the ambient illumination. The reflective LCD intentionally emphasizes diffuse reflectance of the background surface of the LCD to enhance viewability of the dark characters or other symbols.

Diffuse reflectance is often measured with a source of illumination oriented 45° to the normal viewing axis of the display. The light source should be supplied by a regulated DC power supply with regulation better than ±1%. Alternating current may be used from

Figure 5.10. Window pattern used for measurement of contrast degradation due to halation. (After NIDL, ref. 22.)

170 LUMINANCE AND CONTRAST MEASUREMENT

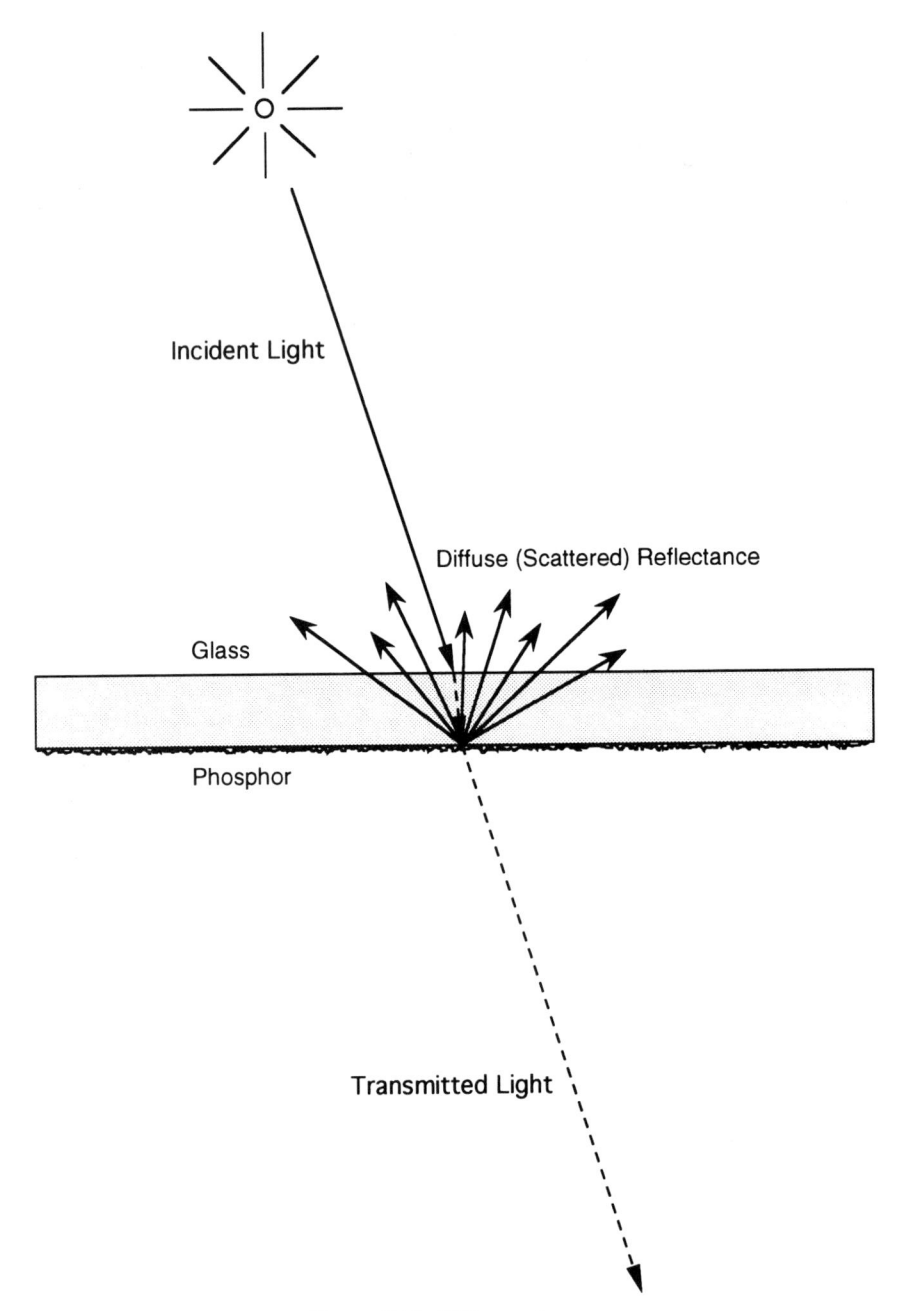

Figure 5.11. Diffuse reflectance.

a constant voltage transformer for either fluorescent or incandescent light sources. A constant-current DC power supply is preferable for incandescent lamps but constant-voltage power supplies are certainly acceptable. For most display applications under fluorescent lighting, a cool-white fluorescent lamp is recommended so that the spectral reflectance of the screen, which may vary somewhat with wavelength, is accurately measured under the

5.9 DIFFUSE REFLECTANCE   171

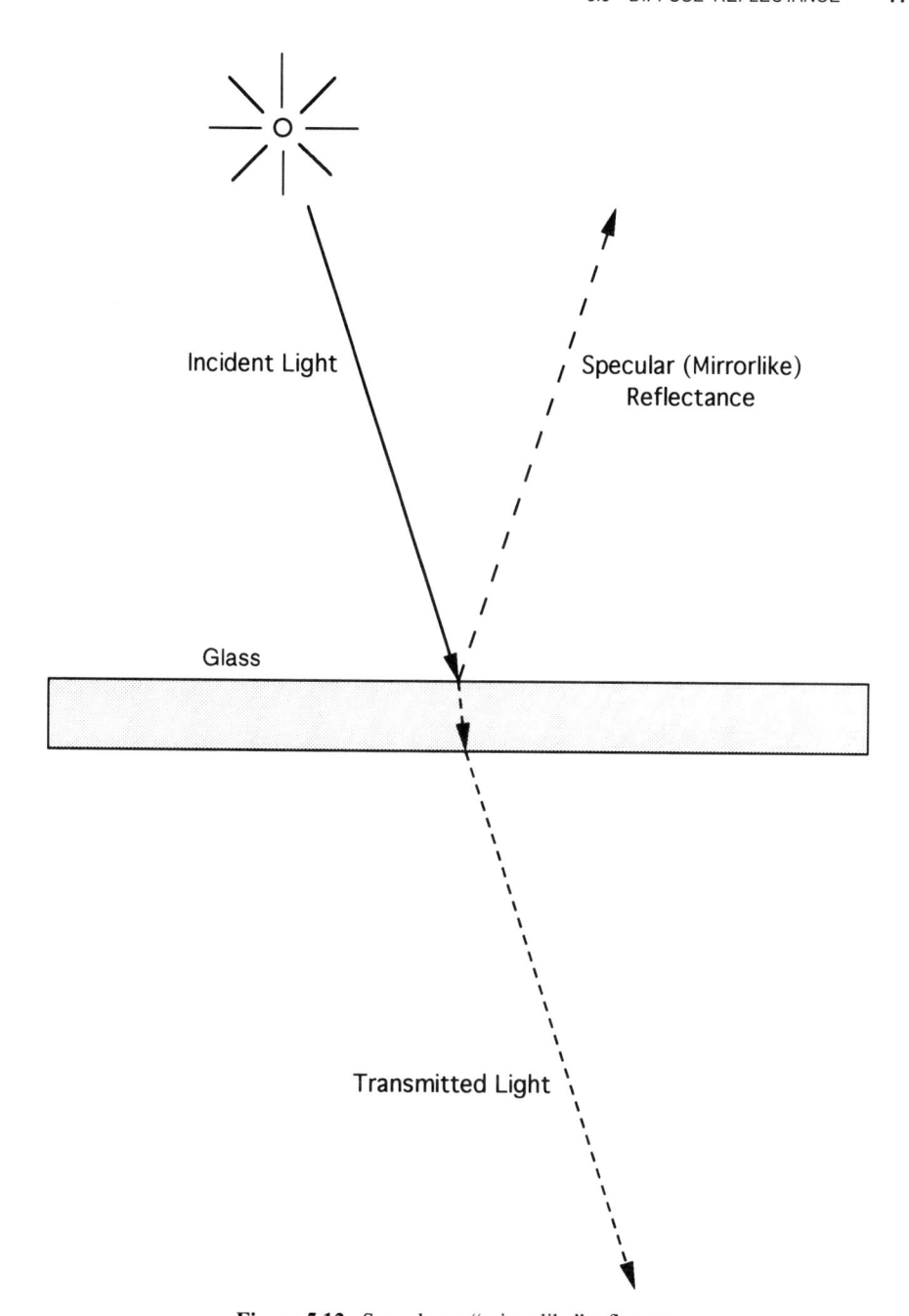

Figure 5.12. Specular or "mirrorlike" reflectance.

expected viewing condition. The actual light level produced by the light source is not critical but should be high enough to avoid errors from any ambient light present. A luminance photometer is set up to view the illuminated area along the viewing axis (Figure 5.13). The acceptance angle of the photometer must be such that it measures only the uniformly illuminated central area of the display. The 45° illumination angle and 0° mea-

172 LUMINANCE AND CONTRAST MEASUREMENT

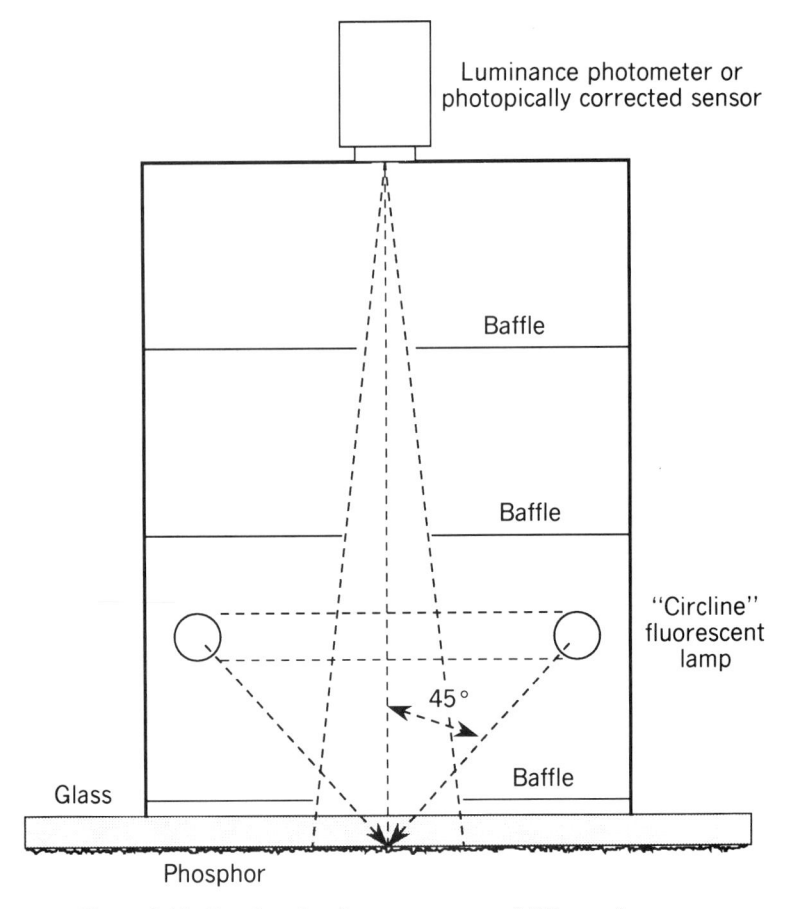

Figure 5.13. Test fixturing for measurement of diffuse reflectance.

surement prevent specular reflections from affecting the readings. Measurement of specular reflectance will be discussed separately. The EIA publication TEP105-12[24] describes a practical measurement method for diffuse reflectance that may be scaled down for smaller displays than the intended television picture tubes.

A luminance measurement is first made of a calibrated standard of diffuse reflectance substituted in the same plane as the display relative to the light source and photometer. The 90% white side of a Kodak R-27 gray card is a readily available reference. The measurement is then repeated for the display itself. It is useful to also place a black velvet or felt sample in the same plane to confirm that the photometer is properly zeroed and that no ambient light is being received by it. To obtain the luminance reading that would represent a perfect white diffuser, divide the reading of the calibrated white reference sample by its reflectance factor (0.9 in the case of the Kodak reference cards). Then divide the luminance reading of the screen under test by the previously calculated luminance representing a perfect white diffuser to obtain the reflectance factor. This may be converted to percent reflectance by multiplying it by 100.

5.10 SPECULAR GLOSS

Specular gloss refers to the mirrorlike reflections produced by polished glass or plastic surfaces. Specular reflections occur when the viewing angle equals the angle of incidence of the ambient light and is worst when the source of light is bright and relatively small in area. Surface treatments such as antireflection coatings are often employed to minimize the annoying reflections that occur at certain angles of ambient illumination with respect to the display screen and viewer. Lower cost treatments, etching for instance, convert the specular reflections to diffuse scattering, which is often not as objectionable.

The EIA publication TEP105-13,[25] ASTM C 346,[26] and ASTM D 523[27] describe procedures to measure specular gloss at angles ranging from 20° to 85° from perpendicular to the surface being measured. These documents are generally designed for measurement of raw CRT panels and flat-panel substrates before being processed into finished devices and, as such, are beyond the scope of this book. The CRT measurement methods proposed by NIDL[28] and ISO 9241-7[29] describe a method of measuring specular gloss using a spot photometer and diffuse light source with each positioned 15° off opposite sides of the perpendicular axis to the display as illustrated in Figure 5.14. By changing the angles to 0° and 45°, respectively, diffuse reflectance may also be measured. Glossmeters and related haze meters are available commercially. These avoid the necessity of fixturing but may not be adequate in all cases. The previously mentioned published standards and instrument manufacturer's instruction manuals should be referred to for the proper measurement techniques.

5.11 PROJECTION SYSTEM MEASUREMENTS

The ANSI prescribes test methods for evaluating data projection equipment (IT7.215[30]). Included are total luminous flux for projectors with a separate screen, luminance for

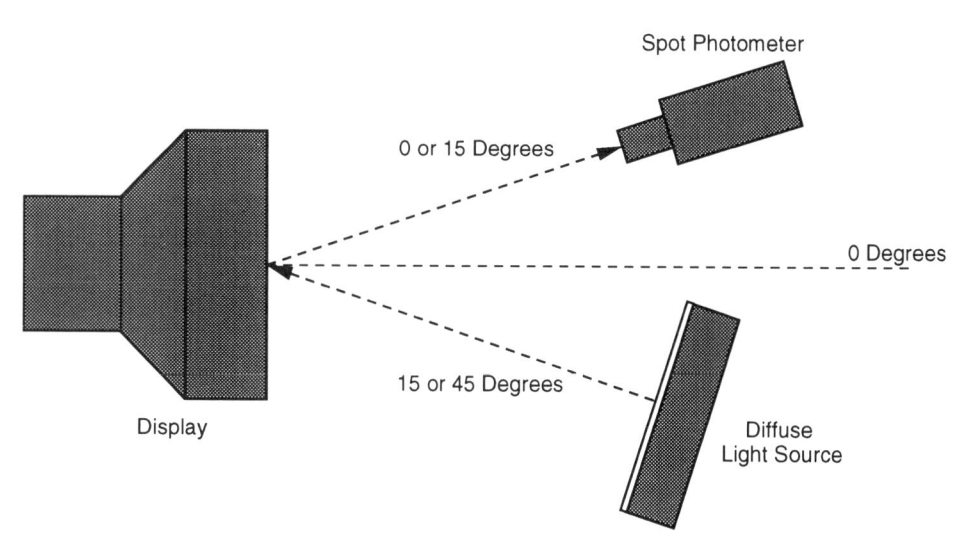

Figure 5.14. Measurement of specular gloss.

174 LUMINANCE AND CONTRAST MEASUREMENT

devices with an integral screen (both are incorrectly called "brightness" by ANSI in IT7.215; other ANSI standards such as ANSI/IEEE 100 and ANSI/IES RP-16 correctly define brightness as a perceived, not measured, quantity), light output uniformity, and contrast ratio.

A special test pattern (Figure 5.15) is used to adjust the 5% level to be distinct from the 0% and 10% patches and the 95% level to be distinct from the 90% and 100% patches to ensure adjustment of the brightness and contrast controls for display of the entire gray scale. Measurements are made at the nine screen locations defined by Figure 5.16 and averaged. For total luminous flux in lumens from a projector with a separate screen, illuminance measurements are made in lux (lumens per square meter) at each of the nine screen locations with the sensor pointed toward the projector. The average of all nine readings in lux is multiplied by the total illuminated area in square meters at the plane of the sensor to obtain the total light flux in lumens. This number will be the same regardless of projection distance since the same number of lumens will be spread over a different area. The light per unit area or screen luminance, however, will decrease by the square of the distance for the same reason. The luminance will also be affected by the screen itself. Obviously, the "whiter" or more reflective the screen is, the greater will be the screen luminance, but even larger increases in luminance are made possible by beaded or "gain" screens that concentrate most of the otherwise wasted light emitted to the sides in the direction of the viewer. Although not described in ANSI IT7.215, the system luminance is measurable using a spot photometer located at the normal viewing position. The position of the photometer is particularly important for gain screens with their somewhat restricted viewing angles. The projection distance at which luminance is measured or the display dimensions must be specified for a meaningful measurement and to be able to equate to other distances and screen dimensions.

Luminance nonetheless is the test specified by ANSI for devices with integral screens. The diameter of the area measured is recommended to be greater than 1% of the smaller screen dimension, either height or width, of the screen or it should encompass at the very least three scan lines. Ten times these specified values would seem to be better to mini-

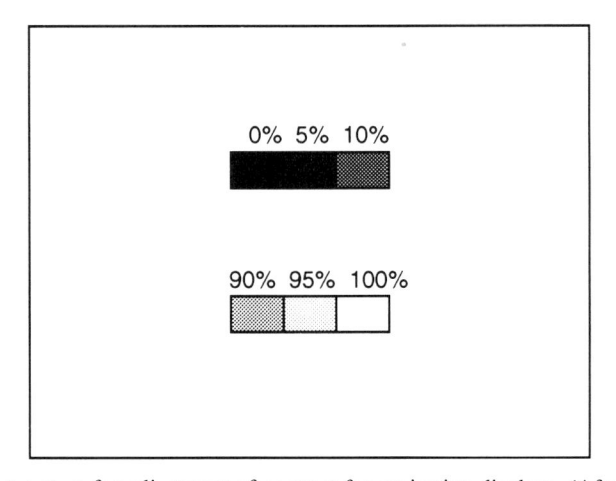

Figure 5.15. Test pattern for adjustment of contrast for projection displays. (After ANSI, ref. 30.)

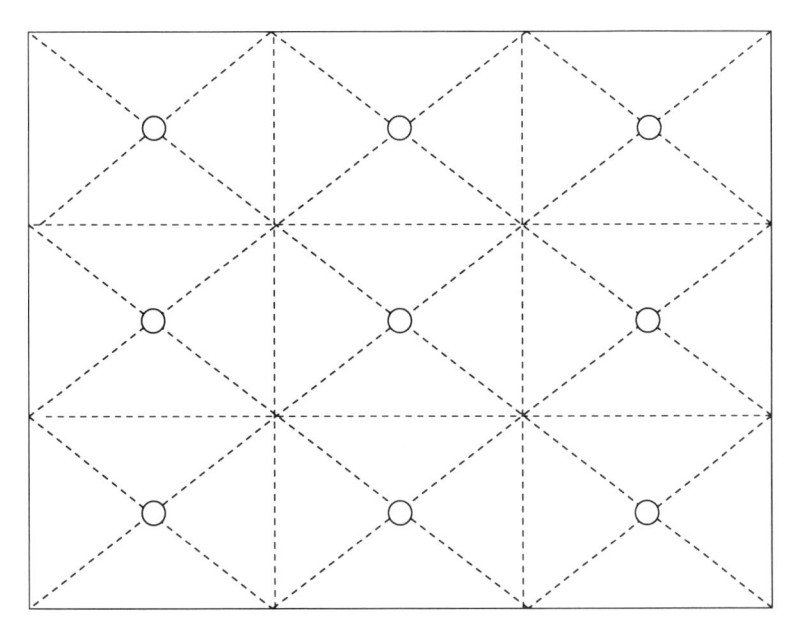

Figure 5.16. Measurement locations for light output of projection displays. (After ANSI, ref. 30.)

mize localized variations in uniformity. Although not mentioned in the ANSI standard, care should be taken to align the luminance photometer with the normal viewing axis if the screen has directional properties. The same adjustment procedure and nine measurement points described for projectors with separate screens is used for this test as well. The ANSI has not followed its own definitions and calls out the deprecated unit, the nit, instead of the more correct equivalent unit, the candelas per square meter.

Luminance uniformity for either type of projection display makes use of the previously measured nine-point data, either in lumens or candelas per square meter. The brightest and dimmest of the nine zones are divided by the average to determine the maximum and minimum deviation from average, and these should be converted to percentages.

A black-and-white checkerboard test pattern (Figure 5.17) is used to measure contrast ratio. Illuminance in lux or luminance in cd/m^2 is measured at the center of each black and white rectangle with the full light level as previously established. The average of the white rectangles is divided by the average of the black rectangles to derive the average contrast ratio.

Another ANSI standard defines test methods for liquid crystal imaging devices (LCIDs) for overhead projectors (IT7.228[31]) After a 10-min period for temperatures to stabilize, the contrast ratio is first adjusted to optimum and measured using a procedure described in the standard. Rather than luminous flux or luminance, this method measures the light transmission efficiency of the LCID itself by comparing it to a test mask (Figure 5.18) using an overhead projector. Measurements are made at the plane of the screen with an illuminance photometer aligned toward the projector.

176 LUMINANCE AND CONTRAST MEASUREMENT

Figure 5.17. Checkerboard test pattern for contrast measurement of projection displays. (After ANSI, ref. 30.)

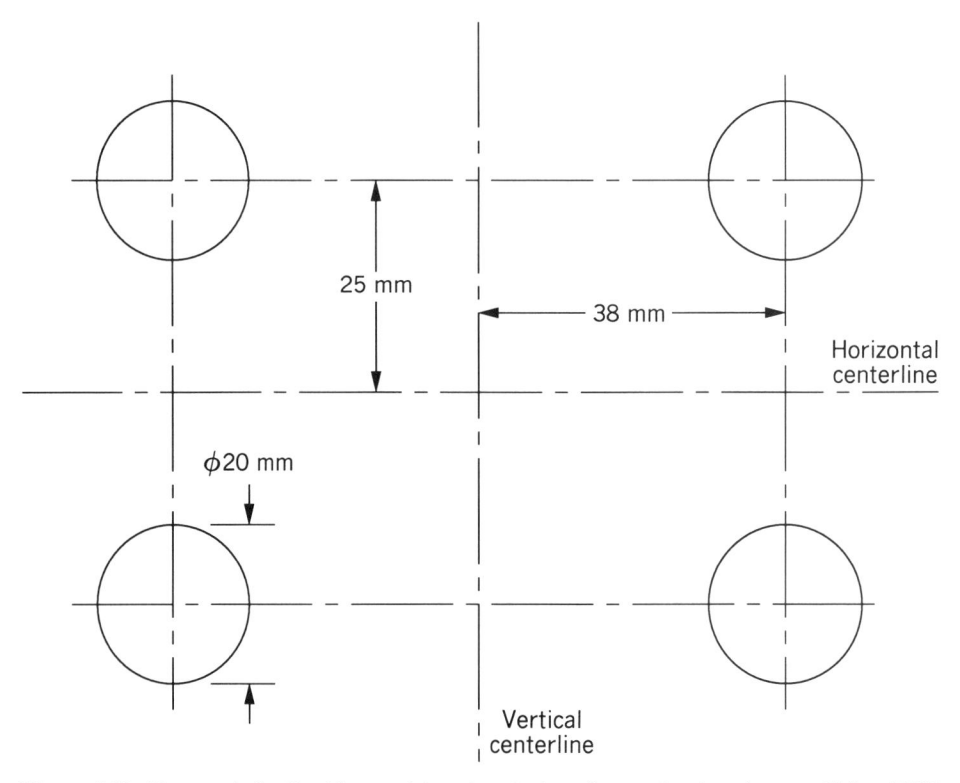

Figure 5.18. Test mask for liquid crystal imaging devices for overhead projectors. (After ANSI, ref. 31.)

5.12 VISIBLE LED OUTPUT

Light-emitting diodes (LEDs) are widely used for status indicators, numeric displays, and to a lesser degree for large matrix-addressed monochrome and limited-color displays. Recently they have begun to replace the more fragile and shorter lived incandescent lamp for emergency exit signs, automobile brake lights, and even traffic signals in multiple series–parallel connected arrays. In these applications addressing of individual LED elements is not used. All devices in the array are on at the same time. Light-emitting diodes are also well suited for backlighting of LCDs because of their low cost, high efficiency, and compatibility with voltages and currents associated with semiconductor technology. The LED displays may be divided into three categories, discrete, alphanumeric, and matrix addressed, each with its own measurement techniques. The techniques will be discussed in the following sections.

Discrete LEDs

Because of their narrow-band spectral output distributions, small size, and often highly directional properties, discrete LEDs can be particularly difficult to characterize accurately. For some reason, after the flurry of work on refining LED measurement methods that coincided with the development of commercial LEDs in the late 1960s and early 1970s, no known standard has ever been published to aid measurement of these devices.

Discrete LEDs are customarily measured in the units of luminous intensity, the candela and the millicandela. The candela is well suited to the geometry of the LED since it determines the amount of light in the center of the LED beam, which is usually the direction from which they are viewed. The intensity at off-axis angles may be estimated from the manufacturer's polar plots of the light distribution (refer to Chapter 2 for examples). High-quality illuminance photometers calibrated in lux or footcandles may be used to determine candelas or millicandelas emitted by an LED providing the distance from the LED to the sensor is accurately measured. At a distance of 1 m, there is a one-to-one relationship between lux and candelas. Likewise, at a distance of 1 ft, there is a one-to-one relationship between footcandles and candelas. The inverse square law allows measurements in either lux or footcandles to be easily converted to candelas for any other distance chosen to fit a particular situation.

The spectral considerations of all types of LEDs are fairly easily met by use of a photometer having accurate spectral correction to the CIE Standard Observer response curve. This is assured by selection of photometers with sensors having CIE f_1' values of 3% or less. This is especially important for the recently developed SiC and GaN blue LEDs, which have maximum emission in a narrow band of wavelengths in the spectral region where the error is worst for lower cost sensors with greater than 3% f_1' specifications. Maximum accuracy may be obtained if the actual spectral curve of the individual sensor being used is available. Just compare its relative response at the peak wavelength of LED emission with the CIE photopic curve and adjust the measured data accordingly.

The LED size and narrow-angle light distribution difficulties are not as easily solved. Here, the key is accurate and reproducible fixturing. First, the LED should be held as perpendicular to the sensor as possible. This is not always easy given the small size of many LEDs and the variety of package shapes in use. Diffuse LEDs with their wider viewing

178 LUMINANCE AND CONTRAST MEASUREMENT

angles are not as critical as those LEDs with transparent packages that have narrow beams. The latter must sometimes have its alignment adjusted for maximum reading on the photometer if the die is not aligned exactly perpendicular to the optical axis or if its mechanical design does not lend itself well to precise alignment. The other side of that argument is that if it cannot be positioned accurately in the test fixturing, it probably will not be any better aligned in its final application and the resulting measurement inconsistency will likely represent the variations that will be encountered in actual use.

The LED-to-sensor distance chosen often involves compromises. Making the measurement at the relatively long distance of 1 m or 1 ft eliminates the computation of candelas from lux or footcandles and makes the distance measurement accuracy less important. The trade-offs are lower sensitivity and the fact that the sensor will intercept a smaller angular sample of the center of the beam than the same sensor at a closer distance. Making use of the inverse square law by using shorter distances will increase the sensitivity significantly. A distance 3.16 times closer will increase the sensitivity by a factor of 10 and $\frac{1}{10}$ the distance will increase the sensitivity by a factor of 100. The angle of the portion of the LED beam sampled will increase linearly as the distance decreases. Thus, a sensor that would intercept a sample of 1° from the center of the beam at a distance of 1 m would intercept about 3° at 0.316 m while a reading of 1 lx would correspond to 0.1 cd or 100 mcd. Use basic trigonometry to calculate the angle subtended by the sensor's active diameter at the selected measurement distance. Then compare that angle to the manufacturer's polar plot of the LED's light distribution to get an idea of whether the sensor is just sampling the peak intensity point in the beam center or is averaging over both the beam center and lower intensity skirts. In the latter case, the measured intensity may be considerably lower than the peak value. It is preferable to use the same sample angle as the manufacturer to obtain best correlation with their specifications. The distance in most situations is less important than the angle since the effect of distance is removed using the inverse square law. If the distance is so short that it becomes difficult to measure that distance with better than 2% accuracy, the measurement accuracy may be degraded. The points used for the distance measurement are the effective plane of the active portion of the sensor and the effective emitting plane of the LED. For a sensor having only a transparent spectral correction filter, the distance is measured to the effective plane of the photosensitive surface. The effective sensor plane will be closer to the LED than the actual physical dimension because of the refractive index of the glass filters on the sensor. The effective filter thickness will be about 65% of the actual thickness for most filters, which usually have refractive indices between 1.50 and 1.55. For sensors having a diffuse surface such as for cosine correction, the distance would be measured to the diffuse surface itself. If the diffuser is domed, an approximation will suffice. Use a point below the highest point on the dome and about 30% of the distance to the lowest point that is exposed to light. The effective emitting plane of the LED depends on whether it is a transparent or diffuse type. The former will be at a point about 65% of the distance from the tip to the die, a result of the refractive index of typical plastic encapsulation being around 1.50–1.55, as with glass filters. Diffuse LEDs will have an emitting plane closer to the tip. In the absence of better numbers from the manufacturer, use the radius of the tip as the dimension for the distance below the tip, or in the case of flat-surfaced LEDs, use the diffuse outermost surface. In either situation, the error will be minor for normal measurement distances, which should be at least 75–100 mm.

The fixturing for the LED and sensor should be in a light-tight enclosure in order to avoid ambient light. The interior should be painted ultraflat black to prevent normally lost

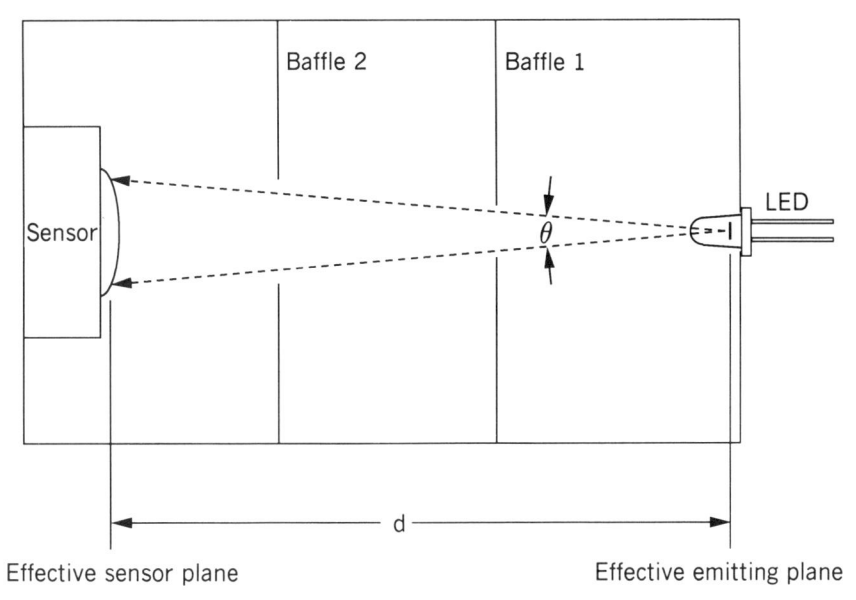

Figure 5.19. Fixturing for LED measurement with an illuminance sensor.

off-axis rays from being reflected toward the sensor. Interior baffling should be used at least at the midpoint of the enclosure for the same reason. The position of the baffle and the diameter of its central aperture should be such that all on-axis rays within the interception angle of the sensor will be passed for the largest LED size to be measured. Two baffles at about one-third of the distance from each end is even better. The inside diameter of the enclosure should be as large as practical to minimize off-axis scattering of light. See Figure 5.19 for one possible method of fixturing. A small number of commercial photometers are available with heads incorporating fixturing for this purpose.

Alphanumeric LEDs

Individual alphanumeric LEDs should be measured in a similar manner except that provision should be made to accommodate larger sizes than for discrete LEDs. Make sure that no direct light from any segment is intercepted by the intermediate baffle(s). Tests may be performed to assure that all segments are of similar light output or all segments may be measured simultaneously to allow matching of light output to other digits that will be used in the same or an adjacent display.

Dot-Matrix LEDs

Dot-matrix LED displays should be measured in a manner similar to other self-emitting displays. Average luminance may be used for this application. The acceptance angle of the luminance photometer should be large enough to include 30 or so LEDs for a representative average luminance value. LEDs are often used for larger scale displays such as message signs. Longer photometer-to-display distances may be used to allow use of a photometer with a measurement angle that is too small for accurate close measurement.

REFERENCES

1. Electronic Industries Association, "Industrial Cathode Ray Tube Test Methods," EIA Publication No. 105-2, Feb. 1981.
2. C. P. Halsted, "Brightness, Luminance and Confusion," *Information Display*, pp. 21–24, Mar. 1993.
3. American National Standards Institute (ANSI), "American National Standard for Metric Practice," Std. 268-1992, 1992.
4. Illuminating Engineering Society of North America (IESNA), "Nomenclature and Definitions for Illumination Engineering," ANSI/IES Recommended Practice RP-16-86, 1986.
5. Commission Internationale de l'Eclairage (CIE), "International Lighting Vocabularly," Publication No. 17.4, 1987.
6. Commission Internationale de l'Eclairage (CIE), "International Electrotechnical Vocabularly," Publication 50, Chap. 845 (Lighting), 1987.
7. National Institute of Standards and Technology (NIST), "Guide for the Use of the International System of Units (SI)," Special Publication 811, 1995.
8. Society of Automotive Engineers (SAE), "Photometric and Colorimetric Measurement Procedures for Airborne Direct View CRT Displays," ARP 1782, Section 6.2, Jan. 9, 1989.
9. National Information Display Laboratory (NIDL), "Display Monitor Measurement Methods Under Discussion by EIA Committee JT-20," Draft, Version 2.0, Part 1, July 12, 1995.
10. National Information Display Laboratory (NIDL), "Display Monitor Measurement Methods Under Discussion by EIA Committee JT-20," Draft, Version 2.0, Part 2, July 12, 1995.
11. National Information Display Laboratory (NIDL), "Display Monitor Measurement Methods Under Discussion by EIA Committee JT-20," Draft, Version 2.0, Part 1, July 12, 1995.
12. National Information Display Laboratory (NIDL), "Display Monitor Measurement Methods Under Discussion by EIA Committee JT-20," Draft, Version 2.0, Part 1, July 12, 1995.
13. Electronic Industries Association (EIA), "Phosphor Linearity Tests Using Illuminance and Microphotometer Detectors," EIA Publication TEP105-16, Aug. 1990.
14. American National Standards Institute (ANSI)/Institute of Electrical and Electronics Engineers (IEEE), "Standard Dictionary of Electrical and Electronics Terms," ANSI/IEEE Std. 100-1977, 1977, p. 281.
15. K. B. Benson and J. Whitaker, *Television Engineering Handbook*, McGraw-Hill, New York, 1992, p. 4.19.
16. Commission Internationale de l'Eclairage (CIE), "International Lighting Vocabulary," CIE Publication No. 17.4, 1987, pp. 58–59.
17. American National Standards Institute (ANSI)/Institute of Electrical and Electronics Engineers (IEEE), "Standard Dictionary of Electrical and Electronics Terms," ANSI/IEEE Std. 100-1977, 1977, p. 133.
18. Electronic Industries Association, "Contrast Measurement of CRTs," EIA Publication No. TEP105-10, Apr. 1987.
19. Illuminating Engineering Society of North America (IESNA), *Lighting Handbook,* 8th ed., IESNA, New York, 1993, p. 927.
20. Electronic Industries Association (EIA), Internal communication, EIA JT33.3 Committee on LCDs, 1988.
21. Electronic Industries Association (EIA), "Contrast Measurement of CRTs," EIA Publication No. TEP105-10, Apr. 1987.
22. National Information Display Laboratory (NIDL), "Display Monitor Measurement Methods Under Discussion by EIA Committee JT-20," Draft, Version 2.0, Part 1, July 12, 1995.

25. Electronic Industries Association (EIA), "Test Method for Specular Gloss," EIA Publication No. TEP105-13, Apr. 1987.
26. American Society for Testing and Materials (ASTM), "Standard Test Method for 45-deg Specular Gloss of Ceramic Materials," ASTM D 346-87, Oct. 1987.
27. American Society for Testing and Materials (ASTM), "Standard Test Method for Specular Gloss," ASTM D 523-89, 1989.
28. National Information Display Laboratory (NIDL), "Display Monitor Measurement Methods Under Discussion by EIA Committee JT-20," Draft, Version 2.0, Part 1, July 12, 1995.
29. International Organization for Standardization (ISO), "Visual Display Terminals (VDTs) used for Office Tasks—Ergonomic Requirements," ISO 9241, Part 7, 1991 (draft).
30. American National Standards Institute (ANSI)/National Association of Photographic manufacturers (NAPM), "Data Projection Equipment and Large Screen Data Displays—Test Methods and Performance Characteristics," ANSI Publication No. IT7.215, 1992.
31. American National Standards Institute (ANSI), "Liquid-Crystal Imaging Devices for Use with Overhead Projectors—Method for Measuring and Reporting Performance Characteristics and Features," ANSI Publication No. IT7.228, 1990.

CHAPTER 6

Color Measurement

6.1 GENERAL BACKGROUND

Chromaticity, or color, is one of the most important measurements for color displays. Poorly adjusted color balance of a single display and mismatched multiple monitors located in close proximity to one another is often objectionable, even to the less sophisticated user. Adjustment of color balance by the "calibrated eye" method is uncertain at best. Surrounding illumination, age of the viewer, the color of objects or other displays viewed immediately prior to the adjustment procedure, and individual color preferences all conspire to preclude accurate color balancing by eye. Instead a colorimeter or spectrophotometer is usually employed for the purpose (Figure 6.1). Even then, accurate color measurement of displays is fraught with pitfalls for the unwary. For a general source of background material on colorimetry of self-luminous displays, see CIE Publication No. 87.[1]

Display manufacturers usually require the most accurate (i.e. expensive) instrumentation for research and design while manufacturing lines do not usually require extreme absolute accuracy. Ruggedness, repeatability, and cost will be the highest ranking factors for manufacturing. Finally, the end user may only require an easy-to-use instrument capable of maintaining consistency of appearance from day to day for a single display. See Chapter 4 for information concerning the choices and trade-offs in selecting a colorimeter.

Chromaticity measurement using the 1931 or 1976 CIE systems is most common and provides good results. An accurately measured pair of chromaticity coordinates and a luminance value provide a unique color appearance that can be exactly matched by the same values on another display. Correlated color temperature measurement does not provide a unique appearance since the chromaticity coordinates may be at any point along the isotemperature line, which may have a tint ranging from greenish to purplish. Some filter colorimeters display correlated color temperature in Kelvin although the accuracy may be suspect. This is due to the fact that there is no simple and convenient algorithm to convert color coordinates to color temperature. A detailed look-up table in the microprocessor ROM is the solution, although one is often better off by acting as the look-up table his or herself and plotting the color coordinates on an expanded correlated color temperature chart.

The selection of the 1931 or 1976 CIE color space is up to the user, although the 1976 CIE-UCS chromaticity probably has the edge. Coordinates in either system may be converted to the other using the equations in Appendix F, and most modern colorimeters provide data in both systems. Selection of the best system may often depend on what units are written into a particular standard, manufacturer's specification, or contract requirement.

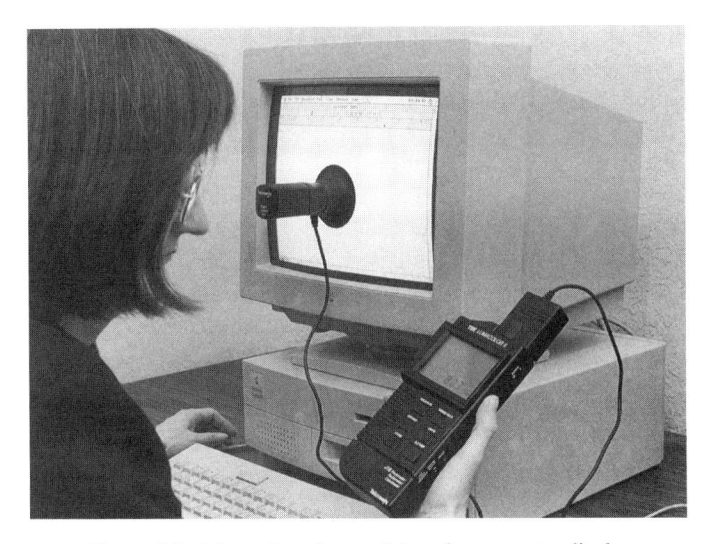

Figure 6.1. Measuring chromaticity of a computer display.

For monochrome displays, simple colorimetric measurements usually suffice. Only the color coordinates need to be measured for normal operating conditions. This is usually of concern to the manufacturer of the CRT or display for quality assurance purposes. The user has no control over the display color and thus has little need for color measurement other than to confirm the use of the proper phosphor or filter. Medical uses and multiple display installations are the exception since uniform appearance may be an important factor for these users.

Color CRT displays require adjustment of color balance, and to achieve proper "tracking" of the color balance from low-drive to high-drive conditions to avoid color shifts as the intensity changes. Interaction of the color balance adjustments between high- and low-drive levels usually makes such adjustment an iterative process. Color uniformity across the entire display is another important measurement since the eye can detect color nonuniformity much easier than luminance nonuniformity.

6.2 PURITY

Before any other adjustments or measurements of a shadow-mask color CRT display are made, the color purity, also known as beam-landing error, should be verified and adjusted if necessary. Adequate warm-up of the display should be allowed. Twenty or thirty minutes will usually be sufficient. The display should be positioned in its normal operating location or the same compass orientation with respect to *magnetic* north to reduce the effect of the earth's magnetic field and nearby electromagnets or permanent magnets if at all possible. This may be especially important for lower cost displays where less magnetic shielding is employed. Try to keep other equipment and large ferrous metal objects at a distance of 1 m or more from the display under test. The display should be degaussed with a hand-held degaussing coil prior to purity verification or adjustment.

Purity is checked by displaying a full-screen raster of red only. Red is usually used since beam-landing errors are more noticeable than with either green or blue. The beams from each electron gun should land only on the phosphor of the corresponding color. There should be no areas of the screen with visible discoloration from a uniform red. Plate 8 shows the appearance of poor purity.

Initial adjustment of an uncalibrated display involves movement of the deflection yoke forward and aft to produce as large a red area as possible in the screen center and then rotation of the purity magnets on the CRT neck to make the screen uniformly red. Green and blue fields are then displayed individually to verify that their purity is also satisfactory. If not, slight readjustment of the purity magnets should be made for the best compromise for purity of all three primary colors.

6.3 CHROMATICITY

The CIE Publication No. 15.2 is the foundation for color measurement.[2] Many standards based on it exist for display chromaticity measurement. Some of these are ASTM E-1336,[3] EBU D-28,[4] R-23,[5] and 3213-E[6]; EIA TEP105-11A[7]; IEC 441[8]; ISO 9241-8[9]; the proposed NIDL monochrome and color display measurement procedures[10,11]; SAE ARP1782,[12] ARP1784,[13] ARP4067,[14] ARP4256,[15] and ARP4260[16]; SMPTE RP71,[17] RP145,[18] and RP167[19]; and the unnumbered VESA "Display Specifications and Test Procedures"[20] standard. It is strongly advised that the pertinent standards be purchased for the detailed procedures for the specific display type or application.

Generally, a color temperature of 6500 K (CIE Illuminant D_{65}) with chromaticity coordinates $x = 0.3127$ and $y = 0.3290$ is recommended for television studio monitors and critical computer displays, although bluer displays, up to 9300 K, are used for home entertainment and less stringent computer applications.

Adequate warm-up of the CRT, emissive flat panel, or LCD backlight should be allowed. Twenty minutes or so will usually be sufficient. For color CRT displays, positioning and degaussing should be as described previously under Purity. All other intended adjustments including size, geometry, linearity, and convergence, should have been performed as described in Chapter 9. Ambient light should be excluded from the measurement area. This is one measurement where residual light cannot be subtracted to correct the readings. While the display is warming up, take the time to clean the screen, especially if it has been in service for some time.

The brightness and contrast controls should be set to their preset positions if the chromaticity of an already calibrated display is just being verified. Procedures for adjustment of white point adjustment will be covered later under Color Tracking. Chromaticity measurements are usually made with a full-screen white field at 100% drive. This may be derived from a video pattern generator, or for some computer applications the blank white page obtained with word processing software is often satisfactory. The page may have to be moved or expanded for measurement of points other than the screen center.

Measurement should be made at least at the screen center. Tristimulus filter colorimeters often use a suction cup to secure the sensor assembly to the screen during a series of measurements or adjustments. A typical example is shown in Figure 6.1. Unless the lips of the suction cup are dampened, the duration that it will stay in place will likely be short. Keep a damp sponge nearby and press the suction cup lips to it before placing it on the screen. Because of their larger size and weight, spectroradiometers should be mounted on a heavy-duty tripod or *x–y* positioner in front of the display. The lower the center of gravity of the tripod, the better since tripod legs are easily tripped over in the dark environment required for the measurement of chromaticity. (The sound of a $20,000 instrument hitting the floor will make a lasting memory!) The spectroradiometer distance should be far enough that an area encompassing at least 30 full pixels is measured to avoid localized color variations. Alternately, some instruments have provision to change lenses to adjust the measurement field.

Liquid crystal displays require consideration of their directional properties. Their optimum viewing direction may be in a direction other than perpendicular to the display. The LCD color varies significantly with viewing angle. For these, the colorimeter or spectroradiometer should be aligned at the intended optimum viewing angle. See the manufacturer's technical specifications to determine that angle. Measurements may be made intentionally at other angles to determine the color coordinate shifts versus viewing angle to further characterize the LCD.

Make at least three measurements to check for repeatability of the color coordinates. Small variations are often normal due to rounding errors in the chromaticity computations and slight electrical drift and/or noise present in the display and/or measuring instrument. Larger errors may be caused by instability of the display video amplifiers, too low a light level, or inadequate integration of rastered displays. For the latter, some colorimeters require synchronization with the display raster refresh rate. For these, follow the manufacturer's instructions for proper electrical connections. Other colorimeters use longer integration times, which may require multiple readings to be made before stabilizing to the correct value. Each time the light intensity or color balance is changed will require allowance of adequate time for the readings to stabilize, but this is usually easier than having to provide electrical connections between the display and colorimeter for synchronization. The CIE recommends that unsynchronized measurements should integrate at least 10 frames.[21] Scanned CCD spectroradiometers adjust for low light levels by increasing their integration time to obtain measurable signal levels.

Chromaticity measurement of a 100% white field is fairly straightforward. The colorimeter or spectrophotometer is positioned to measure the screen center, enough time is allowed for stabilizing and making the measurement, and the readings are read from the instrument. It is when readjustment of chromaticity and/or color tracking is necessary that things get more complex.

6.4 COLOR TRACKING

"Color tracking" refers to the ability of a display to produce the same color at high and low drive levels and is primarily a consideration for CRT displays where light output is a nonlinear function of video drive signals. Usually two luminance levels, high and low, are specified for tracking adjustments and the white balance is adjusted to produce the same color coordinates for each luminance. Approximately 60–100 cd/m^2 (20–30 fL) is usually specified for the highlight white point. For best accuracy with typical colorimeters, the low light luminance should be selected to be 7–14 cd/m^2 (2–5 fL) or so. Not only will this allow more accurate adjustment, but also it is in a range where the eye is more apt to notice color errors. Selecting too low a level increases the risk that the mid-level color will

186 COLOR MEASUREMENT

track improperly because of differing degrees of bowing of the intensity versus drive curves of the three guns. Errors in the mid-levels will be very noticeable. Any residual ambient light will also have less effect on the measurements if these levels are used.

The standard SMPTE RP71 contains the procedures for setting the white point and color-tracking adjustments for color CRT monitors. The document describes procedures using both a visual color comparator and a colorimeter for initial adjustment and to later reestablish the same conditions as the display drifts or ages. It begins by using a white window pattern at 100% video level (100 IRE units, which is +1.0 V) to set the *brightness* and *contrast* controls. The black level is set at the point of extinction using the brightness control for the reference black level (7.5 IRE units for NTSC television, where 0 IRE units, or 0 V, is the blanking level). For computer displays, the black level is often set to extinction using the internal *screen* adjustments that set the CRT first anode or "G2" voltages. With the colorimeter positioned in the screen center with a 100-IRE window pattern, adjust the red, green, and blue video amplifier *gain* controls for the desired chromaticity coordinates and luminance value. These values are $x = 0.313$, $y = 0.329$, and 103 cd/m² (30 fL) for the SMPTE, EBU, and ISO recommended D_{65}.[22-24]

To adjust the lowlight levels, the process is repeated using about 23 IRE units and adjusting the contrast, blue screen, and green screen controls for the same chromaticity coordinates at 7 cd/m² (2 fL). Note that the control names and even functions may be different for some monitor manufacturers. Refer to the manufacturer's instruction manuals for details pertaining to the particular monitor type.

6.5 COLOR UNIFORMITY

Color uniformity characterization consists of several chromaticity measurements of a white full field made at prescribed locations, usually either five or nine, on the display screen. A proposed method for ISO 9241 Part 8[25] specifies the center and four corners as defined by the following points:

0.1 H, 0.1 W
0.1 H, 0.9 W
0.5 H, 0.5 W
0.9 H, 0.1 W
0.9 H, 0.9 W

where H is the height of the viewing area and W is the width. The color difference in the 1976 CIE UCS system is computed for the center and the position with the greatest color deviation from the center using the following formula:

$$\Delta u'v' = [(\Delta u')^2 + (\Delta v')^2]^{1/2}$$

Note that this differs from the CIE recommendations for color difference, which are specified as ΔE^*_{uv} and includes luminance as well as chromaticity. Since displays usually have some luminance nonuniformity and this is handled separately as part of the luminance measurement, it is better if not included in display color uniformity evaluation. It is best to separate luminance and contrast to permit analysis of each since the color nonuniformity is usually more objectionable than luminance nonuniformity and they have different causes.

The EIA standard TEP105-18[26] is very similar to ISO 9241 Part 8, except that the choice of using either the color difference equation ($\Delta u'v'$) or the CIE ΔE^*_{uv} equation is permitted.

A NIDL-proposed color uniformity method for VESA[27] is by no means complete at this time, but it appears similar to the ISO method except that nine points are used. The additional points are

0.1 H, 0.5 W
0.9 H, 0.5 W
0.5 H, 0.1 W
0.5 H, 0.9 W

It certainly does not hurt to measure the additional points, although the five-point method will usually catch most color uniformity problems that, at least for CRTs, are likely to be in the corners. Also, if other areas of color nonuniformity are visibly noticeable, these probably should be measured as well or the display color readjusted if necessary.

6.6 COLOR GAMUT

The color gamut is the range of colors that may be displayed by the display. It is recommended that a spectroradiometer be used to determine color gamut since it involves measurement of the color primaries, which are at a distance from the white calibration point used for tristimulus filter colorimeters. Calibration errors will be greater as the distance from the calibration point increases.

To determine the color gamut, a window pattern or full field should be displayed. One hundred percent red, green, and blue signals are individually applied and the x, y or u', v' chromaticity measured for each in the screen center. The three sets of chromaticity values are plotted on the appropriate CIE chart. The triangle formed by the points indicates the color gamut (Figure 6.2). Any color within the triangle may be produced by mixtures of the red, green, and blue signals. Any ambient light present will dilute the purity of the three color primaries by shifting them toward the white point of the ambient light. The color gamut triangle area is reduced accordingly.

6.7 SPECTRAL OUTPUT

When using a spectroradiometer to measure chromaticity, the spectral output curve of the device is an intermediate step in the process of producing the chromaticity coordinates. It may usually be selected for examination separately. It is not as useful by itself, although it contains valuable information that allows the individual phosphors, other light sources, or filters to be identified as to generic type. It is also particularly useful in selecting contrast enhancement filters for optimum results. A typical spectral output curve for a color CRT display screen is shown in Figure 6.3.

188 COLOR MEASUREMENT

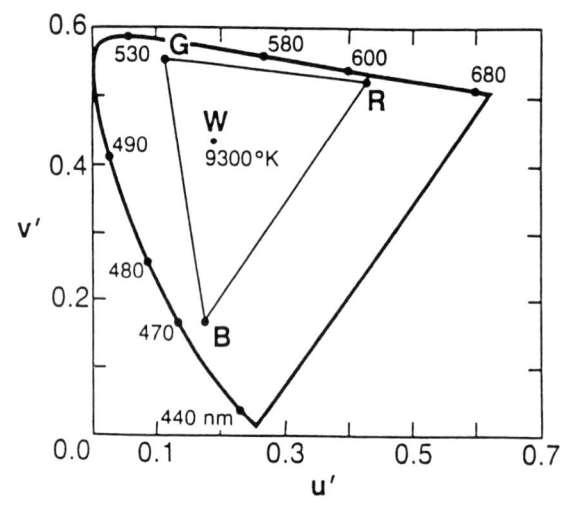

Figure 6.2. Color gamut triangle. The three phosphor primaries are indicated at the points of the triangle by R, G, and B. Any color within this triangle may be produced by adjusting the relative amounts of each of the phosphor primaries.

XXD (P22 Sulfide/Oxysulfide) Phosphors

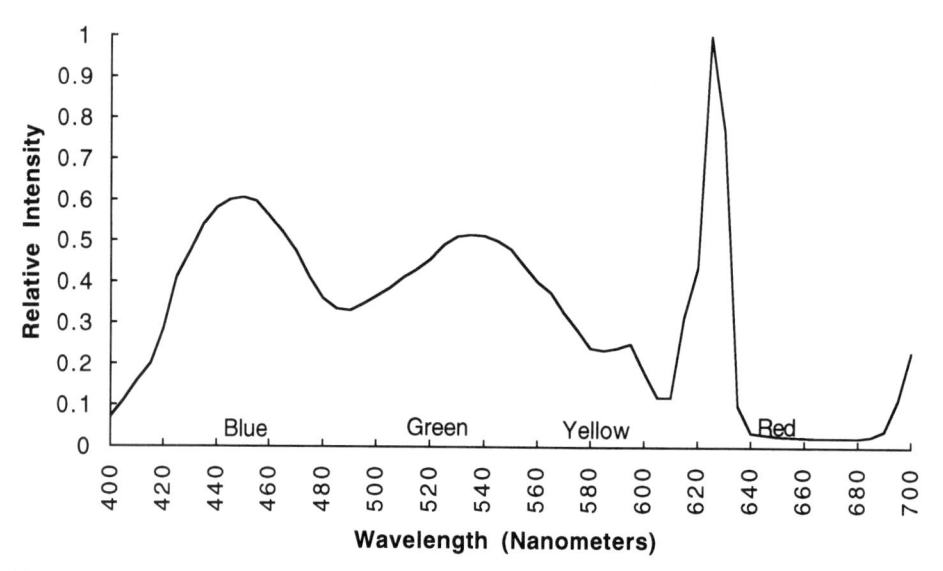

Figure 6.3. Spectral output distribution for a typical color display screen when set for a white color balance. The phosphor screen is EIA type XXD, formerly identified as P22 sulfide/oxysulfide.

6.8 COLOR ANISOTROPY

Color LCDs exhibit differences in luminance and color with viewing angle, or *anisotropy*. A proposed ISO standard based on ISO 9241[28] and proposed VESA standard[29] are directed at measurement and specification of LCD anisotropy. The ISO proposes that the chromaticity should be measured in 1976 CIE chromaticity units over a range of angles and that the following equation should be used to compute the chromaticity variation with those angles:

$$\Delta u'v' = [(\Delta u')^2 + (\Delta v')^2]^{1/2}$$

The LCD manufacturer's specification should be used to determine the rated viewing angles and chromaticity measurements made within those bounds. Otherwise, the measurement is similar to the previously described chromaticity measurement techniques.

Some filter colorimeters contain three separate sensors with no common diffuser mounted in a triangle about the optical axis of the sensor head (refer back to Figure 4.12). Measurement variations of both the chromaticity coordinates and luminance may be observed for some LCDs as the head is rotated in 90° increments about the optical axis. This variation can be as much as ±10% of the readings and is due to nonuniform illumination of the three sensors by the directional light distribution of the LCD. Averaging the four sets of readings will produce the average chromaticity and luminance. The effect may be observed visually by first viewing the display from exactly perpendicular to it. Shift the viewing position a few degrees off-axis above, below, left, and right and certain positions will be found that exhibit slight reddish and bluish coloration. It is to these slight variations that the instrument is responding. The variation will be minimized when the LCD operating voltage (V_{op}) is adjusted optimally for perpendicular viewing. As observed with triangularly mounted sensors, this variation may be employed to aid adjustment of the operating voltage for best chromaticity at the desired viewing angle. Rotate the head about the desired viewing axis in 90° increments and adjust the operating voltage for minimum spread of chromaticity values with angle.

Photometers and spectroradiometers may exhibit polarization effects from mirrors, gratings, or other optical elements. They are calibrated with unpolarized light but will produce readings varying by several percent for various polarizations of light. Since LCDs produce strongly polarized light output, the measurement instrument must have a calibration or correction data for the polarization being measured. When various angles and planes for a LCD are being measured, extensive characterization of the photometer or spectroradiometer is required. The correction data is best computed using a dedicated computer for the system.

REFERENCES

1. Commission Internationale de l'Eclairage (CIE), "Colorimetry of Self-Luminous Displays—A Bibliography," CIE Publication No. 87, 1990.
2. Commission Internationale de l'Eclairage (CIE), "Colorimetry," CIE Publication No. 15.2, 1986.
3. American Society for Testing and Materials (ASTM), "Standard Test Method for Obtaining Colorimetric Data from a Video Display Unit by Spectroradiometry," ASTM E 1336-91, Apr. 1991.

4. European Broadcasting Union (EBU), "The Chromaticity of the Luminophors of Television Receivers," EBU Technical Statement D-28, 1980.
5. European Broadcasting Union (EBU), "Procedure for the Operational Alignment of Grade 1 Colour Monitors," EBU Technical Recommendation R23, 1980.
6. European Broadcasting Union (EBU), "Chromaticity Tolerances for Studio Monitors," EBU Standard 3213-E, Aug. 1975 (reaffirmed 1981).
7. Electronic Industries Association (EIA), "Measurement of the Color of CRT Screens," EIA Publication TEP105-11-A, Dec. 1988.
8. International Electrotechnical Commission (IEC), "Photometric and Colorimetric Methods of Measurement of the Light Emitted by a Cathode-Ray Tube Screen," IEC Publication No. 441, 1974.
9. International Organization for Standardization (ISO), "Ergonomic Requirements for Office Work with Visual Display Terminals (VDT)—Part 8: Requirements for Displayed Colours," Draft, ISO Publication No. 9241-8, 1992.
10. National Information Display Laboratory (NIDL), "Display Monitor Measurement Methods Under Discussion by EIA Committee JT-20," Draft, Version 1.0, Part 1, July 12, 1995.
11. National Information Display Laboratory (NIDL), "Display Monitor Measurement Methods Under Discussion by EIA Committee JT-20," Draft, Version 2.0, Part 1, July 12, 1995.
12. Society of Automotive Engineers (SAE), "Photometric and Colorimetric Procedures for Direct View CRT Displays," SAE Recommended Practice ARP1782, Jan. 1989.
13. Society of Automotive Engineers (SAE), "Design Objectives for CRT Displays for Part 25 (Transport) Aircraft," SAE Recommended Practice ARP1874, May 1988 (Reaffirmed May 1993).
14. Society of Automotive Engineers (SAE), "Design Objectives for CRT Displays for Part 25 Aircraft," SAE Recommended Practice ARP4067, Nov. 1989.
15. Society of Automotive Engineers (SAE), "Design Objectives for Liquid Crystal Displays for Part 25 (Transport) Aircraft," Draft, SAE Recommended Practice ARP4256, 1995.
16. Society of Automotive Engineers (SAE), "Photometric and Colorimetric Measurement Procedures for Airborne Direct View Flat Panel Displays," Draft, SAE Recommended Practice ARP4260, undated.
17. Society of Motion Picture and Television Engineers (SMPTE), "Setting the Chromaticity and Luminance of White for Color Television Monitors Using Shadow-Mask Picture Tubes," SMPTE Recommended Practice RP71, 1977.
18. Society of Motion Picture and Television Engineers (SMPTE), "SMPTE C Color Monitor Colorimetry," SMPTE Recommended Practice RP145, 1994.
19. Society of Motion Picture and Television Engineers (SMPTE), "Alignment of Professional Television Color Picture Monitors," Draft, SMPTE Recommended Practice RP167, Sept. 1992.
20. Video Electronics Standards Association (VESA), "Display Specifications and Test Procedure," VESA Standard Version 1.0, Revision 1.0, Oct. 3, 1994.
21. Commission Internationale de l'Eclairage (CIE), "Guide to Characterizing the Colorimetry of Computer-Controlled CRT Displays," CIE Technical Committee TC 2-26 Final Report, May 1994, p. 10.
22. Society of Motion Picture and Television Engineers (SMPTE), "SMPTE C Color Monitor Colorimetry," SMPTE Recommended Practice RP145, 1994.
23. European Broadcasting Union (EBU), "Chromaticity Tolerances for Studio Monitors," EBU Standard 3213-E, Aug. 1975 (reaffirmed 1981).
24. International Organization for Standardization (ISO), "Ergonomic Requirements for Office Work with Visual Display Terminals (VDT)—Part 8: Requirements for Displayed Colours," Draft, ISO Publication 9241-8, 1992.

25. International Organization for Standardization (ISO), "Ergonomic Requirements for Office Work with Visual Display Terminals (VDT)—Part 8: Requirements for Displayed Colours," Draft, ISO Publication 9241-8, 1992, p. 16.
26. Electronic Industries Association (EIA), "Color Field Uniformity Test Procedures," EIA Publication TEP105-18, Apr. 1995.
27. National Information Display Laboratory (NIDL), "Flat Panel Display Measurements," Draft for VESA, Oct. 6, 1995, p. 17.
28. International Organization for Standardization (ISO), 9241 Part 3 (to become ISO 13406), Proposed Addendum, 1991, p. 10.
29. National Information Display Laboratory (NIDL), "Flat Panel Display Measurements," Draft for VESA, Oct. 6, 1995, p. 18.

CHAPTER 7

Resolution Measurement

7.1 GENERAL BACKGROUND

Information density in displays is increasing rapidly with the phenomenal growth of data and graphics displays. The introduction of color to data and avionics displays required improvement of the shadow-mask CRT, which previously had only limited resolution demands placed on it for television use. The color shadow-mask CRT requires special considerations when measuring resolution in order to avoid anomalies caused by the physical structure of the screen. All of these factors have increased the need for display and CRT resolution measurement. Resolution measurement is required only for CRT displays since other matrix-addressed displays are determined by the mechanical size and fixed spacing of pixels.

Resolution must not be confused with addressability. Resolution is a measure of how much information detail may be viewed. Addressability is the number of data points or pixels that may be individually controlled and is usually specified in both the horizontal and vertical axes, such as 2048 × 1536. Sometimes, addressability is expressed as the total number of addressable pixels, derived by multiplying the horizontal pixels by the vertical pixels and is slightly more than three million pixels for the previous example. Just because that number of pixels is addressable does not mean that they are also resolvable, at least for CRT displays. Cost-cutting measures in electron gun or circuit design may result in poor corner or overall focus. Murch et al. developed a simple metric, the resolution–addressability ratio (RAR) for relating addressability and resolution.[1-5] The equation used for RAR is

$$RAR = W/S$$

where

    W = full width of the raster line profile at half-maximum luminance
    S = peak-to-peak separation between adjacent raster lines

The RAR method was later adopted by EIA as a technical bulletin for CRT displays.[6]

The resolution of data displays and data display CRTs is a most difficult parameter to characterize accurately. Large differences between resolution measurements are often observed, owing to the diversity of display formats, such as raster scan and stroke writing, alphanumerics, and graphics. The large diversity of measurement techniques also produces different results. Differences in test operator perception and skills add to the difficulties. Resolution data should therefore be used with caution. The following are some major sources of resolution measurement error:

Beam current or drive level
Calibration
Magnetic fields
Electrostatic fields
Mechanical vibration
CRT operating voltages
CRT power supply ripple and noise
Resolution measurement method
CRT focus adjustment
Optical focus of measurement device
Phosphor "noise"
Size, shape, and alignment of light-sampling device

When comparing displays or CRTs, the display engineer must be sure that resolution data are comparable. A display from one manufacturer specified as having 0.5 mm resolution at the half-amplitude points with 200 μA beam current may actually have better resolution than a display from another manufacturer specified as having 0.3 mm spot size measured with the shrinking-raster technique at 100 μA.

The resolution of a CRT display may be expressed in any of several ways, each of which has different connotations and is not usually directly comparable with the others, especially if drive or duty cycle differences are considered. Resolution may be measured in an even greater number of ways, as listed in the following:

Visual
Shrinking raster
Vernier line method
Photographic method
Television camera method
Television wedge chart
Radial resolution test chart
Slit/aperture scan
    Mechanical scan
    Electrical scan
    Double-slit scan
Radial scan contour plotting
Dot-matrix scan contour plotting
Image dissector scanning
CCD scanned array
Variable-slit method

194 RESOLUTION MEASUREMENT

Spot photometer with aperture
Video photometer
Modulation transfer function
   Measured
   Calculated

This chapter will present information with advantages and limitations for most of these varied techniques.

Test Patterns for Resolution Measurement

Test patterns for resolution measurement of CRTs and CRT displays include line pairs, dot patterns, and crosshatch patterns. Line pairs are used for visual determination of resolution. They are often available from video signal generators in the form of the SMPTE 133[7] or other similar test pattern. Dots are most useful for analysis with two-dimensional scanning devices such as CCDs that record data from all portions of the dot. Crosshatch patterns are often used for measurements in just one or two axes. Both vertical and horizontal beam profiles often may be obtained by rotating slit scanning devices by 90°. Test pattern generators capable of producing line pairs, dot, and crosshatch patterns are available from several manufacturers. See Appendix K.

7.2 SPOT CHARACTERISTICS

In this chapter, we will define beam profile as the current distribution within the electron beam itself. The spot or line profile is the viewable result on the phosphor screen. The profile of the current distribution in a typical undeflected electron beam at low beam current is usually nearly Gaussian (Figure 7.1). At higher beam current, the electron beam itself may deviate from Gaussian due to electron space-charge effects or cathode damage from bombardment by ions, usually as a result of residual gases within the evacuated CRT envelope. When the electron beam strikes the phosphor screen, the light output dis-

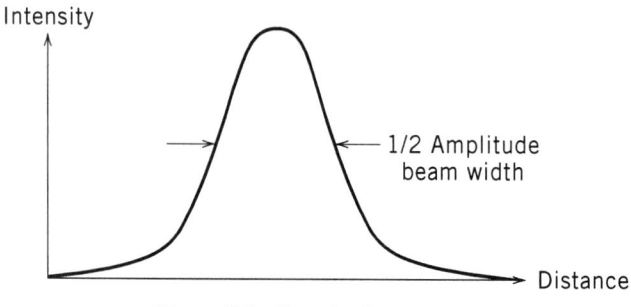

Figure 7.1. Gaussian beam profile.

7.2 SPOT CHARACTERISTICS

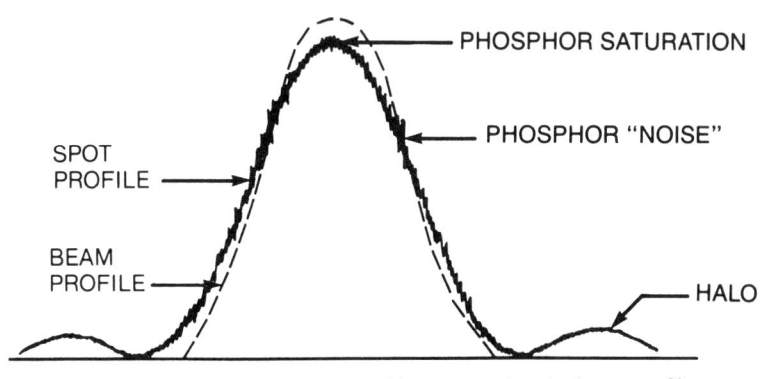

Figure 7.2. Abberations of the spot profile compared to the beam profile.

tribution can differ from the beam current distribution in one or more ways, as listed in the following (Figure 7.2):

- The light output of an undeflected beam may be nonlinear due to phosphor saturation at high current densities, thus rounding the peak.
- Reflections within the CRT faceplate glass may cause spurious light at points well away from the center peak.
- Phosphor graininess, or "noise," causes a more jagged profile shape.
- Etched glare reduction panels broaden the profile by scattering light from the spot.
- Screen structures such as phosphor dots for shadow-mask tubes may cause gross features that limit resolution (Figure 7.3).

The beam or spot may also not be symmetrical in the x, y, or intermediate planes. Astigmatism, coma, and other aberrations may be present that require more than a simple, single-axis profile to analyze it properly.

In addition to the above stationary-beam problems, deflection of the beam also adds coma, astigmatism, and focus aberrations to the spot. In some cases it is even desirable to have an elliptical spot to compensate for video bandwidth limitations. Further, magnetic

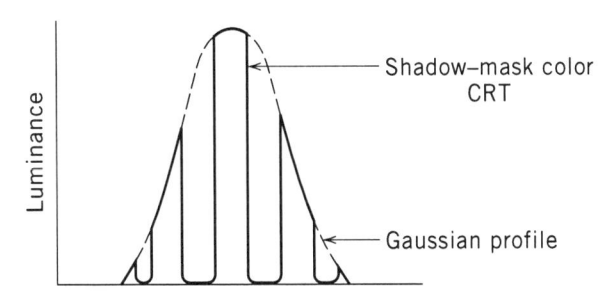

Figure 7.3. Spot profile for a shadow-mask color CRT showing screen structure superimposed on the Gaussian beam profile.

fields, power supply ripple, and mechanical vibration externally influence resolution and its measurement. All of these factors are important, as they determine the ultimate, useful resolution of the display. These factors must be eliminated to characterize properly the CRT alone, but they need to be considered in determining the resolution of a complete display system if they are present in the display itself.

7.3 WAYS OF EXPRESSING RESOLUTION

Resolution can be expressed in any of several ways (Figure 7.4):

- As *spot size,* typically spot width at the 50% luminance points
- As *line width,* at either the width at 50% amplitude or 60.7% if measured by the shrinking-raster method
- As *trace width,* the width at approximately the 5% amplitude point, which is the trace limit as seen by the eye and measured with a microscope (there have been some indications that the visual trace width may correspond more closely to the 1% amplitude under some viewing conditions)
- As the *number of TV lines,* as judged from merging of tapered alternate black and white bars on a video test pattern
- As the *modulation transfer function* (MTF, in cycles per unit distance for a specified modulation depth), which is a measure of the spatial frequency response of a display
- As the *modulation transfer function area* (MTFA) which combines the MTF with the *contrast transfer function* (CTF) of the human visual system
- As the number of *resolvable* pixels in the vertical and horizontal axes

Photographic resolution stated in line pairs is related to television lines by a factor of 2. It takes two television lines, one black, one white, to equal one photographic line pair. Actual television resolution is always less than the number of scanning lines by an amount known as the *Kell factor* because of overlap of the lines. The value of the Kell factor has been determined experimentally, and 0.7 is the generally agreed upon value.[8]

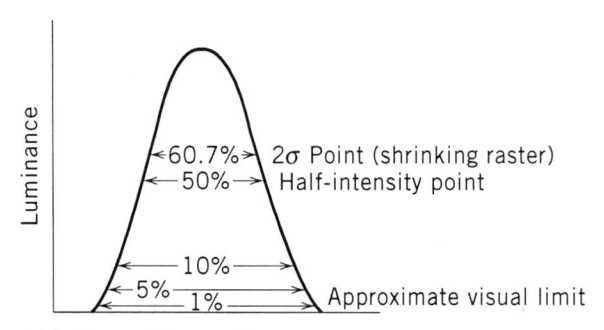

Figure 7.4. Some of the amplitudes at which resolution may be specified.

For a Gaussian spot the width at which resolution is expressed may be converted to other equivalent widths by the following table:

Height (%)	Correction Factor
60.7	1.00
50	1.18
10	2.15
5	2.45
1	3.03

The correction factor for heights other than the 60.7% level associated with shrinking-raster measurements may be calculated using the equation

$$\text{CF} = \left[-2 \ln\left(\frac{h}{100}\right) \right]^{1/2}$$

where

h = height in percent at which the width is measured

While the 50 and 60.7% amplitudes are most commonly specified for resolution, they may not be the most realistic amplitudes to use as a measure of spot quality. A spot having a small core, through CRT design compromise or manufacturing defect, may have low-level haze that will be objectionable particularly for alphanumeric displays. For displayed characters such as "e," the center will tend to fill in, owing to overlapping and consequent addition of light from the skirts of the spot when it is in the several close locations necessary to form the character. It would be difficult for the industry to change from the "small" (50 to 60.7%) spot sizes presently quoted to a more realistic, but larger, 5–10% figure. A possible compromise might be to express resolution at both levels. For example, a spot size might be quoted as 0.21/0.34 mm for 50/5% levels, or better still, it could be quoted as 0.21 mm at 50% with a 50/5% ratio of 0.62. This ratio would be used as a figure of merit; the best spot characteristics would be exhibited by a small 50% spot size and a figure of merit as close to 1.0 as possible.

It is also recommended that as many operating parameters as possible be stated when specifying resolution to ensure similar test conditions for manufacturers and users.

7.4 VISUAL METHODS

Microscope Trace Width Measurement

The simplest and lowest cost visual measurement method utilizes a microscope (Figure 7.5) or magnifier with calibrated reticle to measure spot or line width. This method is highly subjective and results in a width measurement at the 1–5% level of the spot profile; the actual level varies considerably with ambient light and trace luminance. A single dot or line, expanded raster, or circular trace may be used, depending on the ultimate application. The pattern is adjusted to the correct luminance, focused, and the width read directly from the reticle

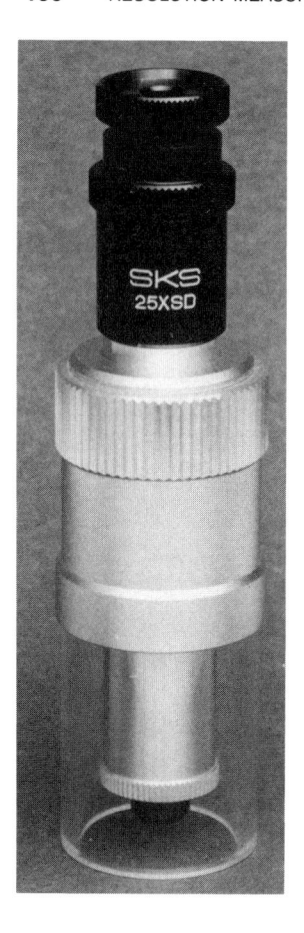

Figure 7.5. A 25 X hand-held microscope with a calibrated reticule suitable for visual resolution measurement.

at the extreme limits of its detectability. The reading should be repeated in the orthogonal axis (90° to the first measurement), if the pattern permits, in order to determine the amount of astigmatism present. The readings should be equal if there is no ellipticity of the CRT spot.

Shrinking-Raster Method

The shrinking-raster method is widely used for CRT testing because it requires minimal instrumentation and is more accurate than the microscope method described above. It also produces "comfortingly" small spot size numbers, since the width is effectively measured at the 60.7% amplitude. With this method, a raster of usually 10–100 lines is first expanded until the lines are easily resolvable. Then the raster is shrunk until the individual lines are just barely discernible, as shown in Figure 7.6. To obtain line width, the just-resolved raster's total vertical size is divided by the number of lines. If the raster axis can be rotated by 90°, the measurement should be repeated to determine the amount of ellipticity of the spot.

Among the drawbacks of the shrinking-raster technique is the subjectivity of judging when the lines merge; ±20–25% variation between observers is not uncommon. It is also difficult to apply it to structured screens such as those used in shadow-mask color CRTs.

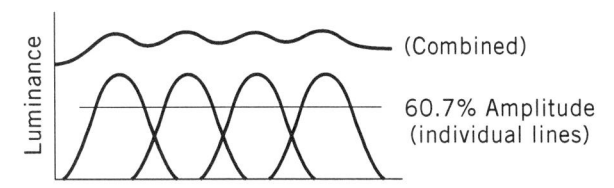

Figure 7.6. Shrinking-raster measurement of resolution showing individual line profiles and the combined light intensity.

A low-frequency signal applied to the axis of the raster orthogonal to the scanning lines may be used to help eliminate effects of the screen structure by allowing the eye to integrate the raster and phosphor pattern. This is practical only in CRT test sets where the flexibility to use scanning other than the normal raster is available.

Vernier Line Method

This less well-known method is said to have the advantage of measuring the total width of the visible trace with greater repeatability than the microscope method while still requiring minimal instrumentation. In the vernier line method, the amplitude of a small-amplitude square wave is adjusted until the top of one segment is coplanar with the bottom of the next on the CRT, as shown in Figure 7.7. At this point, the visual trace width is obtained by measuring the amplitude of the square-wave input to the display and comparing it with the deflection factor for the display. The deflection factor may be obtained by applying a known DC voltage to the input and dividing by the distance through which the trace shifts. This method is most useful for oscilloscope displays since the ability to apply a square wave to the vertical axis and a time base to the horizontal axis exists within the instrument. The measured line width is at the 1–5% amplitude, as in the microscope method. The advantage of the vernier line method lies in using the eye for adjustment of the square-wave amplitude only and not for the actual measurement of line width with a reticle, as in the microscope method.[9]

Photographic Method

The photographic method is simple, provides qualitative indication of astigmatism, coma, and halation, and is especially useful for comparing the effects of gun design, alignment, focus voltage, beam current, and so on. It is less useful for quantifying spot size, owing to differences in the sensitivity thresholds and gamma (linearity) characteristics of film and the eye, which preclude spot size measurements from the photograph. With this technique, a series of photographs using a constant exposure is taken of the spot (pulsed to avoid phosphor damage or saturation) at various currents or focus voltages, as shown in Figure 7.8. The film or spot is moved slightly for each exposure, so that an entire series of

Figure 7.7. Vernier line resolution measurement. The amplitude of a square-wave amplitude is reduced until the top of the negative portion is just level with the bottom of the positive portion.

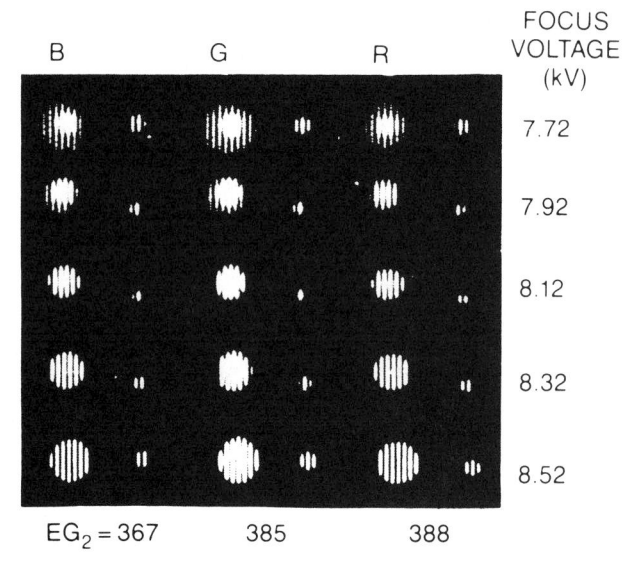

Figure 7.8. Photographic resolution measurement. Each beam is photographed at two beam currents and shows the effect of varying the focus voltage.

exposures appears on one print. The photographic method is most suited for CRT manufacturers performing electron gun design studies.[10,11]

Television Camera Method

This method started out similar to the photographic method, except that a television camera was used to display a highly magnified image of the spot or trace on a monitor. While easier for viewing, it was generally not very useful for quantitative measurements, owing to the limited dynamic range and because TV system linearity (gamma) and threshold sensitivities differed greatly from those of the eye.

It was not until the advent of the computer-controlled video photometer that the basic television camera method became practical as a quantitative resolution measurement technique. The video photometer (also known as a display analysis system), which makes use of the excellent linearity and stability of the CCD array, ultimately made the television camera technique highly successful. Measurements with this device will be covered later under Diode Array Systems.

Television Test Pattern Method

Since the beginning of commercial television, resolution has been determined by a test pattern having wedge-shaped alternate black-and-white bars with a calibrated scale, as shown in Figure 7.9. Standardized test charts have been developed by both EIA and SMPTE,[12–15] and every television station at one time had an RCA "Indian head" test pattern or a customized test pattern with their identification. The test pattern may be produced by a monoscope (a camera tube containing an integral target with an etched test pattern), televising a test chart with a television camera, or electronic generation. The bars may be either vertical or horizontal, and the calibration is usually in number of TV lines.

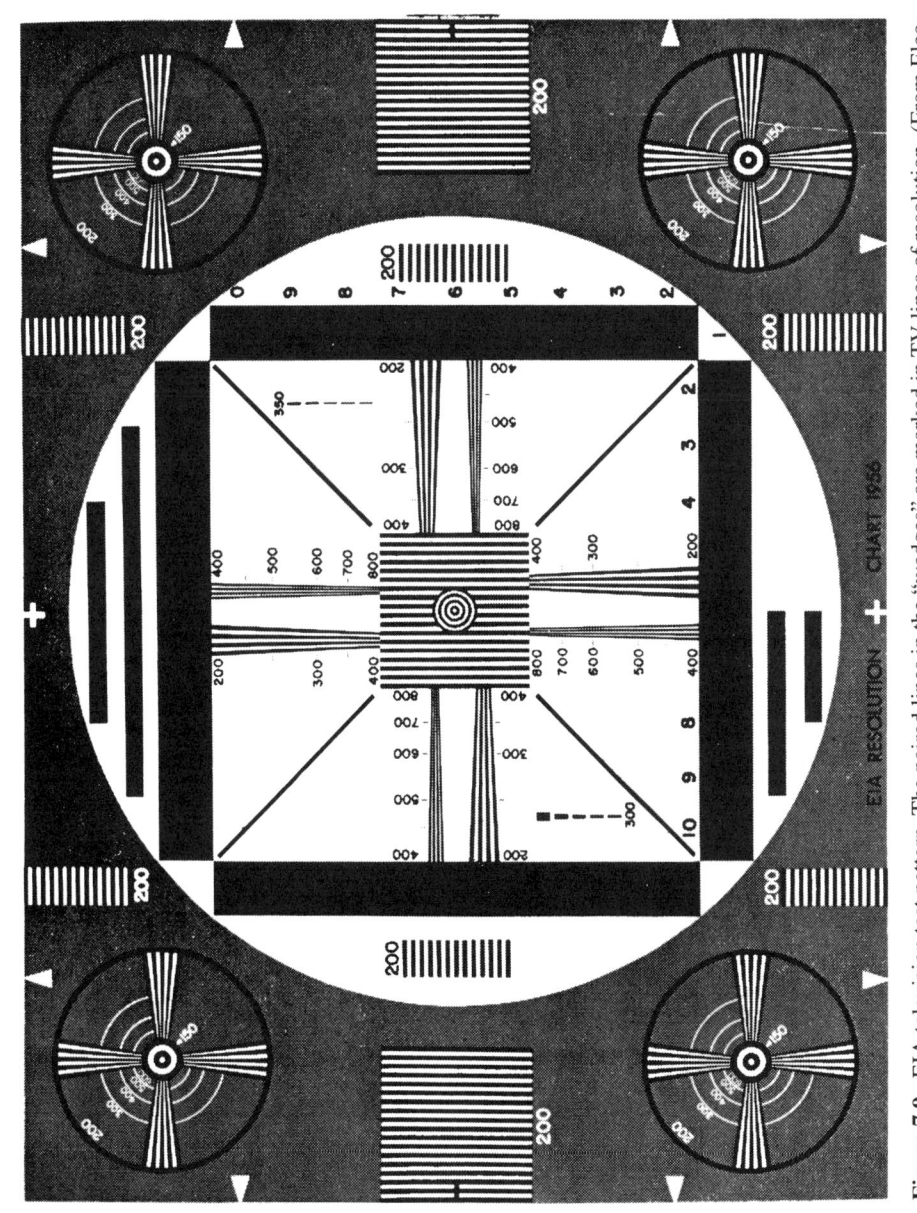

Figure 7.9. EIA television test pattern. The paired lines in the "wedges" are marked in TV lines of resolution. (From Electronic Industries Association Resolution Chart, 1956. Reprinted by permission.)

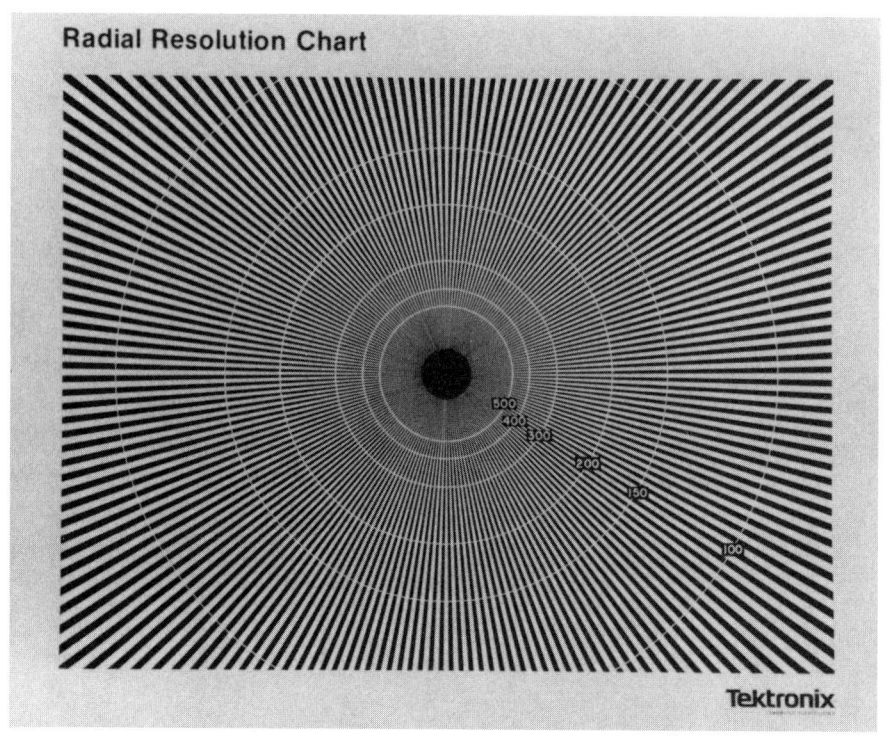

Figure 7.10. Radial resolution test pattern. (Tektronix, Inc. All rights reserved. Reproduced by permission.)

Two TV lines of resolution are equivalent to one *photographic line pair*.[16] The vertical lines are sometimes calibrated in megahertz of video bandwidth for determining system performance. The resolution is visually determined by the point at which the individual lines are just discernible on the monitor or receiver under test. Two improved variations of this method use the "zone plate" with concentric circles of decreasing thickness[17] and the radial resolution test chart[18] shown in Figure 7.10, which allow resolution to be evaluated at all angles rather than just vertical and horizontal.

As with other visual techniques, there is a certain amount of subjectivity in the TV test pattern method. Furthermore, unless the associated electronics performance is significantly superior to that of the CRT, the result will be a measure of performance of the system rather than the CRT alone. This is not necessarily a problem since more users need to know system performance rather than CRT performance.

7.5 PHOTOMETRIC METHODS

Aperture/Slit-Scan Methods

Aperture/slit-scan methods have greater potential for accuracy than the visual methods. However, these methods are more complex and require greater operator knowledge and skill to avoid errors. Most of the slit-scan methods either electrically scan the image of the beam past a stationary detector having a finite area, as shown in Figure 7.11, or me-

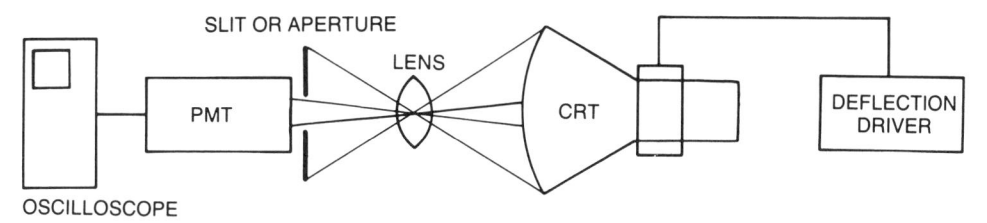

Figure 7.11. Electrical scanning of a slit or aperture.

chanically scan the detector across the image of a stationary spot or line, as in Figure 7.12. The resultant beam profile is displayed on a CRT or plotter as a curve of light intensity versus position.

Both slits and round apertures are commonly used. Each has distinct advantages. An aperture has the advantage of allowing a beam to be scanned past it in any direction to produce a profile of the beam in that axis. A slit is best where the profile of a scanned line is necessary since it gathers more light but alignment of the slit exactly parallel to the line is required to prevent the line width from appearing larger than it actually is. The slit is normally oriented 90° to the direction of the scan (Figure 7.13). This is easy to position, since the beam may traverse anywhere along the length of the slit.

The double-slit method is similar except that a second slit is spaced a known distance from the first. This distance is used to calibrate the horizontal scale of the displayed spot profiles; the display then shows a pair of profiles whose centers are separated by the effective slit spacing, as shown in Figure 7.14, and may be scaled accordingly.

Ideally the slit(s) should have the narrowest possible width, or the aperture should have the smallest possible diameter, because the measured spot or line size may be increased by as much as the slit width or aperture diameter. In practice, the minimum slit or aperture size is usually determined by signal-to-noise considerations. Usually a slit or aperture of about 10% of the beam diameter is adequate to avoid significant errors, although it has been reported that up to 40% is acceptable for truly Gaussian beams.[19] Usually the scanned line is imaged at the plane of the slit or aperture using a lens or microscope. The light passing through the slit or aperture is detected with a photomultiplier tube, and the output is displayed on a PC, oscilloscope, or plotter.

As an alternative to the aperture or slit methods, a fiber-optic probe may be placed at the image plane of a microscope and the output coupled to a photomultiplier tube via a fiber-optic cable, as shown in Figure 7.15. The advantage of this method is that the fiber position may be viewed through an eyepiece to aid correct positioning. The fiber-optic probe may be in either a circular or slit configuration. The photomultiplier tube is

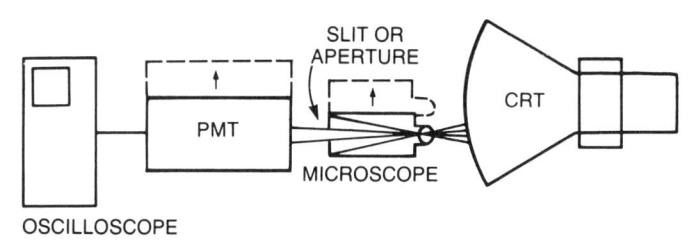

Figure 7.12. Mechanical scanning of a slit or aperture.

204 RESOLUTION MEASUREMENT

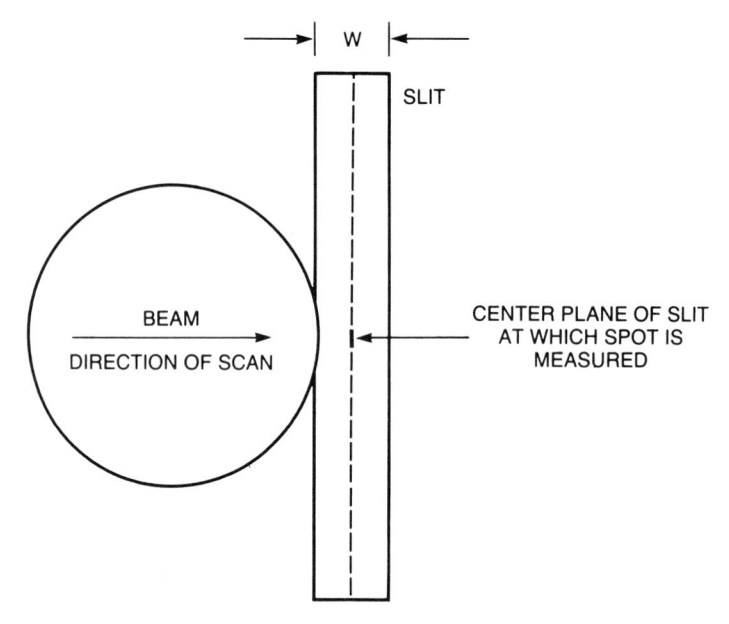

Figure 7.13. Relationship of slit and CRT beam.

separate from the microscope, reducing the mass and increasing the mechanical stability of the microscope. Mechanical stability is of key importance when spot sizes of 0.1 mm or less are to be measured. The microscope must be rigidly supported so that no motion relative to the CRT screen can occur. Local AC magnetic fields and electrical noise on the deflection system must also be minimized. Both of these considerations apply to other resolution measurement methods as well.

Several scan methods are currently in use. Mechanical scanning of the microscope assembly across a stationary spot or trace is widely used in aperture/slit scanning methods. Although this method is slow because of mechanical constraints, in many cases it allows a display to be measured using its own circuits without application of external electrical scanning signals. It is also relatively free of phosphor decay effects, an important consid-

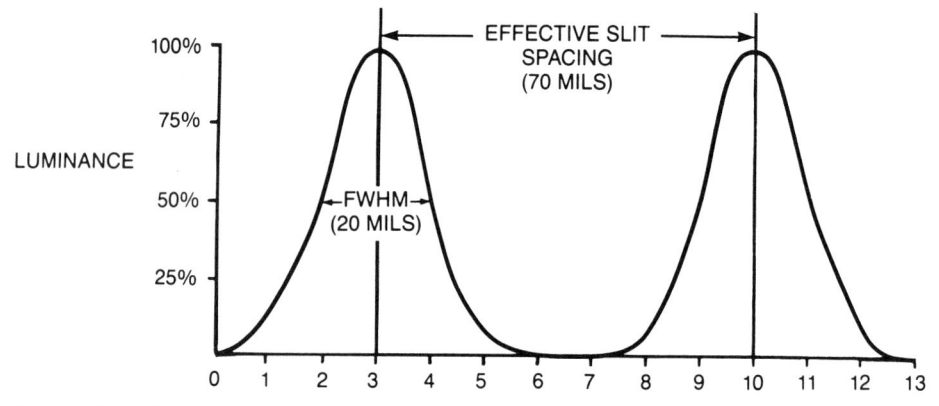

Figure 7.14. Display of spot profiles using the double-slit method. The scale is established by the known spacing of the slits.

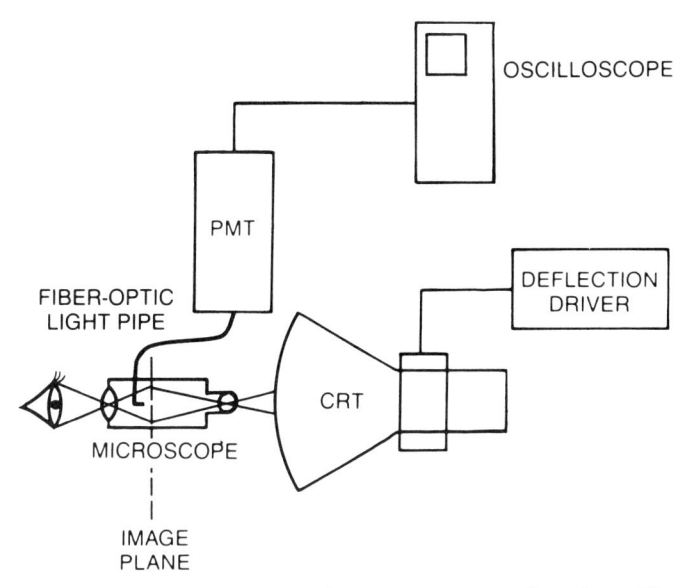

Figure 7.15. Diagram of setup for scanning of a spot or beam profile using a fiber-optic pickup mounted in a microscope.

eration for long-decay phosphors such as type GRA (formerly P39). Even "faster" phosphors such as type WWA (formerly P4) will show some decay problems when using 60-Hz electrical scanning. The presence of decay effects may be verified by using triangular scanning waveforms instead of a sawtooth. Any nonsymmetry between the superimposed beam profiles is usually caused by decay effects, as shown in Figure 7.16.

Electrical scanning of the CRT spot past a fixed microscope is the most practical way to profile shadow-mask CRTs without the screen structure affecting the profile. Such scanning also eliminates the effects of phosphor "noise," or graininess, on spot profile smoothness by allowing the system to remain focused on the same phosphor particles for the entire test.

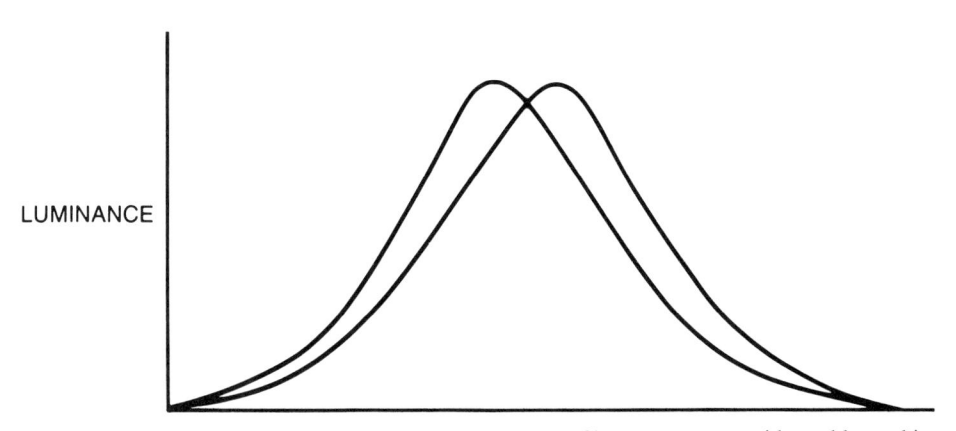

Figure 7.16. Effect of phosphor decay on spot or beam profile symmetry as evidenced by making two scans in opposite directions.

Either line scan on a single axis or a fast line scan on one axis with a slower mechanical or electrical scan on the other axis (perpendicular to the slit) may be used for measurement of CRTs themselves. Two-axis electrical scanning avoids phosphor burning by reducing current density. Z-axis (intensity) duty cycling may be used to hold the average beam current density to a safe level while the beam profile is measured at high peak current densities with either scanning method.

Variable-Slit-Width Method

The variable-slit-width method uses a calibrated, adjustable-width slit aligned parallel to a single scanned line. The line is imaged and centered on the slit by a means of lens with known magnification, and the light passing through the wide-open slit is measured with a photomultiplier tube. The slit width is then reduced until a reading of $\frac{1}{2}$ is obtained. The width of the reduced slit divided by the optical magnification is a measure of the width of the line. Note that the width of a Gaussian spot at which a reading of $\frac{1}{2}$ light output is obtained is not equal to the 50% amplitude of the line but is closer to 80%. A related approach uses a spot photometer with multiple apertures of known sizes instead.[20]

Image Dissector Method

This method is distantly related to the aperture scanning technique. A spot or line is imaged on the photocathode of an image dissector tube by a lens. The image dissector may be thought of as a photomultiplier tube having a very small effective area or aperture at its photocathode. This small aperture may be electrostatically or magnetically scanned across the photocathode (and thus across the image of the CRT spot or line) using conventional CRT scanning techniques. The output of the image dissector is then displayed on a CRT or x–y plotter.[21] Figure 7.17 shows a block diagram of such a system. This is an obsolete technique today but was an intermediate step leading toward diode array resolution measurement equipment.

Spot Contour Plotting Methods

Two methods of electrical scanning offered further measurement advantages but increased system complexity by requiring a desktop computer for control and data manipulation. Both methods produced a two-dimensional isointensity (isophote) plot of the spot. The plot shows the spot intensity contours, making beam aberrations such as coma and astigmatism easily visible. These aberrations are often not readily apparent in conventional x- and y-plane profiles.

The first method uses a series of radial scans in 30° increments through the spot center. The light from the screen is sampled by a microscope with a small fiber-optic pickup and fed to a photomultiplier tube (PMT). The PMT output signal is converted to digital form and processed by a desktop computer system to normalize, interpolate, compute percentages, and plot the intensity contours at preselected percentage levels. Figure 7.18 shows a block diagram of the system and Figure 7.19 is a typical printout of profiles, contour plots, and spot diameters at various percentages of peak intensity. The advantage of contour plotting is that it provides the CRT design engineer with more quantitative and qualitative data than conventional single-axis or two-axis beam profiles. Changes in spot shape and size with operating voltages and currents are readily demonstrated.[22]

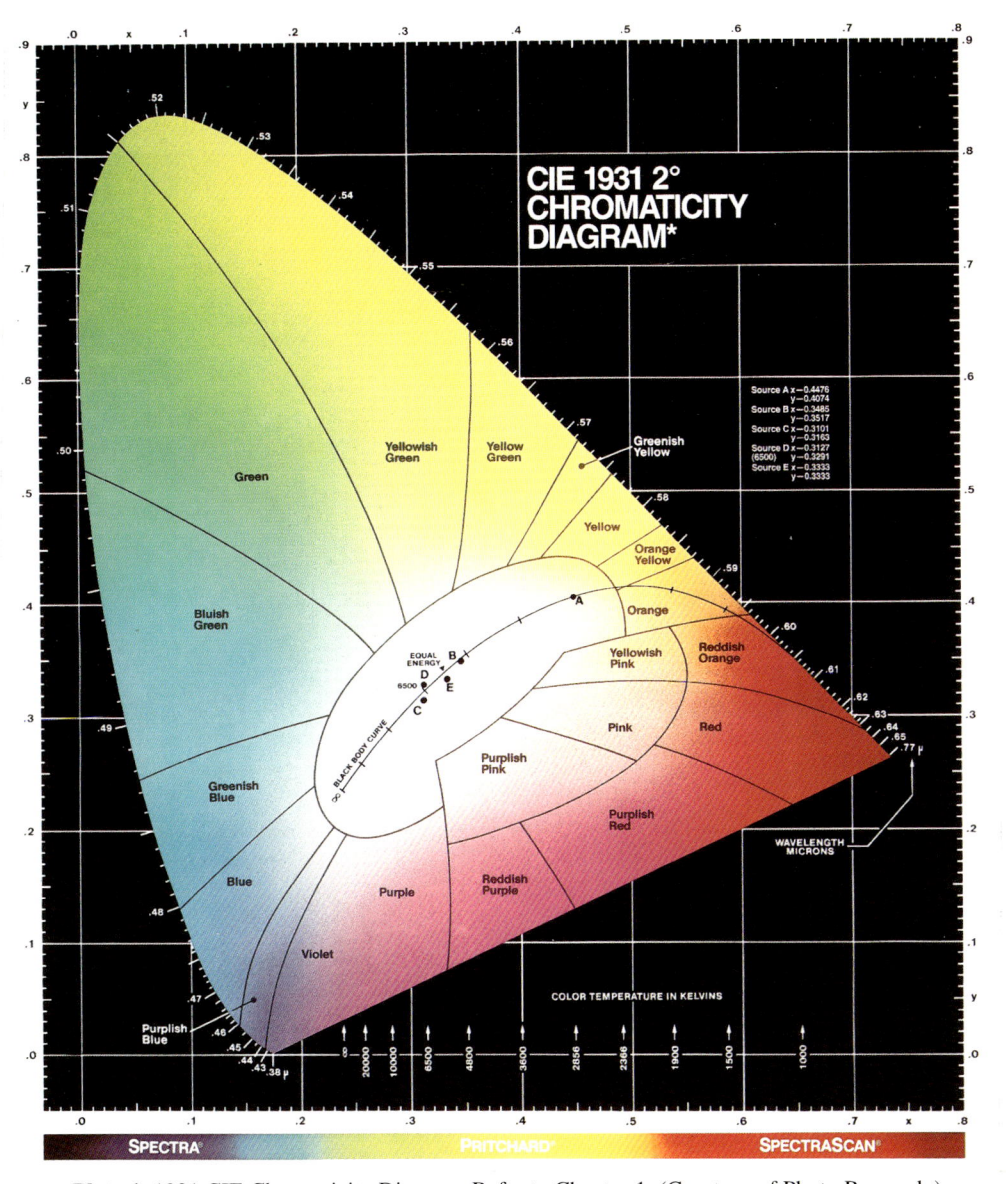

Plate 1 1931 CIE Chromaticity Diagram. Refer to Chapter 1. (Courtesy of Photo Research.)

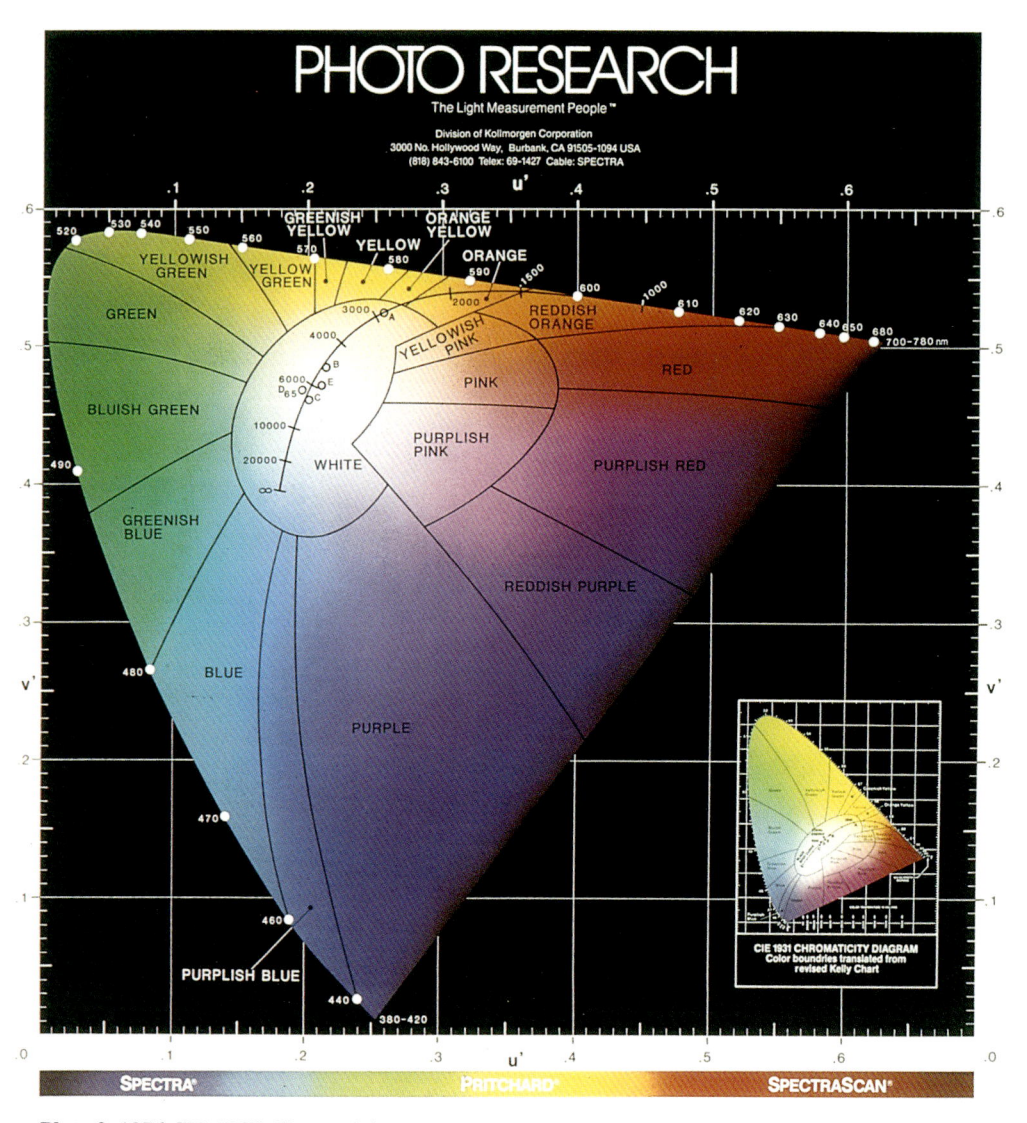

Plate 2 1976 CIE–UCS Chromaticity Diagram. Refer to Chapter 1. (Courtesy of Photo Research.)

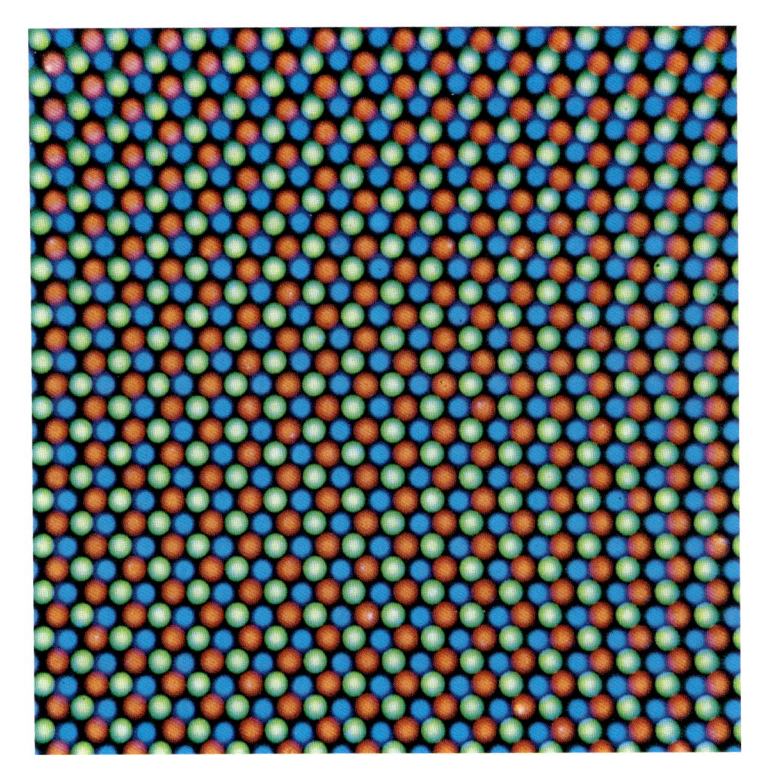

Plate 3 Color dot triad pattern screen. Refer to Chapter 3.

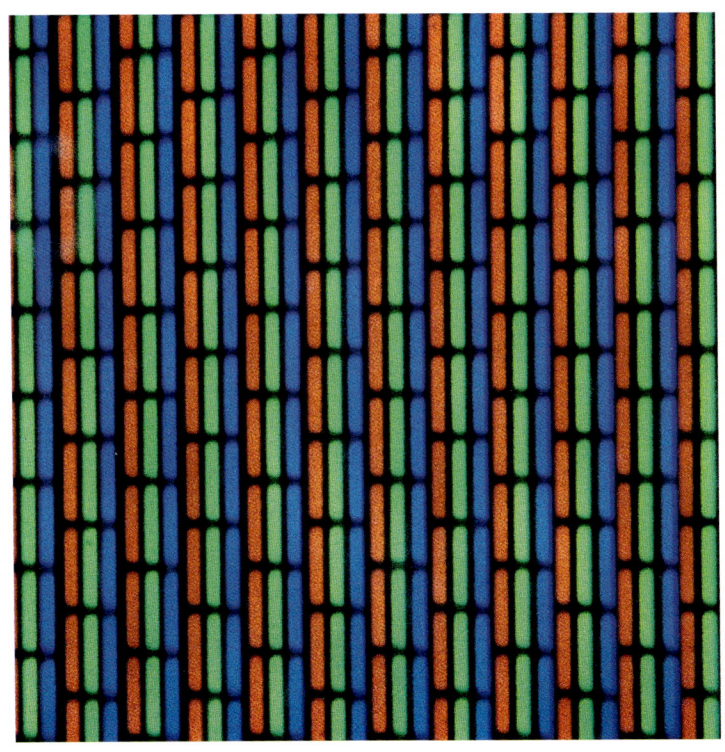

Plate 4 In-line pattern screen. Refer to Chapter 3.

Plate 5 Trinitron screen pattern. Refer to Chapter 3.

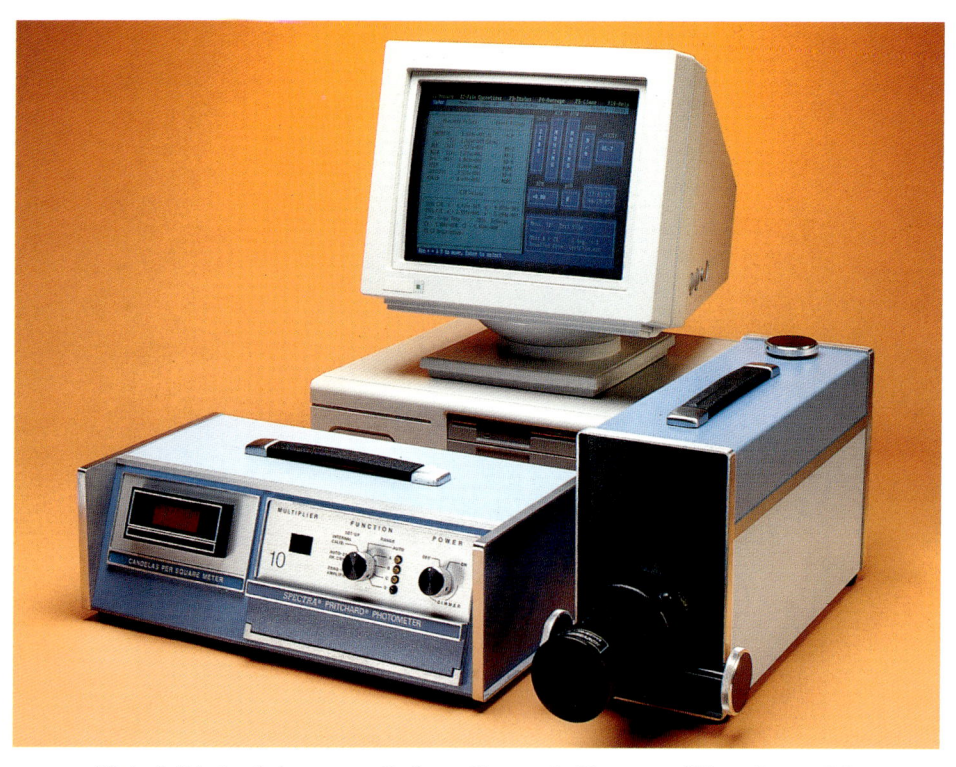

Plate 6 Pritchard photometer. Refer to Chapter 4. (Courtesy of Photo Research.)

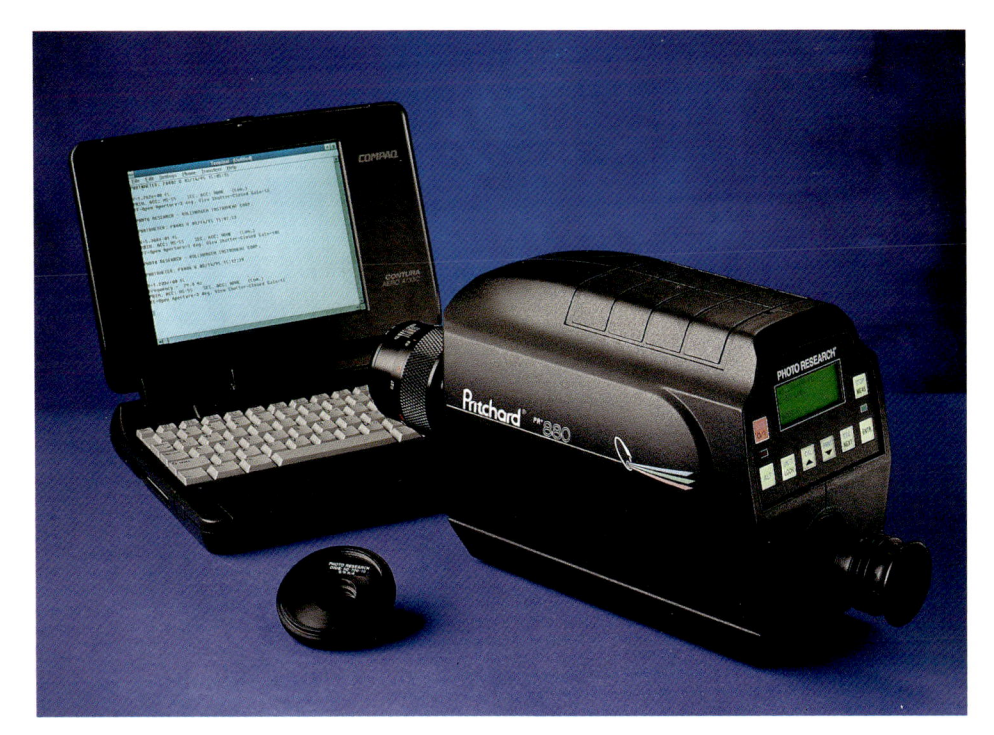

Plate 7 Portable scanning colorimeter using a CCD array. Refer to Chapter 4. (Courtesy of Photo Research.)

Plate 8 Misadjustment of color purity. Refer to Chapter 6.

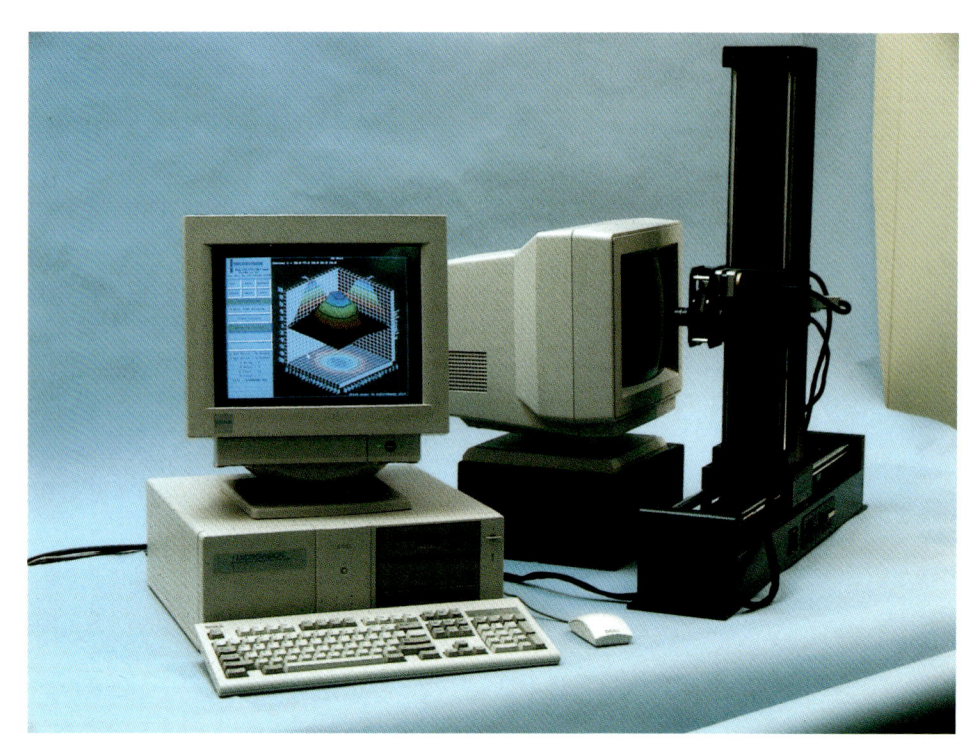

Plate 9 Display analysis system. Refer to Chapter 7. (Courtesy of Microvision.)

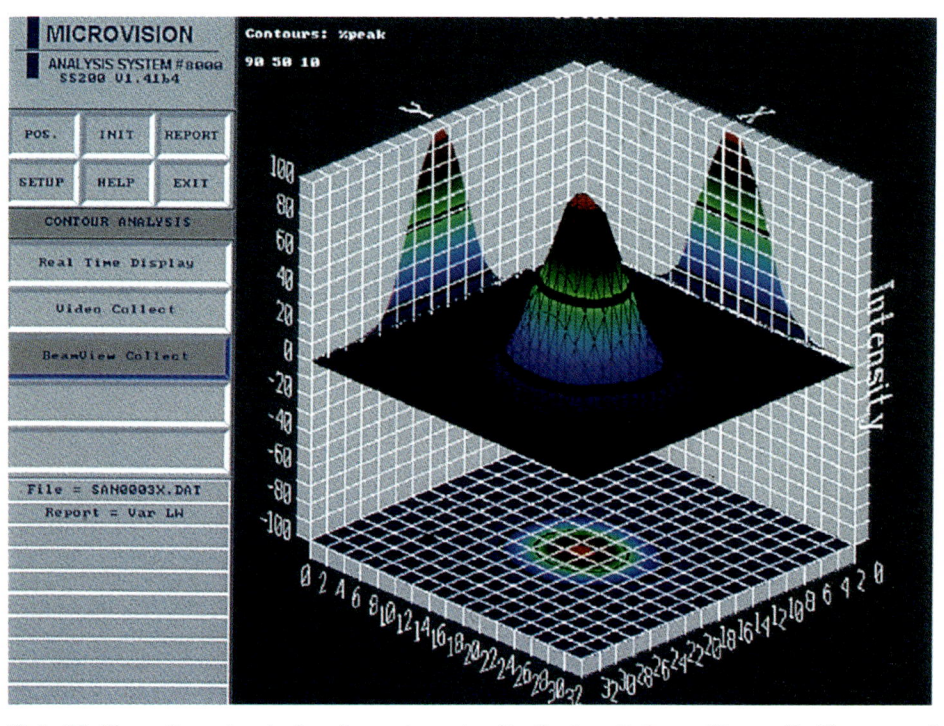

Plate 10 Three-dimensional plot of spot intensity distribution. Refer to Chapter 7. (Courtesy of Microvision.)

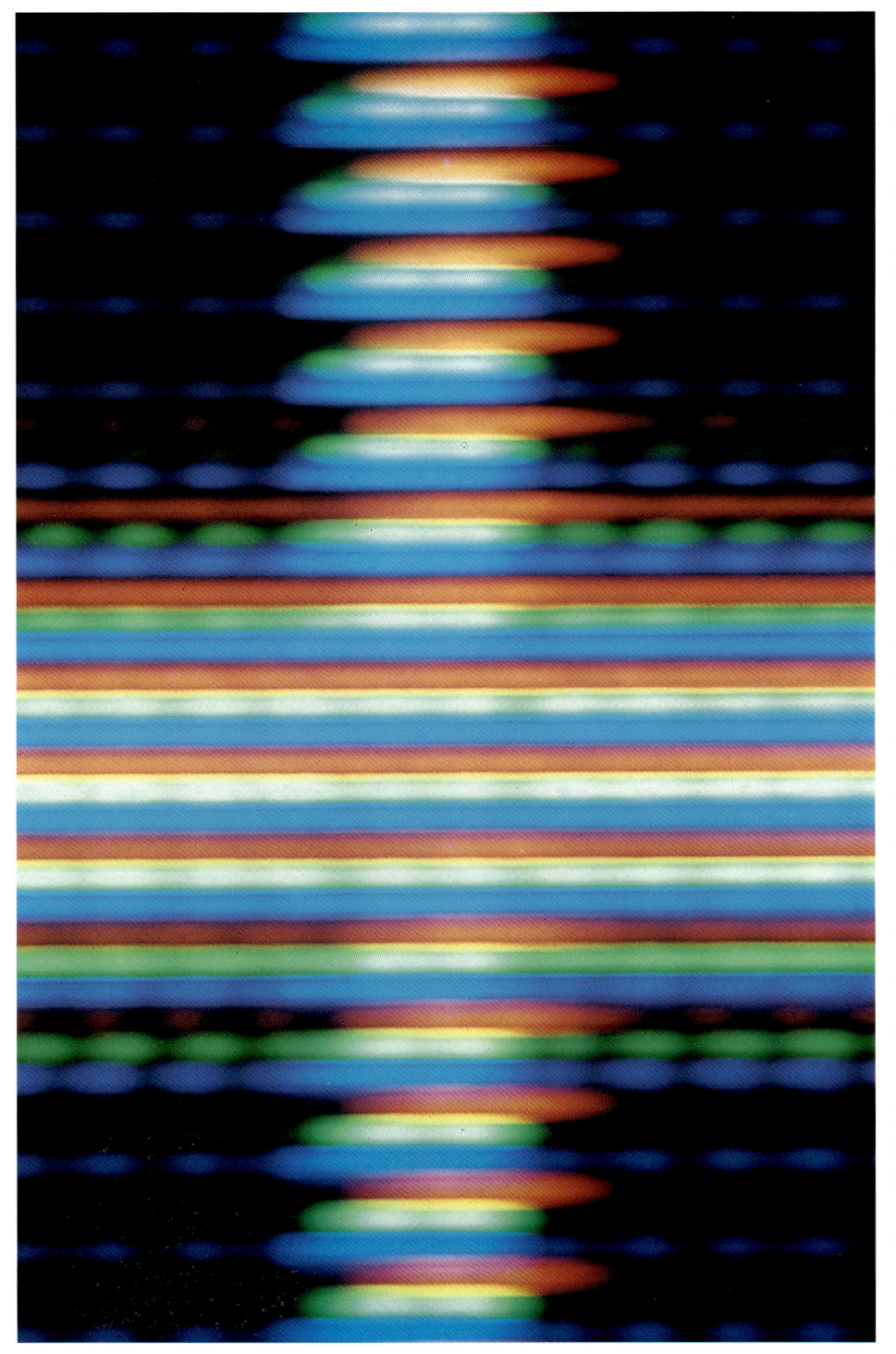

Plate 11 Highly magnified view of intersection of a vertical and a horizontal line showing misconvergence. Refer to Chapter 8.

Plate 12 Convergence test pattern. The red, green, and blue line segments should be adjusted to be in a straight line. Refer to Chapter 8. (Courtesy of TEAM Systems, Inc.)

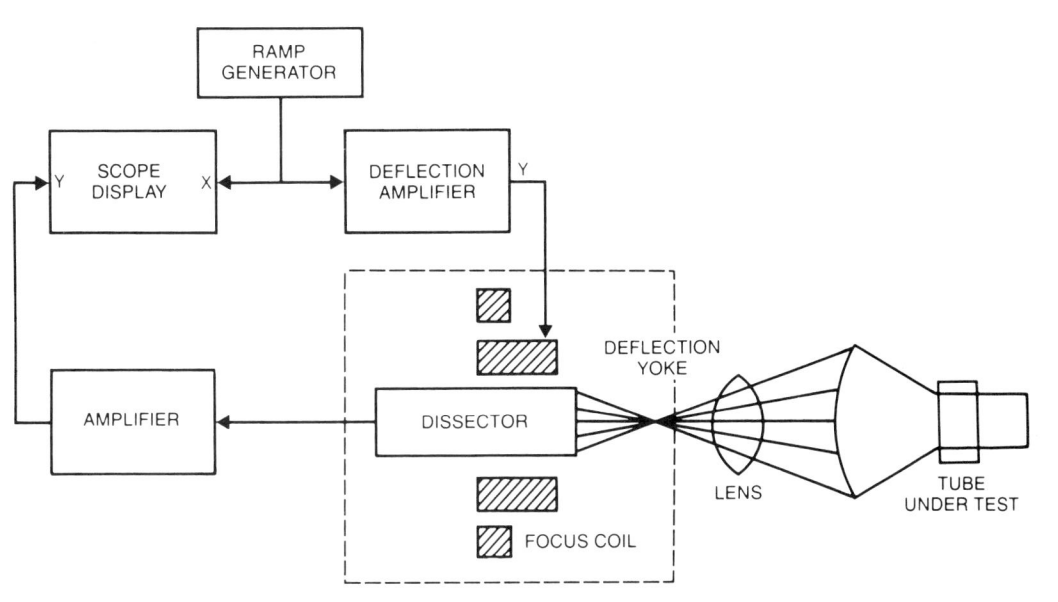

Figure 7.17. Resolution measurement with an image dissector tube. While obsolete today, this was an evolutionary step preceding the CCD.

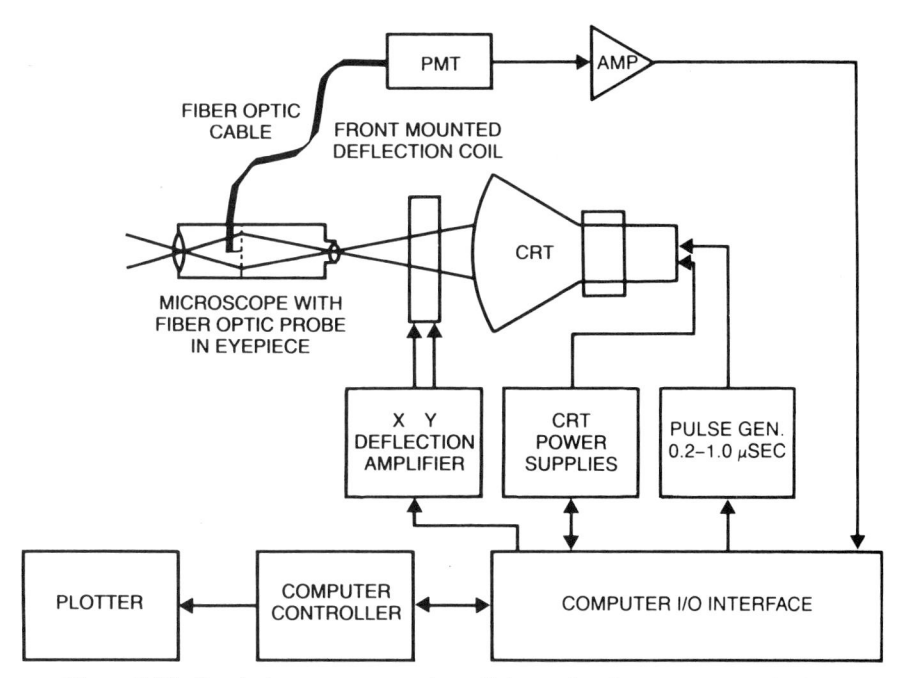

Figure 7.18. Resolution measurement by radial scanning for spot contour plotting.

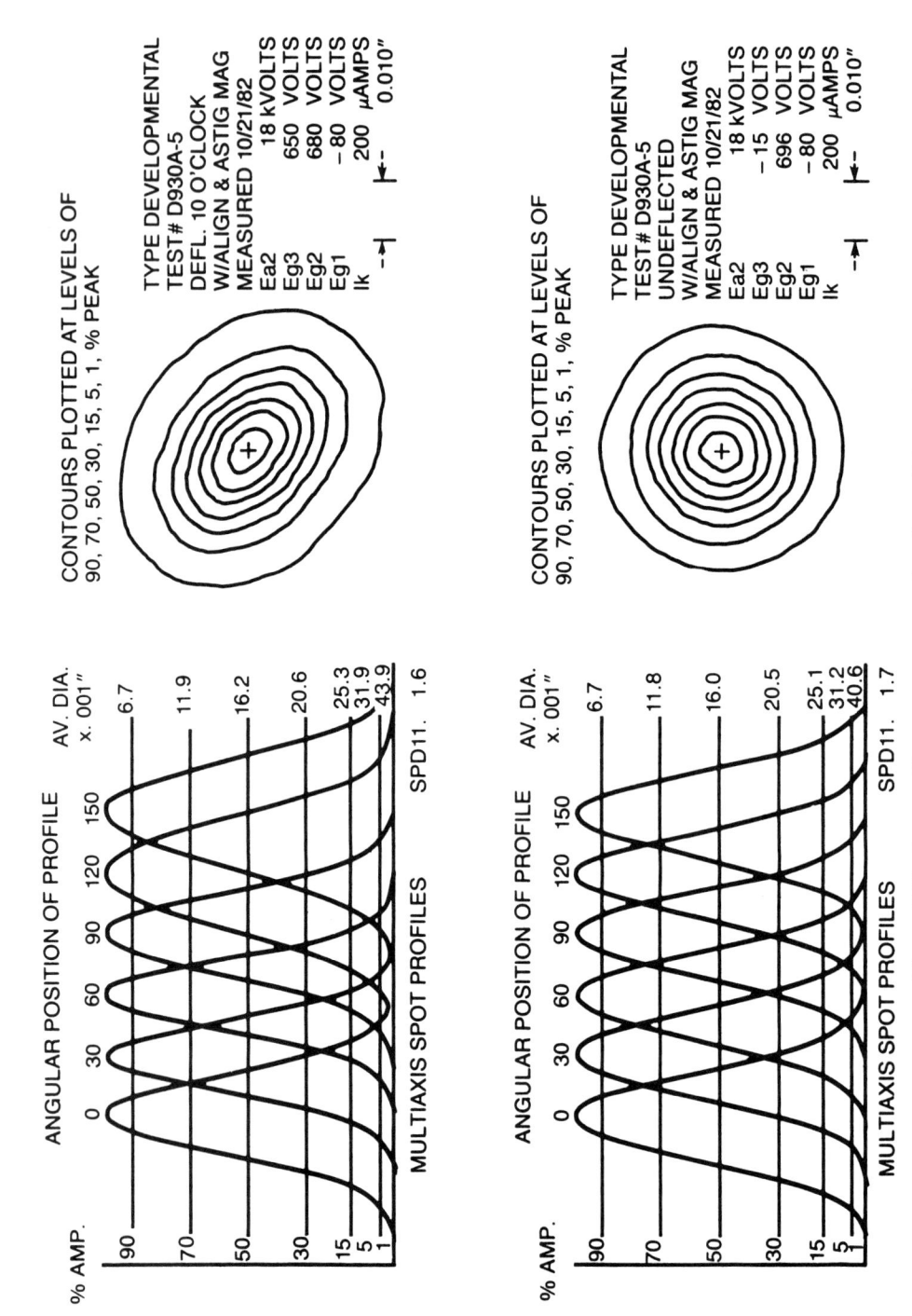

Figure 7.19. Spot profile measurements made using radial scanning.

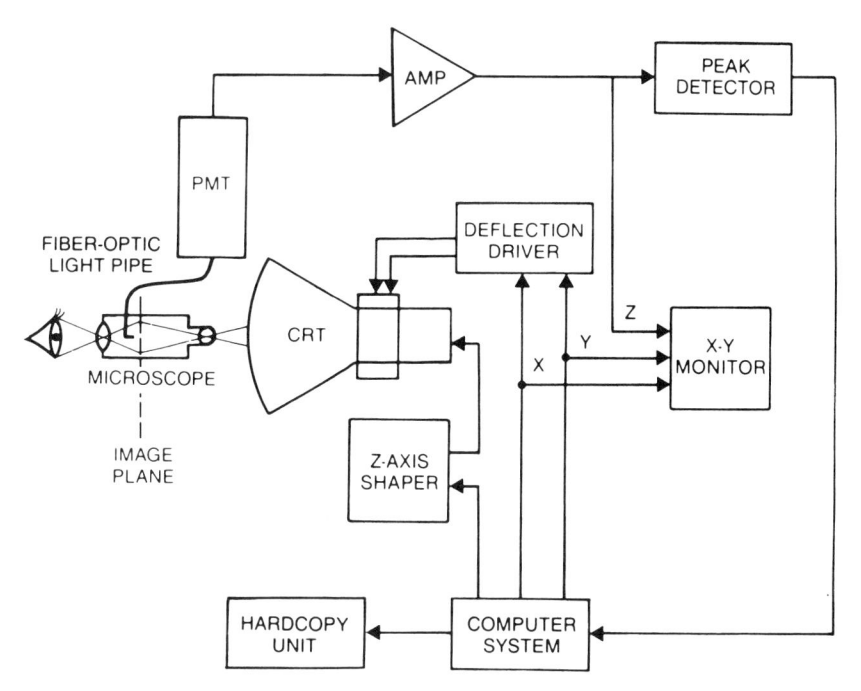

Figure 7.20. Dot-matrix scanning for spot contour plotting.

The second method, shown in Figure 7.20, is similar except that a dot-matrix scan format is used. Dot-matrix scanning offers the advantages of uniform spacing of data points over the measurement area, faster operation, and processing of the data in array format. The data stored in the array may be easily accessed to plot conventional spot profiles, compute MTF, or convolve with system scan rates and video bandwidth to derive system performance. Figures 7.21–7.25 are examples of typical CRT spots measured with the dot-matrix spot contour plotting system.[23,24] These demonstrate a typical well-focused spot and some of the common spot abberations that may be encountered.

Both spot contour plotting methods are usable for structured screens, such as color shadow-mask CRTs and monochrome CRTs, because the microscope position on the CRT remains fixed and the beam is scanned past it to obtain successive samples of the spot. The effects of phosphor graininess are also eliminated.

The dot-matrix scanning method has evolved into the faster and more powerful diode array scanning methods.

Diode Array Scanning Methods

Diode array (CCD) scanning methods are a relatively recent advance and represent the current state of the art for resolution measurement. Several instruments utilizing diode arrays are now commercially available. Two methods of diode array scanning are used. One uses a linear diode array (one single row of sensor elements) to scan a slice of the spot or line and may read out linewidth directly on a digital display (Figure 7.26). This method is simpler and better suited for use in manufacturing environments by relatively unskilled personnel. The trade-off is that beam profile data are only obtained in a single axis. The sensor must be reoriented to make measurements in the other axis.

OPERATOR
 DIANNE
DATE:
 10/17/83
TUBE TYPE:

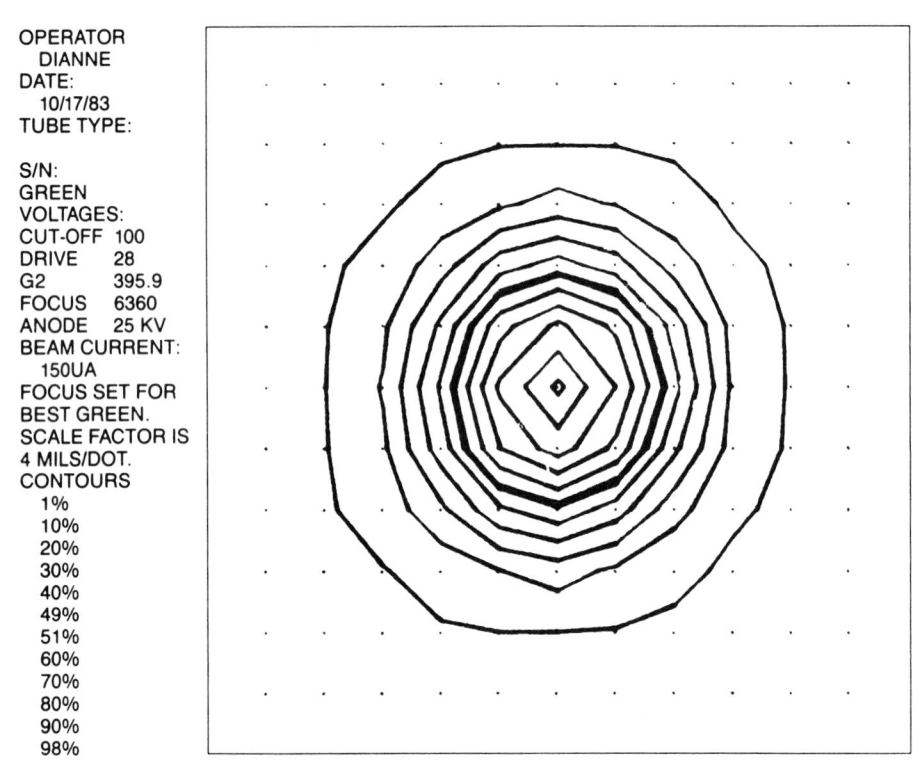

S/N:
GREEN
VOLTAGES:
CUT-OFF 100
DRIVE   28
G2   395.9
FOCUS   6360
ANODE   25 KV
BEAM CURRENT:
 150UA
FOCUS SET FOR
BEST GREEN.
SCALE FACTOR IS
4 MILS/DOT.
CONTOURS
 1%
 10%
 20%
 30%
 40%
 49%
 51%
 60%
 70%
 80%
 90%
 98%

Figure 7.21. Contour plot of a good spot.

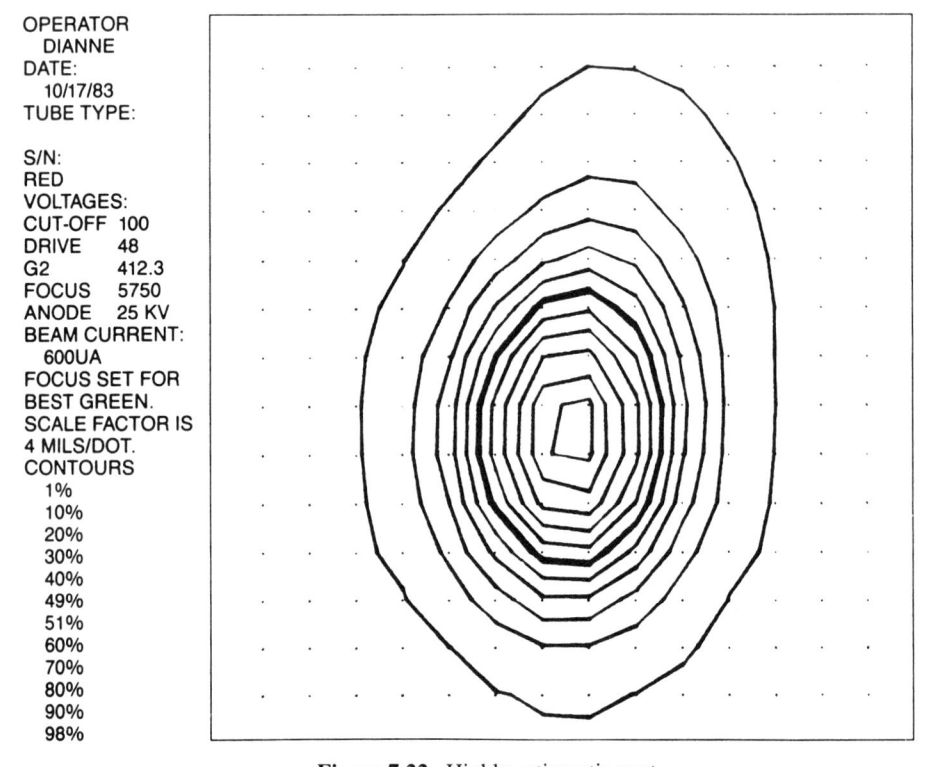

OPERATOR
 DIANNE
DATE:
 10/17/83
TUBE TYPE:

S/N:
RED
VOLTAGES:
CUT-OFF 100
DRIVE   48
G2   412.3
FOCUS   5750
ANODE   25 KV
BEAM CURRENT:
 600UA
FOCUS SET FOR
BEST GREEN.
SCALE FACTOR IS
4 MILS/DOT.
CONTOURS
 1%
 10%
 20%
 30%
 40%
 49%
 51%
 60%
 70%
 80%
 90%
 98%

Figure 7.22. Highly astigmatic spot.

OPERATOR
  D.L.
DATE:
  7-29-83
TUBE TYPE:

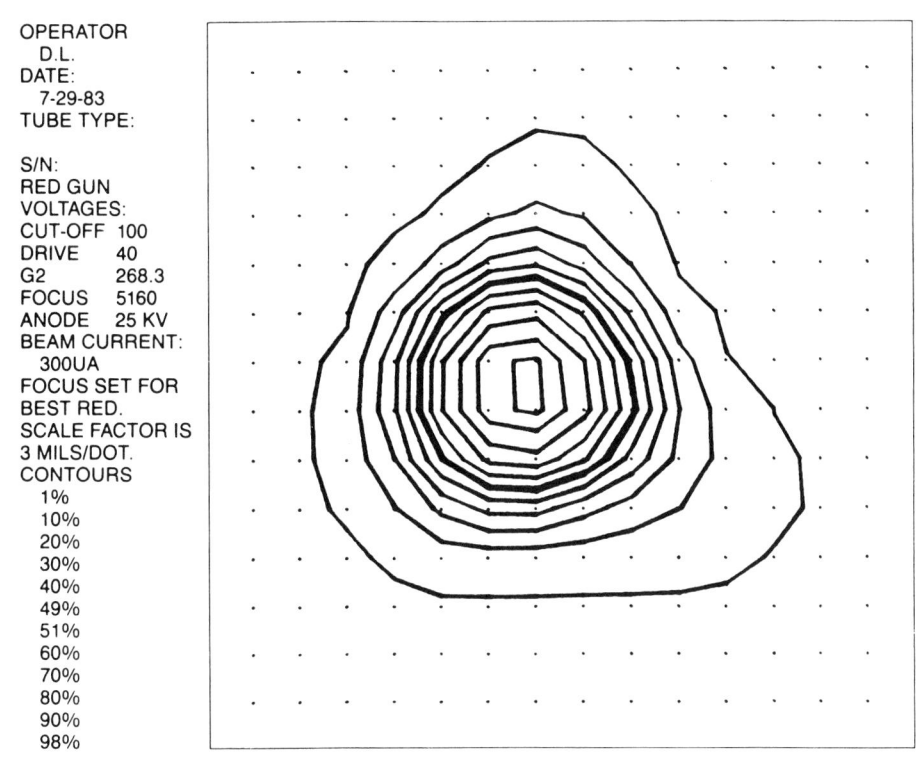

S/N:
RED GUN
VOLTAGES:
CUT-OFF 100
DRIVE   40
G2   268.3
FOCUS   5160
ANODE   25 KV
BEAM CURRENT:
  300UA
FOCUS SET FOR
BEST RED.
SCALE FACTOR IS
3 MILS/DOT.
CONTOURS
  1%
  10%
  20%
  30%
  40%
  49%
  51%
  60%
  70%
  80%
  90%
  98%

Figure 7.23. Spot aberration of a delta-gun color CRT at 300 μA.

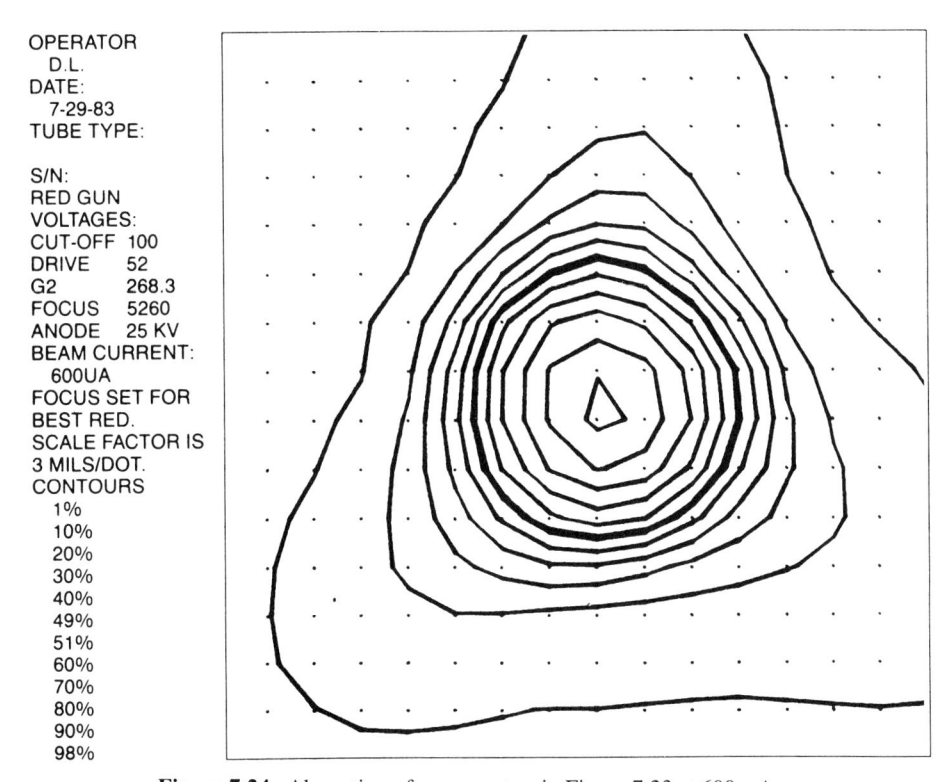

OPERATOR
  D.L.
DATE:
  7-29-83
TUBE TYPE:

S/N:
RED GUN
VOLTAGES:
CUT-OFF 100
DRIVE   52
G2   268.3
FOCUS   5260
ANODE   25 KV
BEAM CURRENT:
  600UA
FOCUS SET FOR
BEST RED.
SCALE FACTOR IS
3 MILS/DOT.
CONTOURS
  1%
  10%
  20%
  30%
  40%
  49%
  51%
  60%
  70%
  80%
  90%
  98%

Figure 7.24. Aberration of same spot as in Figure 7.23 at 600 μA.

212 RESOLUTION MEASUREMENT

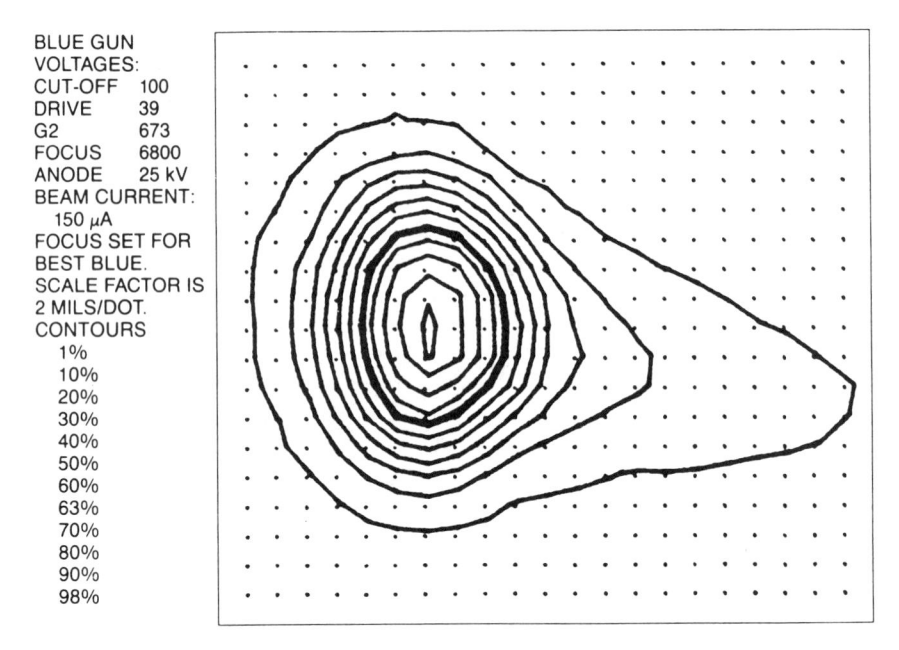

BLUE GUN
VOLTAGES:
CUT-OFF   100
DRIVE   39
G2   673
FOCUS   6800
ANODE   25 kV
BEAM CURRENT:
  150 µA
FOCUS SET FOR
BEST BLUE.
SCALE FACTOR IS
2 MILS/DOT.
CONTOURS
  1%
  10%
  20%
  30%
  40%
  50%
  60%
  63%
  70%
  80%
  90%
  98%

Figure 7.25. CRT spot profile having good central core and severe coma at the 1–10% contours.

Two-dimensional diode arrays allow a complete beam profile to be obtained in one operation. The Photo Research PR-900 video photometer (Figure 7.27) and Microvision SS-200 display analysis system (Plate 9) are representative examples of such a system. Both units are highly versatile and permit a number of display parameters to be evaluated. They consist of optical heads containing the CCD arrays that are essentially video cameras, mechanical means to accurately position the head at the desired location on the display screen, and a personal computer to control and display the measurements. The relatively high cost of such systems makes them best justified for display laboratories, design engineering, and high volume production lines. A still more elaborate system, the Photo Research PR-950 video photometer uses multiple video cameras for fully automated test-

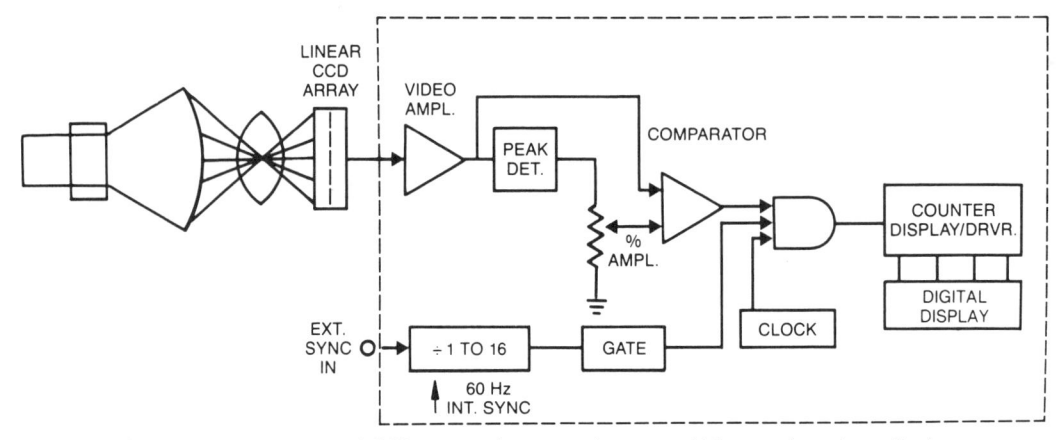

Figure 7.26. Line-scanned CCD system for measuring spot width at a selected amplitude.

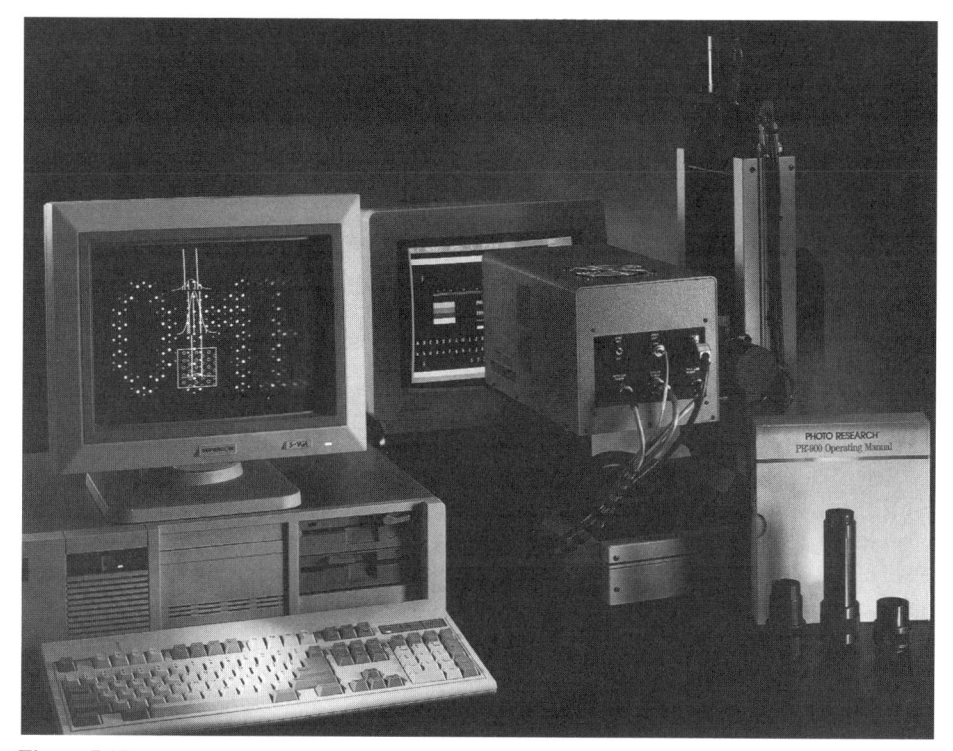

Figure 7.27. Typical video photometer using a two-dimensional scanned CCD array. (Courtesy of Photo Research.)

ing at several screen locations simultaneously. Computer processing of the array of two-dimensional luminance data is sometimes used to depict the intensity distribution of the beam in a three-dimensional plot, as shown in Plate 10.

Since the diode array scanning methods use a stationary CRT spot or line while in effect moving the detector, measurement of beam profiles for structured screens such as color shadow-mask CRTs require special software techniques to derive the profile. The Microvision system uses an optional test pattern generator to move the CRT spot in small increments during the measurement for computer reconstruction of the beam profile, in essence, the same approach as with the previously described dot-matrix scanning method.

7.6 MODULATION TRANSFER FUNCTION

The MTF was originally introduced as a measure of information transfer for optical and photographic systems.[25,26] This soon was adapted to analysis of television and, finally, data and imaging displays.[27,28] The MTF for displays indicates the ability of the CRT to display a constant-amplitude sine wave modulation at various spatial frequencies (cycles per unit distance rather than per second). It is usually expressed as percent of modulation versus spatial frequency in cycles per millimeter (preferred), centimeter, or inch. Figure 7.28 shows the form of an MTF curve. The MFT is roughly analogous to familiar amplifier gain–bandwidth curves, or *Bode* plots. The MTF curves of each element in a system may be convolved to obtain the combined system MTF curve.

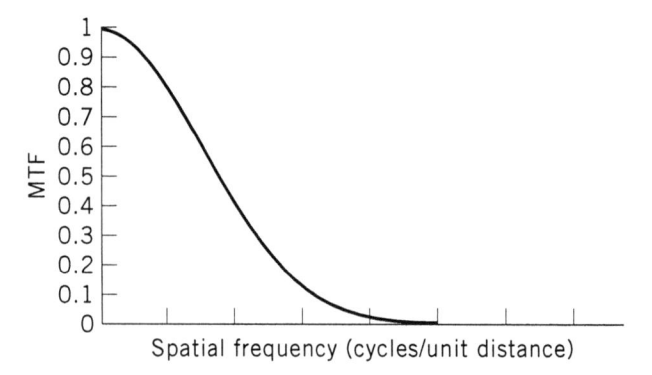

Figure 7.28. MTF curve format.

Some time ago, the International Electrotechnical Commission (IEC) recommended that the spatial frequency at the 60% modulation depth be adopted as the standard reference level for MTF comparisons.[29] This level has the advantage of ease of use at the expense of not fully describing the spot characteristics. Other levels have been suggested, such as 50%.[30] At 60% spatial frequency depth, the 60.6% amplitude may be calculated using the equation:

$$f_{60(\sin)} = \frac{1}{2\pi\sigma}$$

where

$f_{60(\sin)}$ = sinusoidal spatial frequency for 60% modulation depth
$2\pi\sigma$ = width of the spot at the 60.6% amplitude

This equation applies only to Gaussian spot distributions.

Note the two minor inconsistencies present in the IEC recommendation. The 60.7% height is more correctly that of the 2σ amplitude, and 60.7% rather than 60% is also the corresponding modulation depth for a Gaussian spot. Considering the accuracy of resolution measurements, these discrepancies are not serious.

The integrated area under the entire MTF curve after correction for the CTF of the human visual system is often used as an index of image sharpness and is referred to as MTFA (Figure 7.29).[31] The CTF is used since it describes the average resolving capability of the human eye at normal viewing distances. It is of little value to have a visual display capable of displaying more information than the eye can resolve. The MTFA is more useful than the single-point computation of the spatial frequency at the 60% modulation level previously described, but it is still only an approximate indication of spot quality, as is any single-value index. Its greatest value is for comparing near-Gaussian spots. The steepness of the MTF curve and other distortions present in non-Gaussian spots are not obvious in these indices. Generally speaking, the higher the MTFA index, the greater the range of spatial frequencies that may be displayed. The ANSI/HFS standard for display ergonomics uses the MTFA as the recommended measure of display resolution.[32] A minimum MTFA of 5 at each of five screen locations (Figure 7.30) is specified for monochrome displays. A rather lengthy set of computations is required to derive the MTFA in this document. No criteria are specified for color displays since at the time of publication (1988) no accepted method was available.

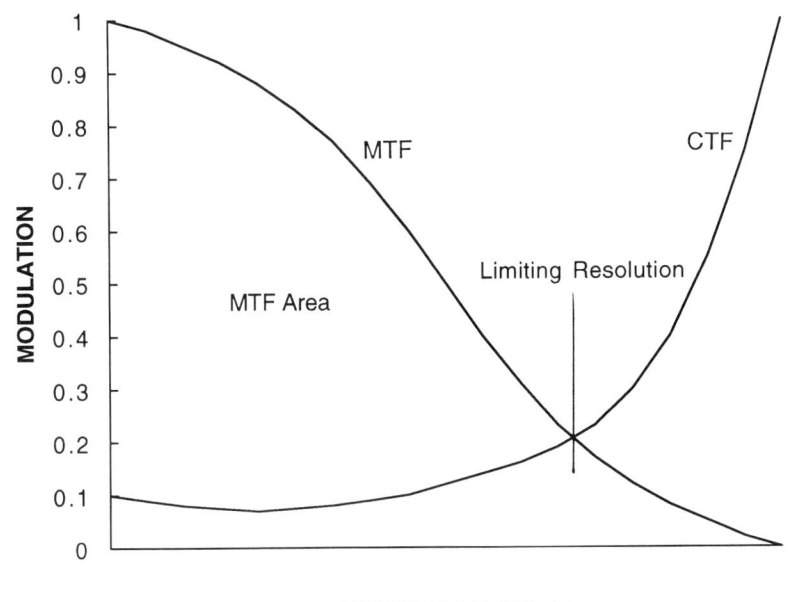

7.6 MODULATION TRANSFER FUNCTION 215

Figure 7.29. MTFA concept.

A variation on the MTFA is obtained by dividing the area under the MTF curve by the integral of the curve of a "perfect" spot having 100% modulation depth for all spatial frequencies. Obviously a practical cutoff spatial frequency based on what information is resolvable by the viewer must be specified to prevent the integral of the perfect spot from reaching infinity.

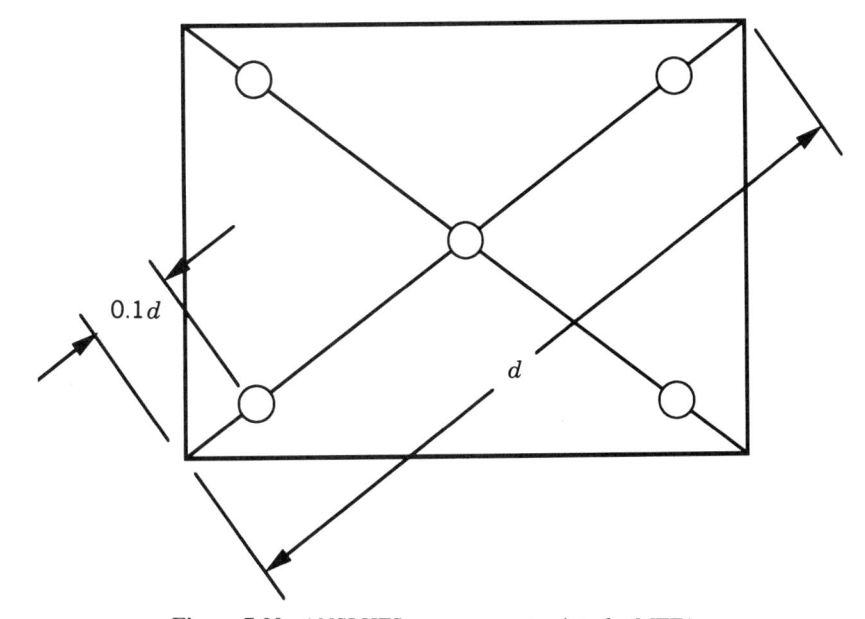

Figure 7.30. ANSI HFS measurement points for MTFA.

Another metric of resolving ability of a CRT display is the *square root integral* (SQRI) method proposed by Barten.[33,34] This method along with the intermediate *integrated contrast sensitivity* method by van Meeteren[35] was derived to better adjust for nonlinearities in response of the human eye than with the MTFA and was based on previous work by Carlson and Cohen.[36,37] Luminance is factored into the equations and used to compensate for the visual nonlinearities. These advanced techniques are beyond the scope of this book and are mentioned to provide references for anyone requiring further information about them.

The entire MTF curve is of course the most useful for the designer, as is spot profile rather than a single spot size number. The MTF curve contains much information about the quality of the spot and its aberrations. Ideally, the MTF should be measured on at least both major axes, since asymmetrical spots are often encountered. Carrying this concept to its logical conclusion would suggest complete two-dimensional MTF contour plotting to quantify beam aberrations such as coma.

The EIA Publication No. 105-17 describes the use of fast Fourier transform (FFT) techniques for computation of MTF curves from CRT line or spot profiles. These techniques are based on work by Bergland and Dolan.[38] The publication includes software to provide an easy means for the computations by merely "plugging in" the numbers from the profile data. Both tabular and graphical results may be derived.[39,40]

The MTF is not usually used for color shadow-mask CRTs without applying special techniques because of the structured screen. The phosphor pattern has the same effect as coarse phosphor graininess, thus masking higher spatial frequencies. To be properly evaluated for the MTF, the screen must have spatial invariance (uniformity). It is possible (but not easy) to reduce or eliminate the screen structure effects by making a series of measurements at slightly different positions and averaging the results or applying curve smoothing.

Square-wave response may be used for convenience instead of sine wave response. The square-wave response in cycles per unit distance may be used by itself. The television test pattern resolution wedges and other line–pair patterns such as SMPTE 133 are prime examples. And it is usually easier to produce the gratings used for direct spatial frequency measurement by using step functions. Resolution measured by the square-wave response will always appear better than by the sinusoidal MTF. Figure 7.31 illustrates the relationship between sine wave and square-wave responses. The response to a square wave may be converted to the equivalent sine wave response, assuming a Gaussian spot profile, using the graph shown in Figure 7.32. At the reference 60% modulation depth, the square-wave spatial frequency would be divided by 1.21 to obtain the sine wave spatial frequency.

Several methods are used to derive the MTF. The first uses conventional spot profile measurements and Fourier transform of the profile data, as in EIA TEP105-17. Either the spatial frequency for a single fixed level of a Gaussian spot may be computed as described previously or more extensive computations may be used to compute the entire MTF curve. There is no requirement of a Gaussian spot shape for the complete Fourier transform while a single-point computation should be attempted only for near-Gaussian profiles. Measurement systems are available that contain PCs and perform the FFT computations with included or optional software.

Of the other methods, the most basic uses a fixed grating, or *Ronchi ruling*, as shown in Figure 7.33, with a lens to image the CRT spot on the grating. The lens magnification and grating spacing are chosen to result in the desired spatial frequency using the equation

$$f = \frac{m}{2s}$$

7.6 MODULATION TRANSFER FUNCTION 217

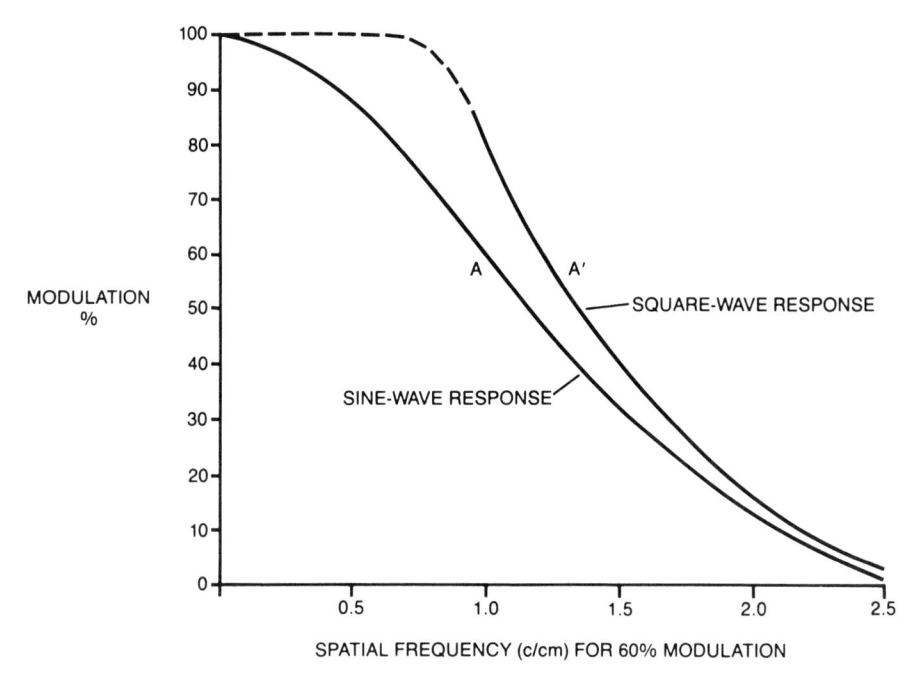

Figure 7.31. Sine wave and square-wave MTF curves of a sample display.

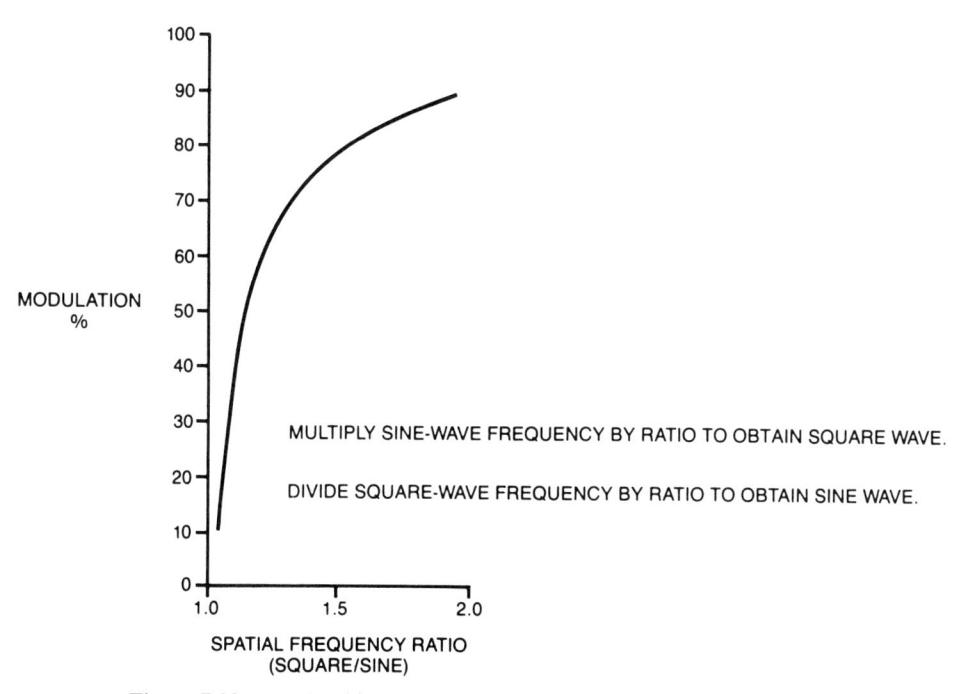

Figure 7.32. Relationship between sine wave and square-wave MTF curves.

218 RESOLUTION MEASUREMENT

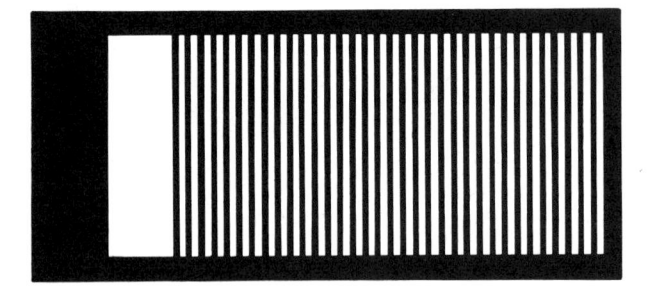

Figure 7.33. Grating used for MTF measurement.

where

m = magnification of the lens
s = spacing of the grating lines

The scanned beam is imaged on and scanned perpendicular to the grating lines. The light through the grating is detected with a PMT, as shown in Figure 7.34. The photomultiplier output signal is displayed on an oscilloscope. The depth of modulation as the beam traverses the closely spaced portion of the grating is compared with the modulation of the wide-spaced area, which is used to establish 100% modulation. Alternatively, the photomultiplier output may be electrically filtered to remove the higher frequency harmonics of the square wave while leaving the fundamental intact, thus yielding the sine wave modulation. The spatial frequency being measured may be changed by rotation of the grating, changing the lens magnification, or replacing the grating with one having a different line spacing or even several different line spacings on one grating, as shown in Figure 7.35. The latter allows several data points to be measured at one time, as shown in Figure 7.36. A closely related technique uses a stationary CRT spot with mechanical scanning of the grating across the spot, as shown in Figure 7.37.

Single slits instead of gratings may also be used for measurement of the MTF. One approach to this method is to employ an electronically generated sine wave bar pattern on the CRT and mechanical scanning of the pattern with the slit, as shown in Figure 7.38. Another similar approach makes use of a fixed slit and electronically moves the image of the sine wave pattern past the slit, as shown in Figure 7.39. Both of these methods are interpreted in a manner similar to the grating techniques.

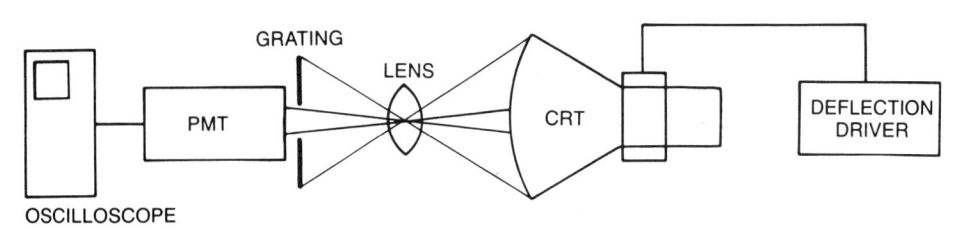

Figure 7.34. Configuration of equipment for MTF measurement.

7.6 MODULATION TRANSFER FUNCTION   219

Figure 7.35. MTF grating with variable spacings.

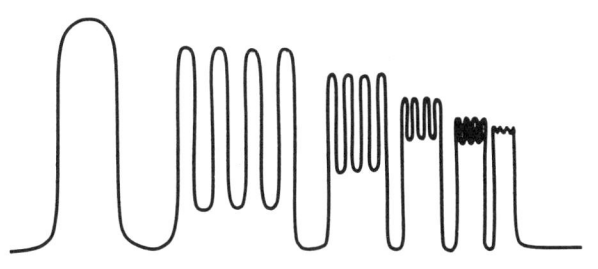

Figure 7.36. Scanned display of multispacing grating.

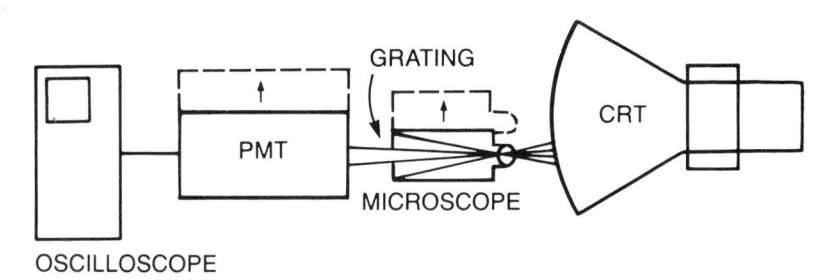

Figure 7.37. MTF measurement with mechanical scanning of a grating across a stationary spot.

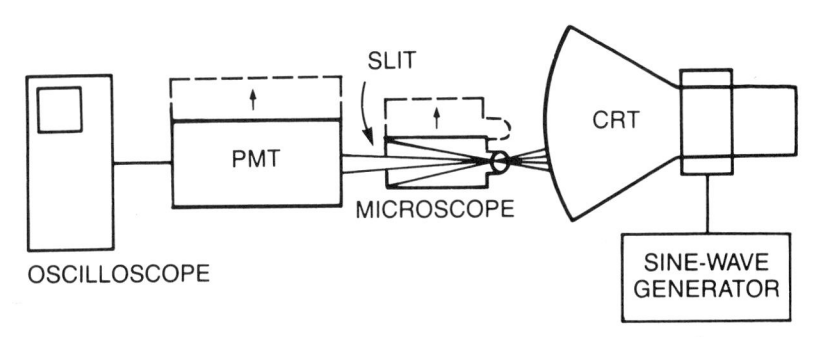

Figure 7.38. MTF measurement with mechanical scanning of a slit across a sine wave pattern.

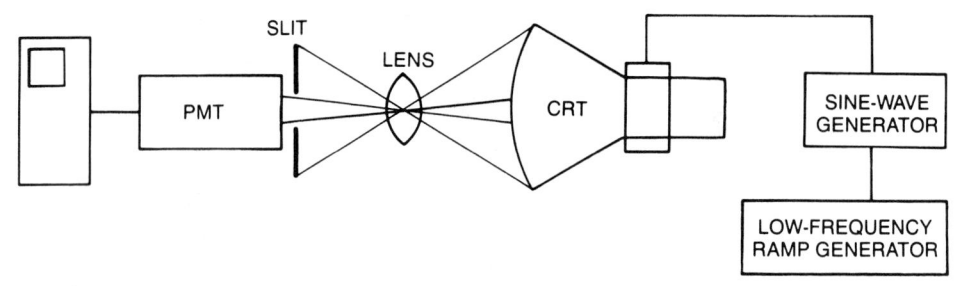

Figure 7.39. MTF measurement with electrical scanning of a sine wave pattern across a slit.

7.7 STANDARDS

A number of standards concern themselves with resolution measurement. Some of these are listed below to aid in selection of appropriate ones to the task at hand:

ANSI/HFS 100-1988[41]	Line luminance profiles, MTFA
EIA TEP105-7-A[42]	Monochrome line profiles
EIA TEP105-8[43]	Modulation depth of 10 and 50% with a shrinking raster
EIA TEP105-9[44]	Line profiles for shadow-mask CRTs
EIA TEP105-17[45]	MTF of monochrome CRTs
IEC 151-14[46]	Radar and oscilloscope CRT visual and slit-scan spot diameter; also MTF
IEC 151-16[47]	Monochrome television picture tube visual spot diameter
ISO 9241-3[48]	Aperture or slit scan measurement of monochrome and color CRT resolution
SAE ARP1782[49]	Line width of monochrome, beam index, and shadow-mask color CRT displays using slit scan and diode (CCD) array scanning techniques
VESA Standard for Display Specifications and Test Procedures[50]	Discusses RAR and line width measured photometrically

In conclusion, there are a great many ways to express and measure display resolution. Data measured with different techniques need to be compared carefully to avoid erroneous conclusions. Whatever the method used, many internal and external factors influence the accuracy of resolution data. While the more sophisticated methods have higher inherent accuracy and repeatability, they also have subtle traps that negate their advantages if not used cautiously. Work is being done within the industry to further develop recommended methods for measuring monochrome and color CRT resolution to minimize future measurement inconsistencies between designers and manufacturers of CRTs and displays and their customers.

REFERENCES

1. G. M. Murch and L. Virgin, "Resolution and Addressability: How Much Is Enough?" *SID Dig.*, Vol. 16, pp. 101–103, 1985.

2. G. M. Murch, L. Virgin, and R. Beaton, "Resolution and Addressability: How Much Is Enough?" *Proc. SID,* Vol. 26, No. 4, pp. 305–308, 1985.
3. G. M. Murch, "Human Factors and Flat Panels Challenge the CRT," *Information Display,* pp. 8–11, Mar. 1987.
4. S. T. Knox, "Resolution and Addressability Requirements for Digital CRTs," *SID Dig.,* Vol. 18, pp. 26–29, 1987.
5. G. M. Murch and R. J. Beaton, "Matching Display Resolution and Addressability to Human Visual Capacity," *Displays,* pp. 23–26, Jan. 1988.
6. Electronic Industries Association (EIA), "Relating Resolution and Addressability," EIA Technical Bulletin TEB27, Sept. 1988.
7. Society of Motion Picture and Television Engineers (SMPTE), "Specifications for Medical Diagnostic Imaging Test Pattern for Television Monitors and Hard-Copy Recording Cameras," SMPTE Recommended Practice RP 133-1986, Jan. 1986.
8. K. B. Benson, *Television Engineering Handbook,* McGraw-Hill, New York, 1992, pp. 4.8–4.9.
9. W. Middlebrook and M. Day, "Measure CRT Spot Size," *Electron. Design,* pp. 58–60, July 19, 1975.
10. D. L. Say, "Picture Tube Spot Analysis Using Direct Photography," *IEEE Trans. Consumer Electron.,* p. 32, Feb. 1977.
11. D. L. Say, "Pulsed Spot Photography," Philips ECG Seminar Notes, June 1982, pp. 7-2-1–7-2-4.
12. "Determining Medical Imaging CRT Resolution, Not a Simple Procedure," *Information Display,* pp. 17–21, June 1986; pp. 10–13, July 1986.
13. Society of Motion Picture and Television Engineers (SMPTE), "Specifications for Medical Diagnostic Imaging Test Pattern for Television Monitors and Hard-Copy Recording Cameras," SMPTE Recommended Practice RP 133-1986, Jan. 1986.
14. J. E. Gray, K. G. Lisk, D. H. Haddick, J. H. Harshbarger, A. Oosterhof, and R. Schwenker, "Test Pattern for Video Displays and Hard-Copy Cameras," *Radiology,* Vol. 154, No. 2, pp. 519–527, 1985.
15. K. B. Benson, *Television Engineering Handbook,* McGraw-Hill, New York, 1992, pp. 21.90–21.93.
16. W. T. Dyall, "Televiewers: What Do You Mean by High Resolution," *Electro-Opt. Sys. Design,* Mar. 1978.
17. J. O. Drewery, "The Zone Plate as an Aid to Testing Display Systems," *Displays,* pp. 209–214, Jan. 1980.
18. Tektronix Inc., "Measuring Television Picture Monitor Resolution," Television Products Application Note 27, Jan. 1981.
19. J. S. Snyder, "Analysis of Resolution Measurements of a Cathode-Ray Tube Spot," 10th Natl. Symp. on Information Display, 1969, pp. 13–26.
20. J. E. Bryden, "Some Notes of Measuring Performance of Phosphors Used in CRT Displays," *Proc. 7th National Conference SID,* pp. 89–97, Oct. 1976.
21. J. J. Hertel, "Image Intensifier MTF Measurement, a Vidissector Camera Application," IT&T Application Note 121, undated.
22. M. E. Carpenter, "Spot Profiling Techniques," Compilation of seminar notes on data display tube characteristics, Philips ECG, June 1982, pp. 7-1-1–7-1-10.
23. P. Burr and B. D. Chase, "Spot Size Measurements on Shadow-Mask Color CRTs," *SID Proc. First European Display Research Conf.,* pp. 170–172, Sept. 1981.
24. B. Baur, "An Advanced CRT Spot Contour Measurement System," *Proc. SID,* Vol. 26, No. 1, pp. 57–59, 1985.
25. O. H. Schade, "A New System of Measuring and Specifying Image Definition," Nat. Bur. Stds. Circular 526, April 29, 1954.

26. O. H. Schade, "An Evaluation of Photographic Image Quality and Resolving Power," *J. SMPTE,* Vol. 73, pp. 81–119, 1964.
27. O. H. Schade, "A Method of Measuring the Optical Sine-Wave Spatial Spectrum of Television Image Display Devices," *J. SMPTE,* Vol. 67, pp. 561–566, 1958.
28. O. H. Schade, "Modern Image Evaluation and Television (The Influence of Electronic Television on the Methods of Image Evaluation)," *Appl. Opt.,* Vol. 3, No. 1, pp. 17–21, Jan. 1964.
29. International Electrotechnical Commission (IEC), "Measurements of the Electrical Properties of Electronic Tubes, Part 14: Methods of Measurement of Radar and Oscilloscope Cathode-Ray Tubes," IEC Publication No. 151-14, 1975, p. 17.
30. P. G. J. Barten, "Spot Size and Current Density Distribution of CRTs," *Proc. 3rd Intl. Display Research Conf.,* Oct. 1983, pp. 280–283.
31. L. E. Tannas, *Flat Panel Displays and CRTs,* Van Nostrand Reinhold, New York, p. 77, 1985.
32. American National Standards Institute (ANSI)/Human Factors Society (HFS), "American National Standard for Human Factors Engineering of Visual Display Terminal Workstations," ANSI/HFS 100-1988, 1988, pp. 17–19.
33. P. G. J. Barten, "The SQRI Method: A New Method for the Evaluation of Visible Resolution on a Display," *Proc. SID,* Vol. 28, No. 3, pp. 253–262, 1987.
34. P. G. J. Barten, "Evaluation of Displays with the SQRI Method," *Proc. SID,* Vol. 30, No. 1, pp. 9–14, 1989.
35. A. van Meeteren, "Visual Aspects of Image Intensification," Thesis, Univ. of Utrecht, The Netherlands, 1973.
36. C. R. Carlson and R. W. Cohen, "A Simple Psycho-physical Model for Predicting the Visibility of Displayed Information," *Proc. SID,* Vol. 21, No. 3, 1980
37. C. R. Carlson and R. W. Cohen, "Visibility of Displayed Information: Image Descriptors for Displays," Report ONR-CR213-120-4F, Office of Naval Research, Arlington, VA, July 1978.
38. G. D. Bergland and M. T. Dolan, "Fast Fourier Transform Algorithms," In *Programs for Digital Signal Processing,* Institute of Electrical and Electronics Engineers, New York, 1979, pp. 1.2-1–1.2-18.
39. Electronic Industries Association, "MTF Test Method for Monochrome CRT Display Systems," EIA TEPAC Publication No. TEP105-17, July 1990.
40. P. Keller and R. J. Beaton, "The EIA Standard for MTFs of Monochrome CRTs," *SID Dig. of Technical Papers,* Vol. 20, pp. 204–207, 1989.
41. Human Factors Society, "American National Standard for Human Factors Engineering of Visual Display Terminal Workstations," ANSI/HFS 100-1988, 1988, pp. 64–66, 85–86.
42. Electronic Industries Association, "Line Profile Measurements in Monochrome Cathode Ray Tubes," EIA Publication No. TEP105-7-A, Jan. 1987.
43. Electronic Industries Association, "Raster Response Measurement for Monochrome Cathode Ray Tubes," EIA Publication No. TEP105-8, Jan. 1987.
44. Electronic Industries Association, "Line Profile Measurements in Shadow Mask and Other Structured Screen Cathode Ray Tubes," EIA Publication No. TEP105-9, Jan. 1987.
45. Electronic Industries Association, "MTF Test Method for Monochrome CRT Display Systems," EIA TEPAC Publication No. TEP105-17, July 1990.
46. International Electrotechnical Commission, "Measurement of the Electrical Properties of Electronic Tubes, Part 14, Methods of Measurement of Radar and Oscilloscope Cathode-Ray Tubes," IEC Publication No. 151-14, 1975, pp. 13–21.
47. International Electrotechnical Commission, "Measurement of the Electrical Properties of Electronic Tubes and Valves, Part 16, Methods of Measurement for Television Picture Tubes," IEC Publication No. 151-16, 1968, pp. 12–15.

48. International Standards Organization, "Ergonomic Requirements for Office Work with Visual Display Terminals (VDTs), Part 3, Visual Display Requirements," ISO 9241-3, 1992, pp. 8–11.
49. Society of Automotive Engineers (SAE), "Photometric and Colorimetric Measurement Procedures for Airborne Direct View CRT Displays," SAE ARP1782, Jan. 9, 1989, pp. 14–25.
50. Video Electronics Standards Association (VESA), "Standard for Display Specifications and Test Procedures," Oct. 3, 1991, pp. 19–20, 37.

CHAPTER 8

Geometry Measurements

8.1 GENERAL BACKGROUND

The geometric, or *spatial,* measurements of a display include the physical size of the entire display, the individual elements that comprise it, active viewing area, defective pixels, scanning distortions, and the nonlinearities and distortions of displayed patterns. Measurements may be expressed in conventional physical units such as millimeters or inches or may be in the number of *pixels* (individual picture elements).

Flat-panel displays are easiest to describe. The display size and pixel size are accurately controlled by fixed processing steps during manufacture and are not subject to any appreciable variation from unit to unit of the same display model. The overall display size and the pixel configuration are therefore the main geometric concern. These are easily discerned using the manufacturer's specification sheet, a ruler, and/or an optical comparator. One other geometric property is unique to flat-panel displays. This is the missing or defective pixel specification. Missing pixels are usually the result of an electrical defect, either broken electrical contact to the pixel or a defective driver for those displays having on-board drivers. A maximum number of missing pixels is specified by the display manufacturer and may also be expressed in the maximum number per unit area to prevent clusters of defective pixels from being concentrated in a small area, which makes them more objectionable.

Flat-panel display devices used in projection systems are another matter. They may exhibit some of the same geometric distortions that are found in CRT displays, although the causes are different. These distortions are caused by optical defects and design compromises rather than the electrical or magnetic aberrations that may be present in a CRT. The distortions may be quantified using similar techniques to those used to measure CRT distortions, however, and will not be discussed separately.

The flexibility of the CRT display, which is what makes it so versatile, makes it prone to several forms of distortion, some relatively constant in nature and others periodic or even momentary. The distortions may be caused internally by its electron optics compromises, electronic driving circuitry, the deflection yoke, or external magnetic fields. Magnetically induced distortions may also be internally produced by transformers or other magnetic components within the monitor. A number of the distortions may be minimized using factory, service, or end-user adjustments provided within the monitor. Each of the common distortions are discussed individually in this chapter. Pertinent industry standards for quantifying the distortions are included. No single standard includes all distortions, so choosing the most useful one is necessary. Instrumentation for CRT display geometry measurement ranges from the human eye and a ruler to fully automated video photometer systems for high-volume production line use. The latter may also provide automatic adjustment of the monitor under test.[1] Many of today's computer displays are adjusted in this manner and few adjustments are provided for the user. Those that are provided may be digitally controlled with menu selections.

8.2 SIZE AND ASPECT RATIO

The size, also referred to as the image area or active display area, is one of the most obvious geometric properties of a CRT display. The maximum physical display size is defined by the CRT and its mask, or *bezel*. For home television receivers, the raster size is usually adjusted to be about 10% larger than the actual viewable screen area. This *overscanning* prevents a black border from becoming visible at extremes of power line voltage variations or as electrical components within the receiver age. Some picture information is lost, but this is usually of little consequence since the action is concentrated in the center of the screen. Conversely, computer displays are usually *underscanned* by 5–10% to make full use of the raster edges where menu bars or other functions are located. Also, CRT electron beam and optical aberrations are worst in the extreme screen corners so underscanning avoids use of these areas. Underscanning results in improved legibility of alphanumeric characters in the corners of the raster.

In any event, the size is often adjustable for user preferences using the height and width, or V size and H size, controls. No special test pattern is necessary for checking or adjustment of the size of underscanned displays. The actual display to be viewed is the best indicator. Note that for multisync monitors, the size can be slightly different in the different scan formats. For overscanned displays, a test pattern such as that used for checking linearity will aid in judging the amount of overscan. Alternately, the display can be adjusted for a few percent underscan and the adjustments moved by twice the amount required to increase the scan to just touch the edges of the bezel or visible CRT phosphor area.

Measurement of screen or display size is straightforward for flat, rectangular screens but is more complex for cylindrical or spherical CRT faceplates with their radiused corners. The VESA Standard for Display Specifications and Test Procedures[2] does not deal with the screen curvature issues and recommends measurement in centimeters, which is a deprecated unit according to ANSI.[3] It is much easier to work in units that are multiples of 10^3 such as the meter and millimeter. The most comprehensive procedure for determining CRT screen size is ANSI/EIA-527,[4] which includes compensation for the slight refractive gain due to the increased glass curvature at the screen edges.

Tests for raster size stability with changes of luminance are described in the proposed NIDL monochrome and color CRT monitor measurement standards for EIA.[5,6] In the NIDL test, a uniform full flat-field raster is displayed and the size measured at 25, 50, 75, and 100% luminance levels. Changes in raster size with luminance are evidence of loss of regulation of the CRT high-voltage power supply, which decreases the electron beam "stiffness" as the voltage drops. This makes the beam more sensitive to the deflection fields and increases the raster size.

A related issue to raster and screen size is the aspect ratio. This is the ratio of the horizontal size to the vertical size, either in millimeters or pixels. Ratios of 4:3 are used for television and many computer displays. Rectangular CRTs have physically similar aspect ratios, and adjustment for a uniform border around the entire raster should come close to the correct aspect ratio of the raster. Small errors in aspect ratio are usually unnoticeable. High-definition television (HDTV) uses an aspect ratio of 16:9. These are now coming

226 GEOMETRY MEASUREMENTS

into use in Japan, Europe, and some demonstration systems in the United States. In some cases, receivers with the 16:9 format use conventional television signals with line doubling and interpolation techniques to provide high quality wide screen displays.

8.3 CENTERING OR POSITIONING

Improper centering is readily observable for underscanned displays. A test pattern containing a centered circle such as shown in Figure 8.1 or other reference pattern must be used for overscanned displays to determine improper centering. Provision for centering adjustment is not as common as it once was due to improvements in the stability of power supplies and deflection amplifiers for CRTs. Where provided, centering adjustments may be either electrical or magnetic. The latter may be in the form of ring magnets on the neck of the CRT. Common nomenclature of the electrical centering adjustments is usually some combination of the H (horizontal) and V (vertical) controls or HOR and VERT with CENT or POS. For magnetic adjustments, refer to the display manufacturer's service manual for identification of the means used and proper adjustment procedure in order to avoid accidental misadjustment of purity or convergence magnets.

8.4 NONLINEARITY AND S DISTORTION

Nonlinearity (Figure 8.2) manifests itself in raster displays as compression of information in certain areas of the screen (often, but not always, near the edge) or elongation of portions of displayed circles. It is not as prevalent or as pronounced as it was in early television receivers because of the steady improvement of deflection circuitry in recent years. If provided at all, linearity adjustment is most likely to be an internal adjustment made during manufacturing. Still, even with the most careful adjustment, slight amounts of nonlinearity will exist in any CRT display due to the physical limitations of deflection yoke winding distribution accuracy and the problems of electrically driving the deflection yoke, which is inductive, with a linearly increasing current.

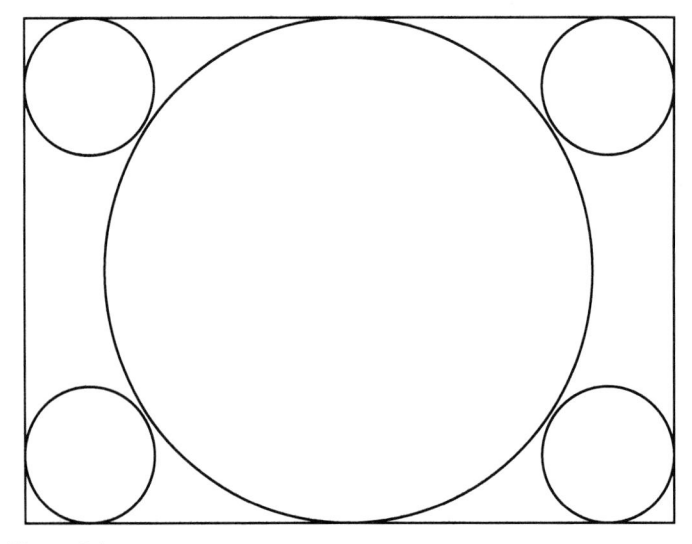

Figure 8.1. Test pattern suitable for adjustment of raster size and linearity.

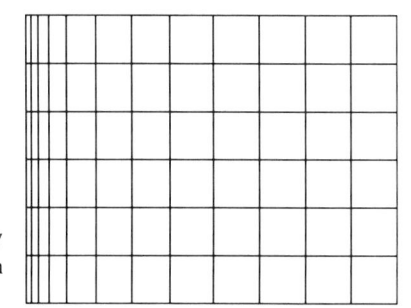

Figure 8.2. Horizontal nonlinearity evidenced by "bunching" at the left side. Nonlinearity may occur in either axis and anywhere on the screen.

A special form of nonlinearity is termed S distortion because of the S-shaped deviation of the linear sawtooth or "ramp" horizontal deflection current waveforms used to produce a uniform rate of travel of the electron beam across the screen. Figure 8.3 illustrates the effect on the sawtooth waveform caused by S distortion. Bunching of the raster will occur on one side of the display with stretching in the corresponding location on the opposite side.

Measurement of linearity errors is described in several display standards and proposals. The NIDL-proposed standards[7,8] use a series of video-generator-produced vertical or horizontal lines of 100% luminance on a black background. The spacing between lines is 5% of the screen width or height respectively. The spacing of the displayed lines is measured along the CRT centerline perpendicular to the lines using a spatially calibrated diode array detector, video photometer, or microscope and calibrated x–y positioner. Nonlinearity is determined by computing the difference in line spacing between the widest and narrowest spacings and expressing the difference in percentage of total screen height. Further use of the data may be made to derive plots of linearity error versus screen location. VESA[9] recommends a square grid test pattern (Figure 8.4) with a minimum of eight squares in the smallest screen dimension, in a similar manner to NIDL, and computes the linearity error using the following equation:

$$\text{Linearity error (\%)} = 100\% \times \frac{L_{max} - L_{min}}{(L_{max} - L_{min})/2}$$

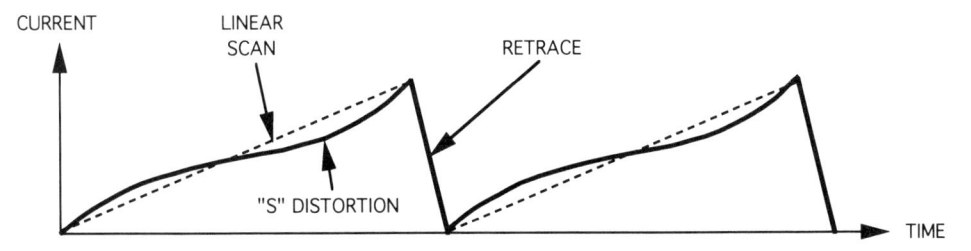

Figure 8.3. Horizontal-deflection sawtooth current waveform with S distortion.

228   GEOMETRY MEASUREMENTS

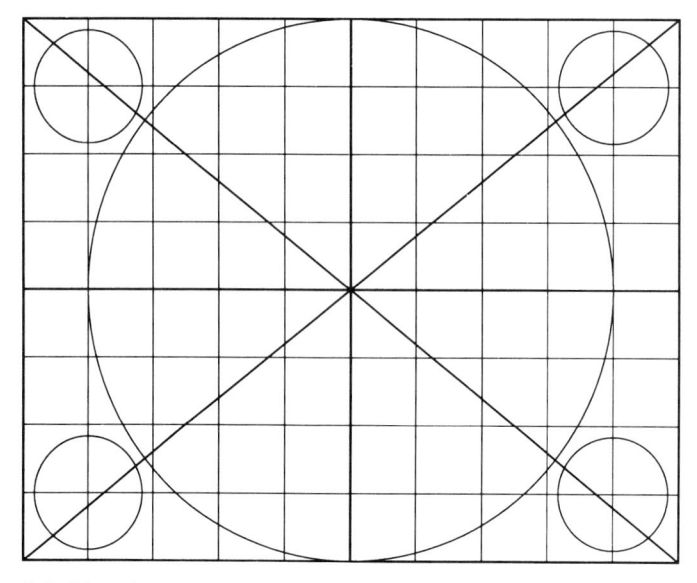

Figure 8.4. Linearity test pattern based on the VESA display standard. (After ref. 9.)

where

L_{max} = maximum square dimension in the axis of interest
L_{min} = minimum square dimension in the axis of interest

The ISO standard 9241-3 uses a different approach oriented toward alphanumeric display users.[10] They specify a maximum of 2% difference in length of a series of the capital letter H in the same row or column.

The SAE standard ARP1782[11] for avionics CRTs combines linearity measurement with keystone, orthogonality, and pincushion/barrel distortion measurement in one comprehensive test. A geometry test pattern (Figure 8.5) is displayed and measurements made at the locations designated by letters. Equations are prescribed by SAE for computing the amount of distortion of each category.

8.5   TILT

Figure 8.6 illustrates *tilt*. Tilt is merely a measure of the amount that the entire pattern is rotated with respect to the CRT mask or faceplate sides and is measured in degrees of rotation from correct alignment. It is usually relatively simple to correct as it is caused by rotation of the deflection yoke on the CRT neck. Provision is made to allow rotational adjustment of deflection yokes for this purpose.

Tilt may also occur in electrostatically deflected rectangular oscilloscope CRTs. In this case, it is caused by misalignment of the electron gun with respect to the CRT bulb. Rotational adjustment may be accomplished magnetically by a coil placed between the deflection plate region of the tube and the screen. A DC applied to the coil will cause rotation of the entire pattern on the screen. The amount of current controls the degree of rotation and the polarity controls the direction of rotation.

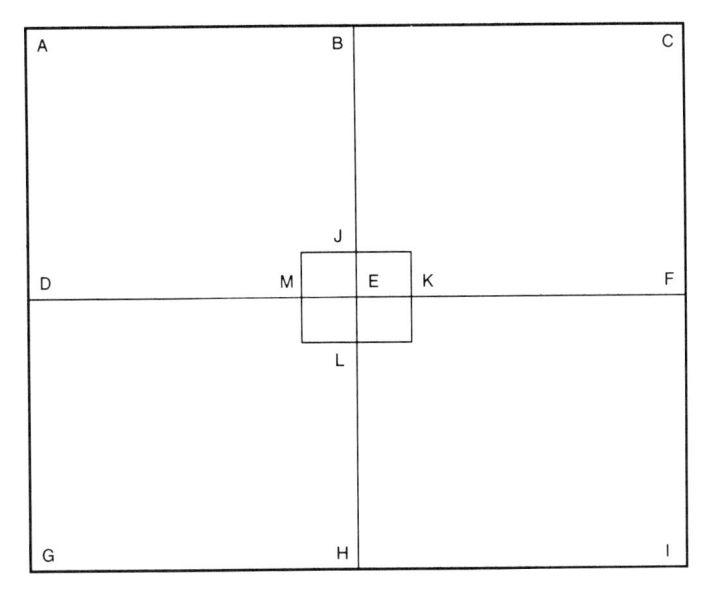

Figure 8.5. Geometry test pattern based on SAE ARP1782. (After ref. 11.)

8.6 ORTHOGONALITY

Orthogonality is the misalignment between the vertical-deflection axis and the horizontal-deflection axis and is expressed in degrees from perpendicular. It is noticeable on a display when vertical and horizontal lines are displayed simultaneously, as in Figure 8.7. The lines should ideally intersect at a 90° angle, but lack of orthogonality will result in a different angle. The cause is usually misalignment of the vertical and horizontal windings in the deflection yoke for electromagnetically deflected CRTs. No adjustment is possible for this defect. Electrostatically deflected CRTs may also exhibit orthogonality. Here, the cause is internal misalignment between the vertical and horizontal plates. Adjustment of orthogonality may be provided by magnetic or electrostatic means in the vertical-deflection-plate region of the CRT for higher performance oscilloscopes.

The VESA test method is the most straightforward method for specifying orthogonality, and the misalignment is simply measured in degrees from perpendicularity.[12] The ISO has another approach to measuring orthogonality, as shown in Figure 8.8.[13] They use the

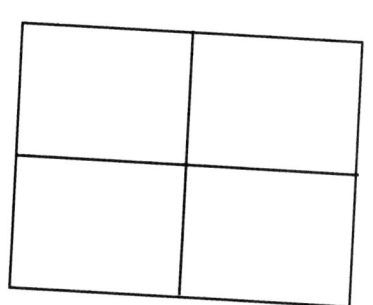

Figure 8.6. Tilt is usually due to rotation of the deflection yoke.

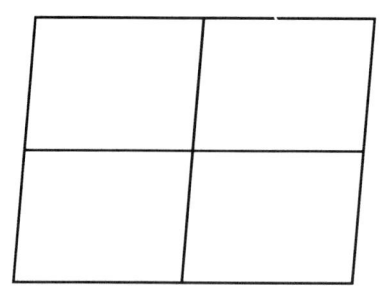

Figure 8.7. Orthogonality distortion as caused by misalignment between the vertical and horizontal deflection coils.

ratio of the difference in length, $D_1 - D_2$, between the extremes of the two raster diagonals and their mean length, $\frac{1}{2}(D_1 + D_2)$. A maximum of 0.04 times the smaller raster dimension (vertical for landscape mode displays and horizontal for page or portrait mode displays) is permitted.

8.7 TRAPEZOIDAL DISTORTION

Trapezoidal distortion, also known as *keystone* distortion, is a difference either in the height of the raster between the left and right sides or in the raster width between the top and bottom that results in a trapezoidal shaped pattern, as shown in Figure 8.9. Actually both may be present in the same display to some degree. The cause of trapezoidal distortion in monochrome CRTs is an imbalance between the pairs of the coils within the deflection yoke for electromagnetically deflected CRTs or misalignment of one or more of the deflection plates in an electrostatically deflected CRT. Replacement of the yoke in the former case or the CRT in the latter are about the only solutions to the problem. Trapezoidal distortion in color CRTs occurs in the red and blue rasters since the red and blue guns are mounted slightly off-axis in the CRT neck. This distortion is corrected for in the deflection yoke design. Projection displays also may exhibit trapezoidal distortion, which is optically produced by the optics being located off-axis to the screen. This distortion is corrected for by applying opposite and equal trapezoidal distortion electronically to the rasters.

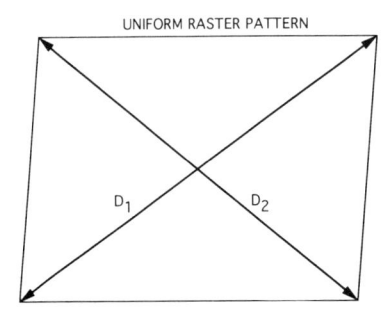

Figure 8.8. Orthogonality test as recommended by ISO; D_1 and D_2 will have unequal lengths if orthogonality distortion is present.

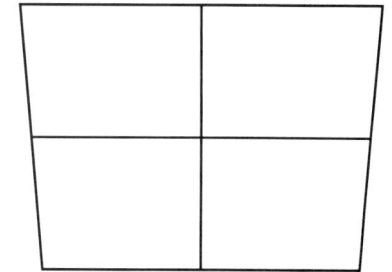

Figure 8.9. Trapezoidal (keystone) distortion.

Both VESA and ISO specify similar trapezoidal distortion measurement methods, although under different terminology.[14,15] VESA calls it parallelogram distortion while ISO includes it as part of their orthogonality test method. In both cases, it is the maximum height (or width) minus the minimum height (or width) divided by the mean height (or width). Just do not mix units of height and width in the same computation. VESA multiplies the result by 100 to obtain percentage of distortion while ISO leaves the result as a decimal with a maximum of 0.02 allowed.

8.8 PINCUSHION AND BARREL DISTORTION

Pincushion ("pin") and *barrel* distortion are illustrated in Figures 8.10 and 8.11, respectively, and are well described by their terminology. Both are a result of projecting an electron beam from a point in the CRT deflection region onto a flat or curved glass faceplate. Analogous distortions occur in optical projection as well. The effects are aggravated by the trend toward minimum overall CRT length. Pincushion distortion is the natural result because of the greater distances that the beam must travel to reach the screen corners. Deflection amplifier gain compensation at higher amplitudes as well as compensation of the deflection yoke windings can minimize the effect. Barrel distortion can result from overcompensation for pincushion correction. Either pincushion or barrel distortion may occur in just one or both axes.

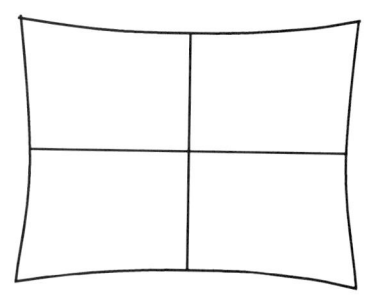

Figure 8.10. Pincushion distortion.

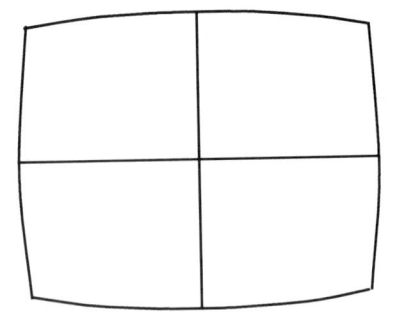

Figure 8.11. Barrel distortion.

The main sources of information for quantifying pincushion and barrel distortion are SAE and VESA. The SAE[16] includes both in the overall raster distortion analysis previously described in the section on nonlinearity and expresses the result in percent. They compare the average of the raster top and bottom width with the width in the center and the average of the height at the left and right sides with the height in the center. The VESA test method[17] uses a rectangular test gauge to determine if the distortion is within specified limits.

8.9 MISCONVERGENCE

For color shadow-mask CRTs, the most widely used display device, convergence and its adjustment are important considerations. It is covered in this chapter rather than with color measurement since it is more a matter of geometric adjustment. Convergence is a measure of how well the red, green, and blue beams are physically aligned with each other to strike the same area of the screen. This section deals mainly with the measurement of misconvergence. If adjustment of convergence is required, refer to the manufacturer's service manual for the exact adjustment procedure for a particular display.

Ideally, the three beams should be aligned to strike the screen at exactly the same point regardless of location within the actively scanned display area. Because the three beams originate from different electron guns that are physically separated from each other, they are deflected to slightly different screen locations. Magnetic adjustments are provided on the CRT neck to allow misconvergence to be minimized. Misconvergence causes sharply defined characters or objects to have colored "fringes," as illustrated in Plate 11. It is most common near the edges of the screen and in the corners. Color fringing may also be caused by different beam sizes of the three guns. For instance, red fringing may result on both sides of the lines of a perfectly converged crosshatch pattern if the red beam has a larger spot size than the green and blue beams. The fact that it is seen as fringes of the same color on both sides of the line is the key to identifying this condition. Misconvergence is indicated by different colors on one or both sides of the crosshatch lines.

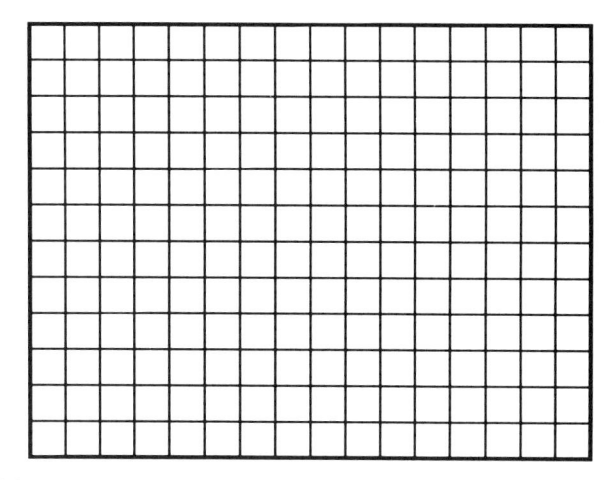

Figure 8.12. Crosshatch test pattern used for convergence testing and adjustment.

Test Patterns

To measure misconvergence, a test pattern of single-pixel-wide white lines or dots is usually used, as shown in Figure 8.12 or 8.13. Both may be combined into a single pattern, as in Figure 8.14. One other recent test pattern is especially designed for convergence testing. It is a crosshatch pattern consisting of a large number of short red, green, and blue line segments (Plate 12), as developed at TEAM Systems, Inc.[18] Adjustment of convergence is reported to be easier using the separate line segments than with the overlapping white lines of conventional crosshatch patterns.

Instrumentation

The human eye is the ultimate consideration for the amount of misconvergence that is tolerable and is most often used for the measurement or adjustment of convergence. Televi-

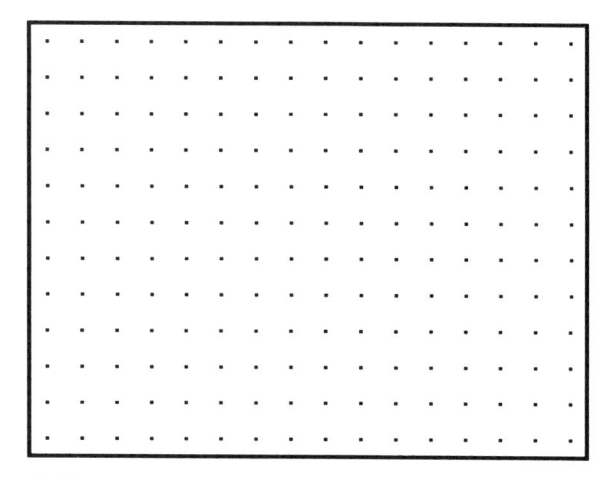

Figure 8.13. Dot test pattern used for convergence testing and adjustment.

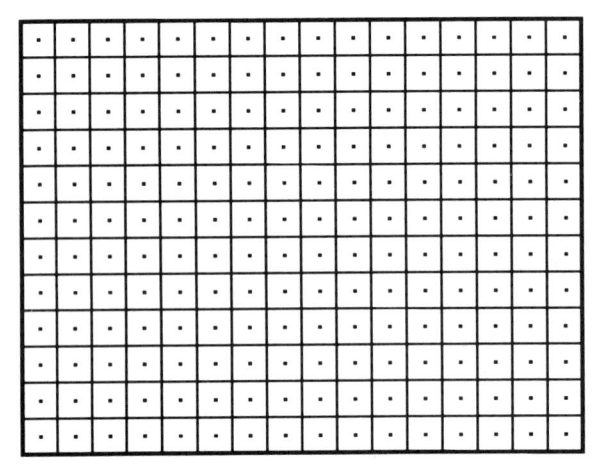

Figure 8.14. Combination crosshatch and dot test pattern for convergence testing and adjustment.

sion receivers, which are usually viewed from distances of 2 or 3 m or more, are less critical for accurate convergence than closely viewed data and CAD displays where the screen corners are of equal importance to the center.

Instrumentation to assist convergence measurement and adjustment ranges from the humble magnifying glass to fully automated systems. The method chosen depends on the quantity of displays involved and the precision required. The simplest measurement method uses a hand-held comparator to measure the degree of misconvergence of the three color primaries at several locations on the screen. Because of the shadow-mask pattern, scanning lines, and the fact that each of the beams is spread across several phosphor stripes or dots, it is difficult to judge the center of each beam, and some interpretation and estimation are necessary to determine the center of each.

The "Klein Gauge" is the next level of sophistication in convergence measurement. This simple-to-operate device (refer back to Figures 4.16 and 4.17) optically splits a white line of the crosshatch into three adjacent segments consisting of the three color primaries that make up the white line. With the two adjustment knobs on the sides of the gauge set to zero, the amount of displacement of each segment is indicative of the physical location of each beam (Figure 8.15). By adjustment of the two knobs, the red and blue segments are made to appear in a straight line with reference to the green segment. The amount of red and blue misconvergence may then be directly read from the scales at the sides. An advanced model using the same concept adds a small monochrome television camera to the gauge to make viewing easier than through the optical eyepiece, but the basic technique remains the same.

Another level of sophistication is the CCD array, such as used in the Quantum Data CG1 portable convergence gauge and the Minolta CC-100 convergence meter. These are digitally based instruments that automatically measure the amount of red and blue misconvergence relative to the green beam. The Minolta measures both the x and y axes simultaneously while the Quantum Data measures one axis at a time but adds a red/blue misconvergence reading. Other CCD-based instruments include some video photometers by Photo Research, Quantum Data, and Microvision. At least one, the Quantum Data CP-1, is combined with a test pattern generator.

8.9 MISCONVERGENCE   235

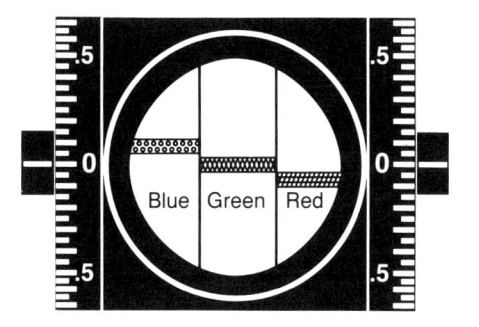

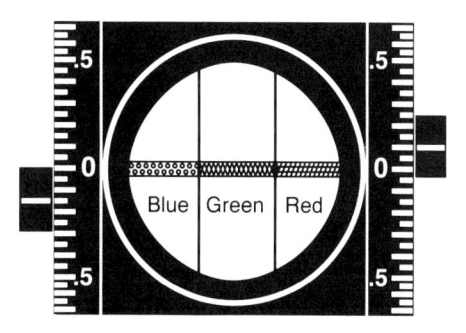

Figure 8.15. Top view of the Klein convergence gauge, showing knob adjustment and green-blue and green-red convergence error readings. (Courtesy of Klein Optical Instruments, Inc.)

Microphotometers may also be used to measure misconvergence, although this technique requires operators experienced in such methods and displays capable of having the crosshatch pattern slowly scanned in the axis to be measured. A photometric microscope is used to record the light output in turn from individual phosphor dots or lines of the three phosphor primaries as the beams are slowly scanned across them. The output is recorded on an x–y recorder or personal computer to show the relative misconvergence of the three beams, as in Figure 8.16. Alternatively the microphotometer may be mechanically scanned across the lines and the results recorded. Some means such as color filters or individual operation of the three electron guns must be used to identify the three beams. The latter is less desirable because of possible shifting of the beams when the other beams are shut off due to less than perfect power supply regulation. The microscope must view an area no larger than the width of one phosphor dot or line. Because of the small amount of light available from such a small area, a photomultiplier tube is best suited as a detector. It may be connected to the microscope via a fiber-optic cable. A single optical fiber with a right-angle bend at the image plane of the microscope serves to

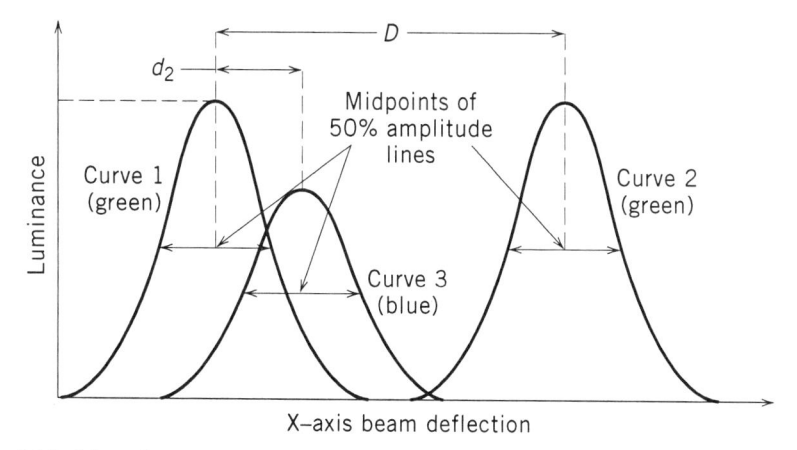

Figure 8.16. Microphotometer scan of red, green, and blue beams showing misconvergence. All three beams should ideally be superimposed. (From "Photometric and Colormetric Measurement Procedures for Airborne Direct View CRT Displays," Aerospace Recommended Practice 1782, Jan. 9, 1989, p. 28. Reprinted with permission from SAE document ARP 1782 © 1989 Society of Automotive Engineers, Inc.)

sense the light from a small area of the screen. Telephotometers such as the Photo Research Pritchard photometer provide some ease of use since they are self-contained instruments capable of viewing extremely small areas. These procedures are best suited for display laboratory use rather than routine testing.

Finally, fully automated video photometer systems that are based on CCD detectors are available for high-volume manufacturing lines. These systems may also include the ability to automatically calibrate the monitor as well as record performance data.

Standards

At least three display standards address the issue of misconvergence measurement. These are VESA,[19] SAE ARP1782,[20] and NIDL (Draft).[21] The VESA standard describes the use of the Klein convergence gauge as described in the section on instrumentation, although it states that electronic misconvergence measurement devices may be used using the same technique. The standard ARP1782 is more comprehensive in its coverage of misconvergence, even describing shifting of the apparent line centers in the horizontal direction by differences in the three phosphor decay curves. A microphotometer is used to measure the three beam profiles, as previously shown in Figure 8.16, using individual phosphor dots or stripes as previously described. Measurements are made in at least nine screen locations in both the vertical and horizontal axes and the misconvergence computed using the midpoints between the half-amplitude widths of each beam profile.

8.10 MOIRE PATTERNS

Moire patterns, akin to *aliasing* in the digital world, are a result of repetitive displayed patterns "beating" against the repetitive patterns of the shadow mask for color CRTs or the pixel pattern of flat-panel displays. The maximum effect occurs when both patterns are of nearly the same spacing. Even with high-resolution displays, the Moire effect can cause a much coarser structure than that of either the pixel pattern or the displayed image to be visible. Moire is often observed in television images of fabrics having a relatively narrow spacing of lines or checks.[22–26] Choice of a display having a different pixel pitch is a possible solution to Moire problems.

Measurement of Moire effects is not often required but consideration must be given to the appearance if fine detail in repetitive patterns is to be displayed. The NIDL apparently has the only measurement method for Moire.[27] They suggest use of a CCD array detector and specify reporting methods.

8.11 JAGGIES

"Jaggies," also referred to as *aliasing*, are an artifact observed in both flat-panel and CRT displays made up of discrete elements.[28] Diagonal and curved lines show a steplike appearance when closely examined. An example is shown in Figure 8.17. Lines displayed exactly vertically and horizontally will not have jaggies. Jaggies are a function of the pixel pitch of the display, and the best way to minimize them is to use the highest resolution display possible since the step size is equal to the pixel and scanning line spacings. Slightly defocusing a CRT beam so that it is larger than the pixel size can also reduce the effect, although at the expense of resolution. This option is not available to the flat-panel user.

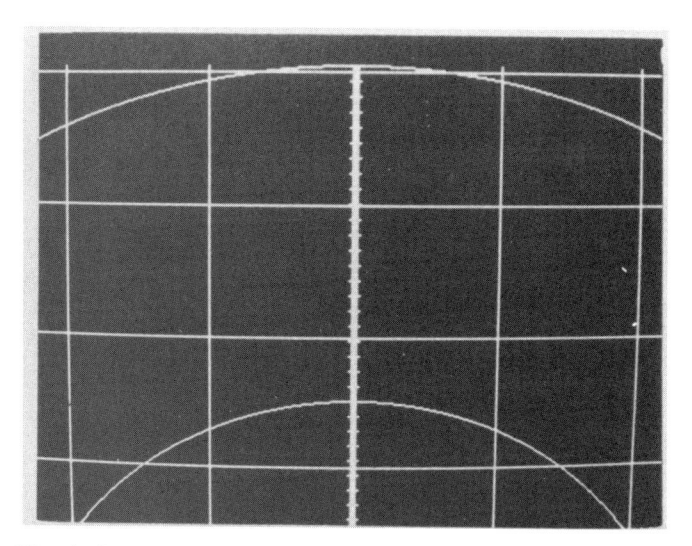

Figure 8.17. "Jaggies" are noticeable on the curved lines displayed due to the discrete steps created by the horizontal scanning lines.

8.12 JITTER

Jitter is the slight variation in position with time of a displayed stationary image element. Jitter may be best observed with a magnifier. The movement may be in either the vertical or horizontal direction or a vector combination of the two. It may be either cyclic or random and may be caused by electrical or magnetic disturbances. Jitter is most likely to be observed in CRT displays.

8.13 MAGNETIC ABERRATIONS

Cathode-ray tubes are prone to disturbance of electron trajectories under the influence of magnetic fields resulting in their displacement on the screen. The magnetic fields may be caused by transformers or inductors within the display itself, nearby equipment or magnets, or even the earth's magnetic field (Figure 8.18).[29,30] The fields may be either AC or DC. Direct current is usually less objectionable since it will primarily cause only a small shift of the displayed pattern screen location, although it can cause a color purity loss in color displays. The AC fields are more troublesome since they can cause *line pairing, waviness,* "*roping,*" "*swim,*" and "*waterfall*" patterns. These names are indicative of their actual appearance.

Line pairing is mainly noticeable in interlaced displays and is caused by one field being shifted in a slightly vertical position relative to the second field, usually by a power line magnetic field. If the vertical frame rate is not synchronized to the power line frequently, the pairing may drift slowly vertically, affecting large areas of the display. Waviness and roping similarly may be the effect of higher frequency magnetic fields. Swim is the slow cyclic motion of characters displayed on the screen and is caused by magnetic fields close to but not the same as the vertical scan rate. Magnetic shielding is usually

238 GEOMETRY MEASUREMENTS

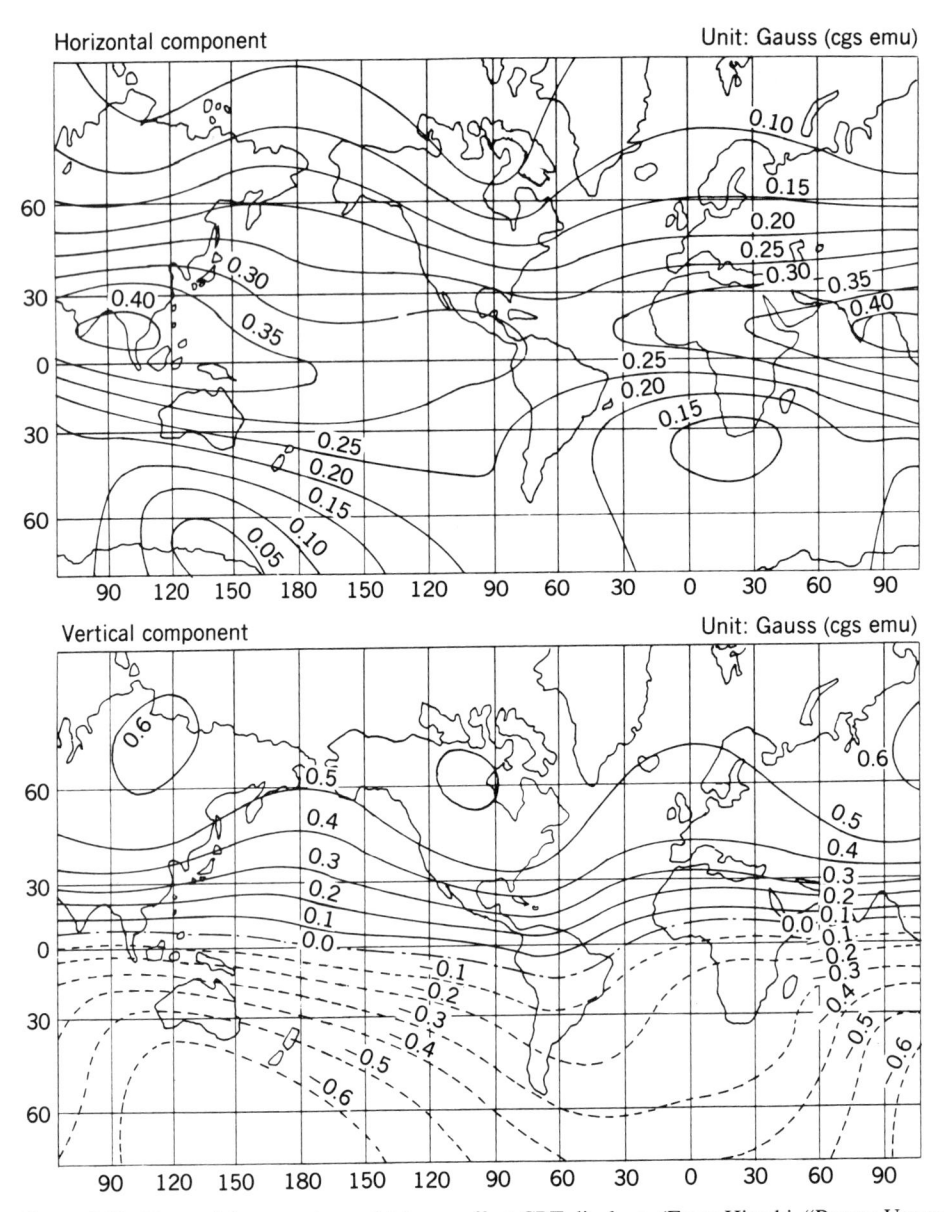

Figure 8.18. Terrestrial magnetism, which can affect CRT displays. (From Hitachi, "Proper Usage of Hitachi CRTs for Display Equipment and TV Receivers," Application Note CE-E590, July 1984, p.11.)

employed in higher quality displays to minimize these effects. The DC effects due to magnetization of the color CRT shadow mask are avoided by use of a degaussing coil located around the CRT screen and activated manually or each time the display is powered up. Measurement of magnetic artifacts and determination of their cause are sometimes difficult and must be handled on a case-by-case basis.

REFERENCES

1. R. Fawcett, "Stereo Imaging and Automated Vision Boost CRT (sic) Production," *Photonics Spectra,* pp. 92–94, Mar. 1995.
2. Video Electronics Standards Association (VESA), "Display Specifications and Test Procedures," Version 1.0, Rev. 1.0, Oct. 3, 1994, p. 9.
3. American National Standards Institute (ANSI)/Institute of Electrical and Electronics Engineers (IEEE), "American National Standard for Metric Practice," ANSI/IEEE Std. 268-1992, Oct. 1992.
4. American National Standards Institute (ANSI)/Electronic Industries Association (EIA), "Screen Definition for Color Picture Tubes," ANSI/EIA-527-1986, June 1986.
5. National Information Display Laboratory (NIDL), "Display Monitor Measurement Methods, Part 1: Monochrome CRT Monitor Performance," NIDL Publication No. 171795-036, Draft Version 2.0, July 12, 1995, p. 78.
6. National Information Display Laboratory (NIDL), "Display Monitor Measurement Methods, Part 2: Color CRT Monitor Performance," NIDL Publication No. 171795-037, Draft Version 2.0, July 12, 1995, p. 37.
7. National Information Display Laboratory (NIDL), "Display Monitor Measurement Methods, Part 1: Monochrome CRT Monitor Performance," NIDL Publication No. 171795-036, Draft Version 2.0, July 12, 1995, p. 73.
8. National Information Display Laboratory (NIDL), "Display Monitor Measurement Methods, Part 2: Color CRT Monitor Performance," NIDL Publication No. 171795-037, Draft Version 2.0, July 12, 1995 p. 37.
9. Video Electronics Standards Association (VESA), "Display Specifications and Test Procedures," Version 1.0, Rev. 1.0, Oct. 3, 1994, pp. 27–28.
10. International Organization for Standardization (ISO), "Ergonomics Requirements for Office Work with Video Display Terminals (VDTs)—Part 3: Visual Display Requirements," ISO 9241-3, 1992, p. 6.
11. Society of Automotive Engineers (SAE), "Photometric and Colorimetric Measurement Procedures for Airborne Direct View CRT Displays," Aerospace Recommended Practice ARP1782, Jan. 9, 1989, pp. 32–36.
12. Video Electronics Standards Association (VESA), "Display Specifications and Test Procedures," Version 1.0, Rev. 1.0, Oct. 3, 1994, p. 29.
13. International Organization for Standardization (ISO), "Ergonomics Requirements for Office Work with Video Display Terminals (VDTs)—Part 3: Visual Display Requirements," ISO 9241-3, 1992, p. 6.
14. Video Electronics Standards Association (VESA), "Display Specifications and Test Procedures," Version 1.0, Rev. 1.0, Oct. 3, 1994, p. 29.
15. International Organization for Standardization (ISO), "Ergonomics Requirements for Office Work with Video Display Terminals (VDTs)—Part 3: Visual Display Requirements," ISO 9241-3, 1992, p. 6.
16. Society of Automotive Engineers (SAE), "Photometric and Colorimetric Measurement Procedures for Airborne Direct View CRT Displays," Aerospace Recommended Practice ARP1782, Jan. 9, 1989, pp. 32–36.
17. Video Electronics Standards Association (VESA), "Display Specifications and Test Procedures," Version 1.0, Rev. 1.0, Oct. 3, 1994, pp. 29–31.
18. T. Van Maldegiam, "A New Convergence Test Pattern," *Information Display,* p. 20, Jan. 1993.
19. Video Electronics Standards Association (VESA), "Display Specifications and Test Procedures," Version 1.0, Rev. 1.0, Oct. 3, 1994, pp. 24–25.
20. Society of Automotive Engineers (SAE), "Photometric and Colorimetric Measurement Proce-

dures for Airborne Direct View CRT Displays," Aerospace Recommended Practice ARP1782, Jan. 9, 1989, pp. 25–29.
21. National Information Display Laboratory (NIDL), "Display Monitor Measurement Methods, Part 2: Color CRT Monitor Performance," NIDL Publication No. 171795-037, Draft Version 2.0, July 12, 1995.
22. K. B. Benson, *Television Engineering Handbook,* McGraw-Hill, New York, 1992, pp. 12.7–12.8.
23. J. C. Whitaker, *Electronic Displays,* McGraw-Hill, New York, 1994, pp. 272–273.
24. K. R. Oakey, "Some Consequences of the Shadow-Mask CRT DOT Structure," *Displays,* pp. 143–148, July 1984.
25. J. P. Wittke, "Moire Considerations in Shadow-Mask Picture Tubes," *Proc. SID,* Vol. 28, No. 4, pp. 415–418, 1987.
26. J. P. Wittke, "Moire Considerations in Shadow-Mask Picture Tubes," *SID Dig.,* Vol. 18, pp. 347–350, 1987.
27. National Information Display Laboratory (NIDL), "Display Monitor Measurement Methods, Part 2: Color CRT Monitor Performance," NIDL Publication No. 171795-037, Draft Version 2.0, July 12, 1995.
28. A. Miller, "Suppression of Aliasing Artifacts in Digitally Addressed Shadow-Mask CRTs," *J. SID,* Vol. 3, No. 3, pp. 105–108, 1995.
29. Hitachi, "Proper Usage of Hitachi CRTs for Display Equipment and TV Receivers," Application Note CE-E590, July 1984, p. 11.
30. A. D. Johnson, "Effects of the Earth's Magnetic Field on CRT Display Geometry," *Proc. SID,* Vol. 30, No. 3, pp. 225–228, 1989.

CHAPTER 9

Time-Related Measurements

9.1 GENERAL BACKGROUND

Occasionally, *temporal,* or time-related, measurements of devices or materials may be required for system design, vendor qualification, manufacturing quality control, or incoming inspection. This may be for displays themselves, peripheral equipment, or even equipment used for processing or testing display devices. Examples include the switching times of color shutters or LCDs, risetime and decay of CRT or EL phosphors, flicker of displays or lighting, infrared and visible light emitting diode response times, xenon flash lamp characteristics, and electro-optical modulator switching characteristics.

Three basic approaches are used depending on the light levels, expected response times, and wavelengths to be measured. The first uses the silicon photodiode in the photoamperic (often called photovoltaic) mode. The second also employs the silicon photodiode but in the photoconductive (biased) mode. The same silicon photodiode may usually be used in either mode, but they are usually optimized for operation in one or the other. The silicon photodiode has the benefits of simplicity of operation, ready availability at low cost, small size, ruggedness, compatibility with voltage and signal levels of today's electronic instrumentation, and good red and infrared sensitivity. It is eminently suited to brighter light sources such as xenon strobe lamps but may be employed for somewhat lower intensities if properly used. The third approach is the photomultiplier tube (PMT). The photomultiplier's principal strength is extreme sensitivity for very low light level measurements, but it must be properly used to obtain accurate measurements and avoid damage to it. Both the silicon photodiode and the PMT are commonly capable of rise- and falltimes in the nanosecond range, and specialized versions are available with picosecond response times. For almost all response time measurements, an oscilloscope is used to display the amplitude-versus-time characteristics of the device or material under test. Figure 9.1 shows the digital storage oscilloscope display of the light intensity versus time for a single pulse of a photographic xenon flash lamp as measured using a photovoltaic silicon diode.

9.2 WITH PHOTOAMPERIC SILICON PHOTODIODES

We begin with a straightforward and easily duplicated temporal measurement, the modulation of the light output by the power line frequency commonly observed with everyday fluorescent lighting. The light output of a fluorescent lamp varies with time since it only emits light during the peaks of the AC applied power line voltage (ignoring the slight

242 TIME-RELATED MEASUREMENTS

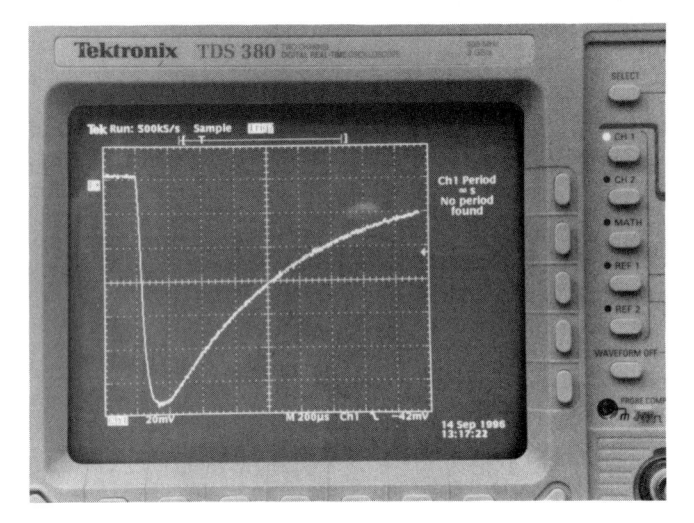

Figure 9.1. Light intensity versus time for a photographic xenon flash lamp as measured with a photovoltaic silicon photodiode. Note that, because of the photodiode internal configuration, the signal is negative in polarity. The zero light level is 3 divisions above center and the peak intensity is about 3.7 divisions below center. Photomultiplier tubes also have negative-going outputs.

phosphorescence or afterglow of the lamp phosphors). There are two peaks for each complete cycle of the AC voltage, one positive and one negative, both of which cause the emission of light from the lamp. Therefore, a fluorescent lamp operated at 60 Hz will exhibit modulation of its light output at 120 Hz, and 50-Hz operation will result in 100 Hz modulation.

To measure the light-output-versus-time characteristics of the relatively slow fluorescent lamp modulation, the configuration shown in Figure 9.2 is used. Virtually any oscil-

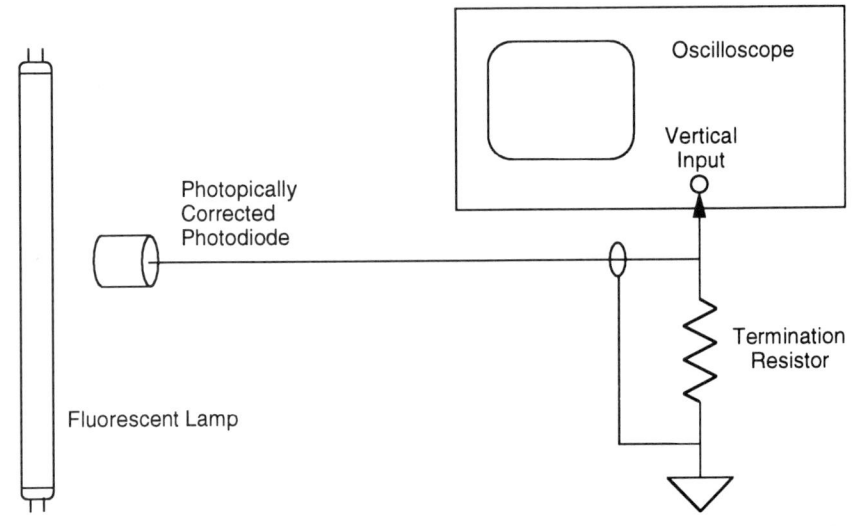

Figure 9.2. Configuration for measuring modulation of light output from AC line operated fluorescent lamps.

loscope may be used in this instance. Trigger the oscilloscope's horizontal time base from the "line" source so that it will always be synchronized with the power line modulation of the lamp. Hold the silicon photodiode near the lamp while viewing the waveform. The modulation of light output will be obvious. As shown in the diagram, the photodiode output is fed directly to the oscilloscope vertical input. Use the DC coupling position of the oscilloscope input switch to provide a ground return for the DC current produced by the photodiode. The 1MΩ input resistance of the oscilloscope provides the DC return path. This is the photovoltaic mode of operation: The photodiode generates a voltage across the 1MΩ input displayed. Under these conditions, the output is semilogarithmic, which compresses the signal and reduces the apparent amplitude (in percentage of peak amplitude) of the modulation, although the signal amplitude is large and easy to work with. Because of this, quantitative results are difficult to achieve without resorting to individual calibrations for the range of amplitudes encountered. The time constant of this arrangement is also relatively long, which limits its use to fairly slow changing phenomena. For the circuit shown, the time constant in seconds is determined using the oscilloscope vertical input resistance and capacitance, the coaxial cable capacitance, and photodiode junction capacitance according to the following equation:

$$t_c = R_{in} \times (C_{in} + (C_c \times l) + C_j) \times 10^{-12}$$

where

R_{in} = oscilloscope DC input resistance, Ω
C_{in} = oscilloscope input capacitance, pF
C_c = coaxial cable capacitance, pF/m or pF/ft (ignore the connector capacitance as it is a minor factor)
l = cable length, m or ft, depending on unit used for C_c
C_j = photojunction capacitance, pF

Typical values would be 1 MΩ for R_{in}, 20 pF for C_{in}, 90–100 pF/m for commonly used RG-58A/U coaxial cable, and 100–200 pF/mm^2 of active area for the photovoltaic photodiode junction capacitance. A convenient and widely used photodiode, the UDT PIN-10DP (Figure 9.3), has a large active area (100 mm^2) for high sensitivity, an integral BNC connector, and rugged housing for durability for general laboratory use. It is specified as

Figure 9.3. Conveniently packaged silicon photodiode with an integral BNC connector.

having about 10,000 pF junction capacitance. From this, it is apparent that the photodiode capacitance is the major factor in the equation. Using the UDT photodiode and a 1-m cable length, the typical time constant might be about 10 ms, not quite enough to display a 120-Hz waveform faithfully. The risetime of the circuit may be computed from the time constant using the equation

$$t_r = 2.2 t_c$$

The PIN-10DP is also available with a photopic correction filter for applications where measurements must be made as seen by the human eye. More about the need for photopic correction will be covered later.

The output of the photodiode in the preceding example may be linearized by changing from photovoltaic to photoamperic operation of the photodiode. This is accomplished by decreasing the load resistance for the photodiode with a resistor across the oscilloscope vertical-input connector. Silicon photodiodes are typically linear over six or more decades when operated into a load resistance of 0 Ω or close to it. Obviously, if we use a 0-Ω load resistor, we will not see any voltage across it. A useful compromise is to use 50 Ω, which is often built into many newer oscilloscopes and is switch selectable. An alternative is to use a 50-Ω BNC termination resistor at the oscilloscope vertical-input connector. Using 50 Ω instead of 1 MΩ will decrease the time constant and risetime by a factor of 20,000 and the linearity will be excellent over a wide range. Of course, the sensitivity will be reduced by a factor of 20,000 also. Depending on light levels and required response times a useful compromise is to use an intermediate resistance value. One thousand ohms is often practical without undue sacrifice of linearity if the signal levels across it are no more than 100 mV or so. Linearity will become significantly poorer as the signal levels approach the beginning of the forward conduction knee for a silicon diode (0.5 V or so).

Photopic correction of the photodiode is usually necessary for time-resolved measurements to avoid excessive sensitivity to the long-wavelength (red) end of the spectrum. A stack of three or four colored glasses ground and polished to specific thicknesses are required to produce a photopic filter for silicon photodiodes. Because of the difficulty of correcting individual photodiodes, the best solution is to use a purchased photopic detector such as one manufactured by one of the companies listed in Appendix K. Of course, the sensitivity of the detector is reduced considerably by the addition of the photopic filter. Usually the better the accuracy of the photopic correction, the greater will be the sensitivity loss.

9.3 WITH PHOTOCONDUCTIVE SILICON PHOTODIODES

The photodiode junction capacitance is the predominant limiting parameter in determining the maximum response time that may be achieved with a given photodiode. The effective capacitance may be reduced considerably by applying a reverse-bias voltage in series with the load resistance across the photodiode, as shown in the circuit in Figure 9.4. Figure 9.5 demonstrates the capacitance reduction versus bias voltage for a typical silicon photodiode operated photoconductively. The effective capacitance reduction is due to the improved collection of the photoelectrons within the depletion region of the silicon wafer. These photoelectrons are generated by exposure of the silicon to light. The response time improves as the applied bias voltage is increased, up to the maximum rated forward

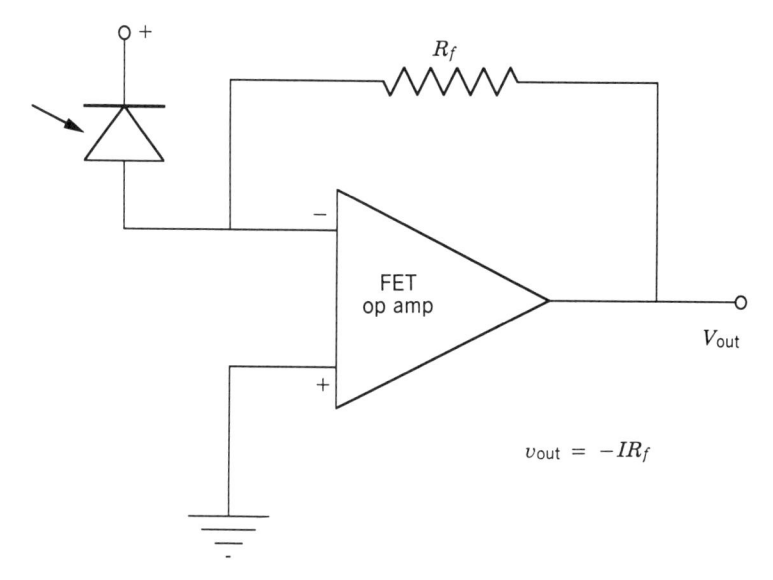

Figure 9.4. Circuit for photoconductive operation of a photodiode for faster response time.

$$v_{out} = -IR_f$$

breakdown voltage of the device. Reduction of response time by 10–100 times is typical, with resultant risetimes in the vicinity of 1–10 ns being typical. The risetime and falltime are not always equal, especially at lower voltages or when operating into an integrating operational amplifier. When operated in the fully depleted condition, which is near maximum bias voltage, the risetime and falltime should be nearly equal.[1-3]

Photodiode Junction Capacitance vs Bias Voltage

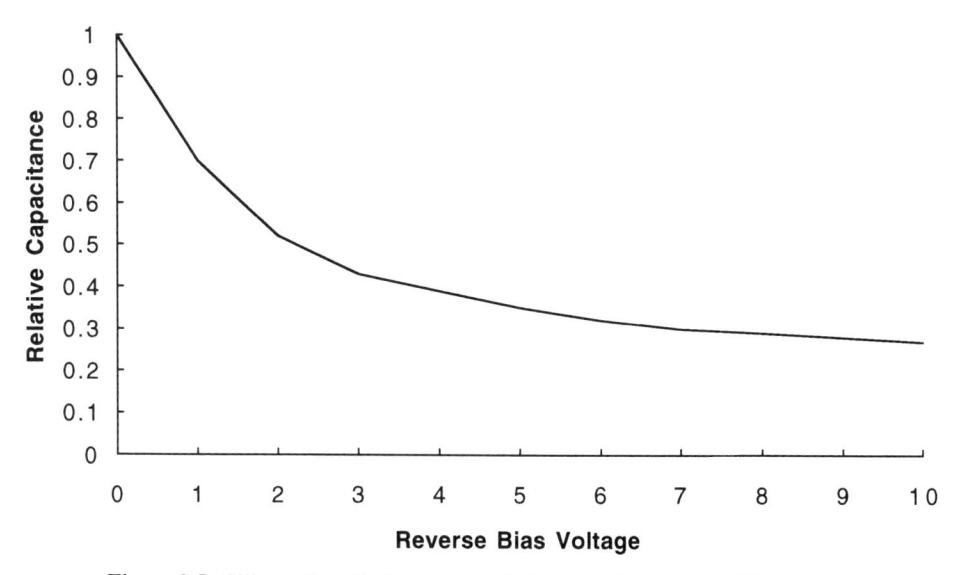

Figure 9.5. Silicon photodiode junction relative capacitance versus bias voltage.

The sensitivity in the photoconductive mode will also be improved by about two times for a given load resistance. At the same time, the spectral response is extended further into the near-infrared region. The latter is not usually an advantage for display measurements since displays produce no useful light output in the infrared and nearby incandescent lamps, which produce copious amounts of infrared radiation, may actually interfere with the display measurement.

"Shot noise," which is not normally a factor in the photoamperic mode, is present in biased operation in the form of *dark current*. Dark current is a residual current even when the photodiode is not exposed to light. The DC component can be canceled out in amplifier circuits by summing an identical current of opposite polarity at the amplifier input or applying an equal current of the same polarity to the inverting input if an operational amplifier is used. Unfortunately, compensation at one temperature will not be the same for another temperature. The dark current, which is a result of bulk resistance of the silicon, approximately doubles for every 10°C increase in temperature.

One other effect of photoconductive operation is the difference in response time with wavelength. Photoconductive operation will result in faster response times at shorter wavelengths.

In most other respects, photoconductive operation is very similar to photoamperic operation. Selection of the proper load resistance for the application as previously described will determine the actual risetime and sensitivity achieved.

The PIN photodiode is often used for its fast response time in the photoconductive mode. In these devices, a deeper undoped depletion or *intrinsic* region is formed between the p- and n-doped regions (Figure 9.6) to reduce the capacitance and allow higher bias voltage without breakdown. Both factors reduce the response time significantly.

One other form of biased operation is worthy of mention, the *avalanche photodiode* (APD), which uses high bias voltage to obtain electron multiplication of the photoelectrons and thus produce larger currents by operating in the breakdown region. The high accelerating voltage causes additional electrons to be generated by impact ionization or *avalanching* within the silicon. Gains of 1000 times are possible. Precise bias voltage and temperature control are required for stability and to avoid damage since the breakdown voltage point is highly dependent on temperature. The APD is not commonly used for general display photometry but is useful where low-intensity fast light pulses must be measured.

9.4 WITH PHOTOMULTIPLIER TUBES

The PMT, while more difficult to apply and more expensive, is the best sensor for fast response time at low light levels, especially in the blue region of the spectrum, where the silicon photodiode has relatively poor responsivity. Risetimes of 1–10 ns are typical for commonly used tubes, and the ability to detect light levels of just a few photons are important advantages of the PMT. The drawbacks include the need for a high-voltage power supply, magnetic and electrostatic shielding requirements, the necessity for avoidance of exposure to excessive light levels, greater fragility, and less stable operation than is possible with silicon photodiodes. These constraints will be discussed individually. Ralph W. Engstrom's classic *Photomultiplier Handbook* is an excellent text on the subject and should be consulted for the theoretical and practical considerations of PMT usage.[4] Originally published by RCA, the handbook has been reprinted[5] by Burle Industries, Inc., which took over manufacturing of the RCA line of PMT products in Lancaster, Pennsylvania.

Planar Diffused Silicon Photodiode

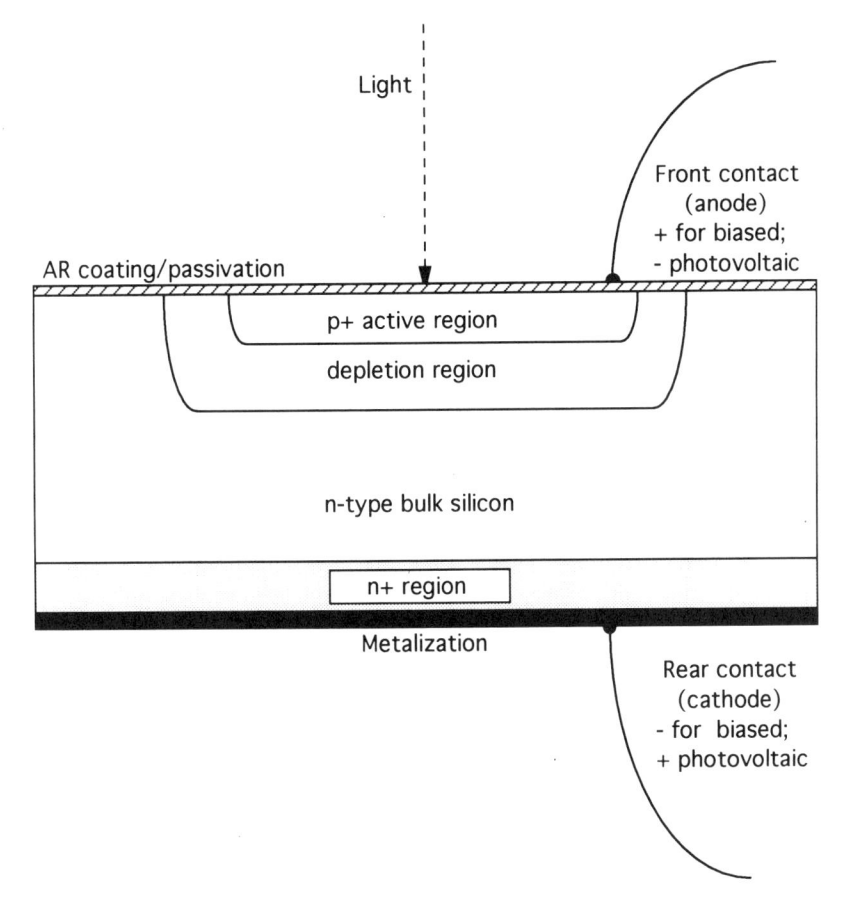

Figure 9.6. Silicon PIN photodiode cross section.

The power supply is usually a negative polarity supply capable of producing 0–1500 V at 1 mA or so for the photocathode. Negative is chosen to provide safe DC coupling from the PMT anode, which is highly positive with respect to the photocathode. The MHV connectors and RG-59/U cable are widely used for the high voltage connection. The MHV connector looks like a slightly longer version of the everyday BNC connector but will not mate easily with the BNC series (Figure 9.7). Do not attempt to use a BNC connector with an MHV connector. Damage to the connector, equipment, or, more importantly, the user (YOU) may result. Good connection of the cable shield to the connector shells at both ends is also very important for safety reasons.

The PMT should be housed in a light-tight housing (Figure 9.8) because of its extreme gain, which may be as much as 1 to 10 million times when operated at its highest rated voltage. Commercially produced housings are available that include the voltage divider chain for supplying all dynode voltages from the single negative high-voltage power supply, an appropriate tube socket, magnetic and electrostatic shielding, and convenient mounting for accessory filters. One additional precaution: Care must be taken to avoid

248 TIME-RELATED MEASUREMENTS

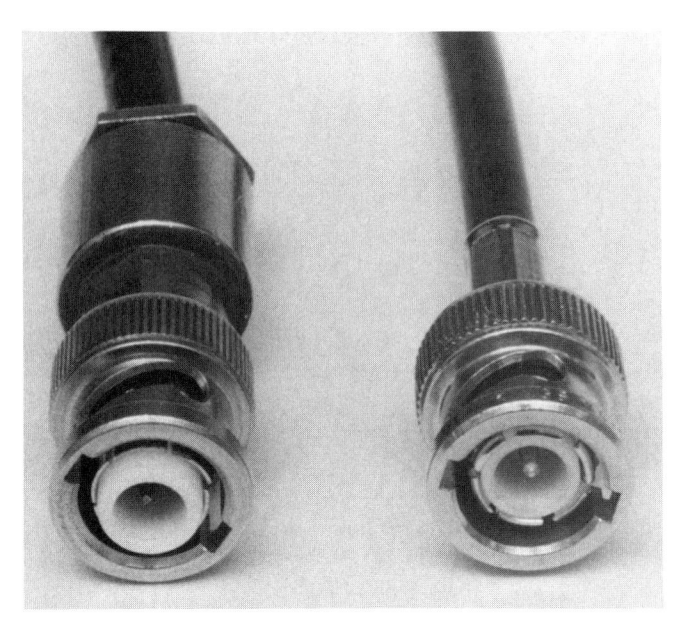

Figure 9.7. Male MHV and BNC connectors showing the extended insulation and additional body length of the MHV type.

nearby grounded objects and even insulated ones from contact or close proximity to the photocathode region of the PMT. The small leakage currents will give rise to the spurious emission of light pulses. These pulses will be detected by the photocathode and will increase the PMT's apparent dark current.

Care must be exercised to avoid exposure of the PMT to ambient light with voltage applied to the tube. Excessive currents will flow, resulting in permanent loss of sensitivity and/or increased dark current. Even though maximum anode current specifications are on the order of 0.1–1 mA, always try to operate them at less than about 1 μA for best life and stability while providing some safety margin for the occasional brief error that seems to occur with even experienced users. When increasing the high voltage for higher sensitivity, remember that the output current increases logarithmically with voltage. Always

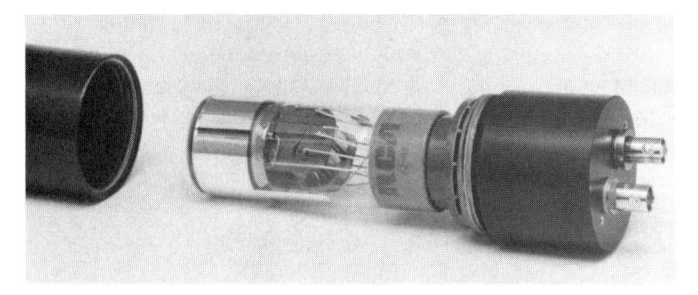

Figure 9.8. Photomultiplier tube housing partially disassembled. The outer light-tight housing, which is lined with a magnetic shield, may be seen at the left. The PMT is installed in its socket assembly, which contains a voltage divider chain, MHV connector for negative high voltage (top), and BNC connector for the output signal (bottom).

start around −300 V and increase it gradually while monitoring the PMT output current until the desired level is reached. Even without applied voltage, exposure of the PMT to fluorescent or other lighting may cause increased dark current for several hours, although permanent damage should not be experienced. This effect is most pronounced for PMTs with the spectrally wide S-20 multialkali photocathodes.

Avoid mechanical shock or mounting stress of the PMT. Vacuum tubes have none of the ruggedness we are accustomed to with today's solid-state electronics!

The stability of the PMT is also somewhat less than that of the silicon photodiode. Short- and long-term gain changes with time or temperature of up to several percent are common. Spectral sensitivity changes of a similar nature may also be encountered. Frequent recalibration is the solution to such variations. Fortunately, absolute sensitivity accuracy is not often a problem for temporal measurements since percentage of light intensity versus time is usually the important characteristic to be measured.

The choice of the PMT itself is an especially important consideration. Two styles are available, the side-window and end-window types (see Chapter 2). The side-window type is exemplified by the old standby type 931-A originally developed in the late 1930s by RCA. It is still widely used today for low-light-level measurements under its original designation and in a number of variants. It has excellent sensitivity and risetime of as little as 1.6 ns and is comparatively compact and rugged as PMTs go. Some variation in sensitivity and dark current will be encountered from tube to tube, and selection or use of a premium version may be required for critical low-light-level applications. The photocathode is about 8×24 mm in size. Half-sized versions of the 931-A have been produced by Hamamatsu in Japan for applications requiring miniaturization.

The end-window PMT was originally developed for nuclear scintillation counting applications where it is coupled to large-area scintillation crystals. Subsequently, it has been utilized in many other applications in versions having photocathode sizes of from about 8 to 460 mm in diameter. The circular photocathode is often easier to incorporate into optical system designs than the side-window types.

The spectral response is determined by the chemical composition of the photocathode. Most have maximum sensitivity in the blue region of the spectrum, just the opposite of solid-state sensors such as silicon. Cesium–antimony (CsSb) is most commonly used and is designated as an S-4 photocathode on a metal substrate as used in side-window PMTs. It is also designated as S-11 when deposited on a glass substrate for end-window PMTs. The S designations are EIA standardized nomenclature, much the same as the older P numbers used for CRT phosphors.[6,7] The S-20 trialkali ($Na_2KSb:Cs$) photocathode has increased red sensitivity compared to the S-4 and S-11 types (Figure 9.9) and is most often used as a sensor for high-sensitivity, wide-range spectroradiometers.

The PMT should be photopically corrected for all but spectrally narrow-band measurements. Photopic correction may be approximated with a Wratten No. 106 filter in front of but not touching the photocathode for S-4 and S-11 surfaces. Without photopic correction, the PMT will respond more to the short-wavelength (blue) components of broadband emissions. The accuracy of photopic correction is somewhat less than can be obtained from corrected silicon photodiodes.

A block diagram showing a typical configuration for common intensity-versus-time measurements is shown in Figure 9.10. The display area to be measured may be optically imaged on the PMT photocathode or a diffuser directly ahead of it or it may be coupled with a fiber-optic cable. The latter allows coupling to a photometric microscope for measuring very small areas such as single pixels, an application at which the PMT excels.

250 TIME-RELATED MEASUREMENTS

Photocathodes

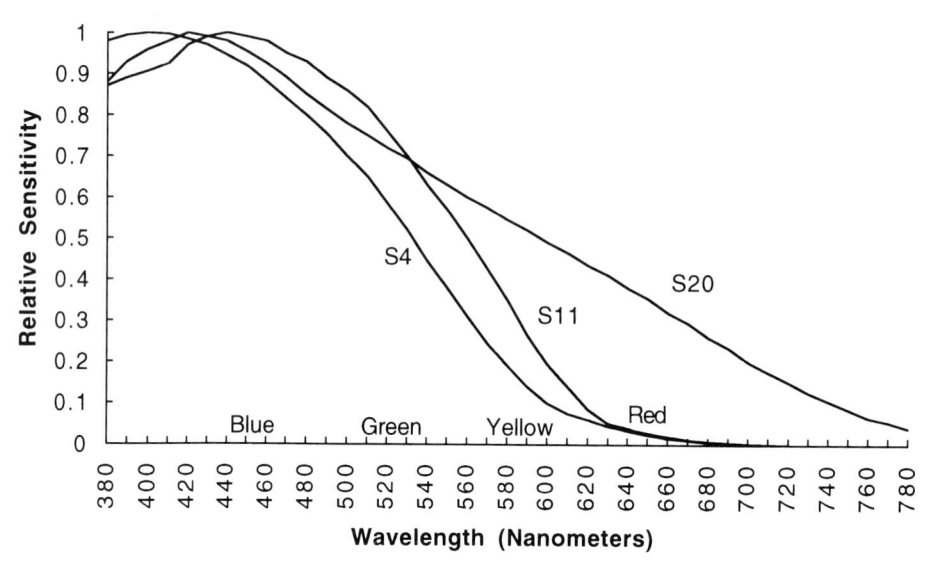

Figure 9.9. Common PMT spectral response curves.

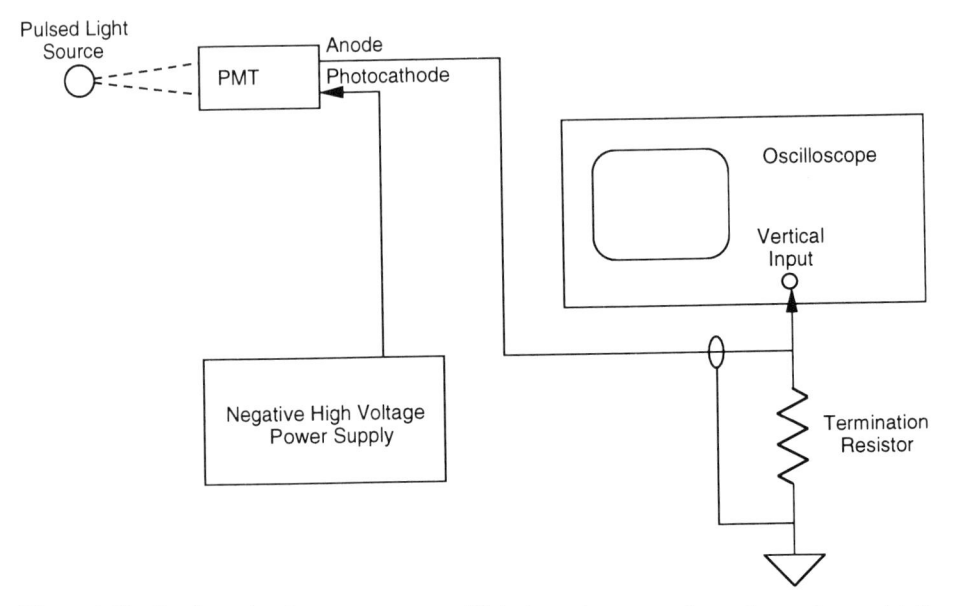

Figure 9.10. Configuration for measurement of light intensity versus time using a photomultiplier tube.

9.5 PHOSPHOR PERSISTENCE MEASUREMENT

Several methods of phosphor persistence or *decay* measurement exist. Unfortunately, different techniques often yield widely different results. In the past, this has led to many disagreements and to comparisons of different phosphors measured with different methods. This may cause the wrong phosphor to be selected for a particular application. As a result, the EIA JT-31 Committee on Optical Characteristics of Display Devices has developed standardized persistence and risetime measurement procedures[8] for use in registering phosphors under the Worldwide Type Designation System (WTDS). Complete data, including persistence curves, for all registered phosphors is available in EIA TEPAC Publication TEP116-C, "Optical Characteristics of Cathode-Ray Tube Screens."[9]

The standardized measurement methods help to reduce the errors caused by excitation of the phosphor under different drive conditions. The widely used zinc sulfide phosphors are especially prone to difference in measured decay values under different drive levels and excitation durations and require considerable measurement care in order to obtain valid results. On the other hand, zinc orthosilicate, one of the earliest CRT phosphors under the old EIA designation of P-1, is one of the most stable materials. Measurements of zinc orthosilicate with almost any technique will produce similar results unless there are nonlinearities present in the detector or associated instrumentation.

Phosphor persistence is usually specified at 10% of initial light output. At this level, phosphors are available having persistence of nanoseconds to milliseconds. Under subdued ambient lighting conditions, persistence is visible and useful out to the 1 or 0.1% level and may exist for up to 1 min or more for radar and related applications. The EIA TEPAC Publication 105-14 classifies phosphor decay and the appropriate measurement methods for each as shown in Table 9.1.

With all three methods, it is customary to graphically plot the decay level versus time. Two fundamental forms of decay are encountered. The first is exponential decay, such as the silicate and rare-earth phosphors exhibit, and is plotted on a semi-log graph in Figure 9.11. The other is the power law decay characteristic of the sulfide phosphors, as shown in Figure 9.12 plotted on a log-log scale graph. Some phosphors may exhibit both characteristics or may show different slopes at the high- and low-light-level portions of the curves. The following test methods are based on TEP105-14.

TABLE 9.1 Phosphor Decay Classifications

10% Decay Time	Description	Method
1 s or more	Very long	Pulsed raster
100 ms–1 s	Long	Pulsed raster
1–100 ms	Medium	Pulsed line or pulsed raster
10 μs–1 ms	Medium short	Pulsed line or pulsed spot
1–10 μs[a]	Short	Pulsed spot
Less than 1 μs	Very short	Pulsed spot

[a]Erroneously listed as 10 s in TEP105-14.

252 TIME-RELATED MEASUREMENTS

Exponential Decay

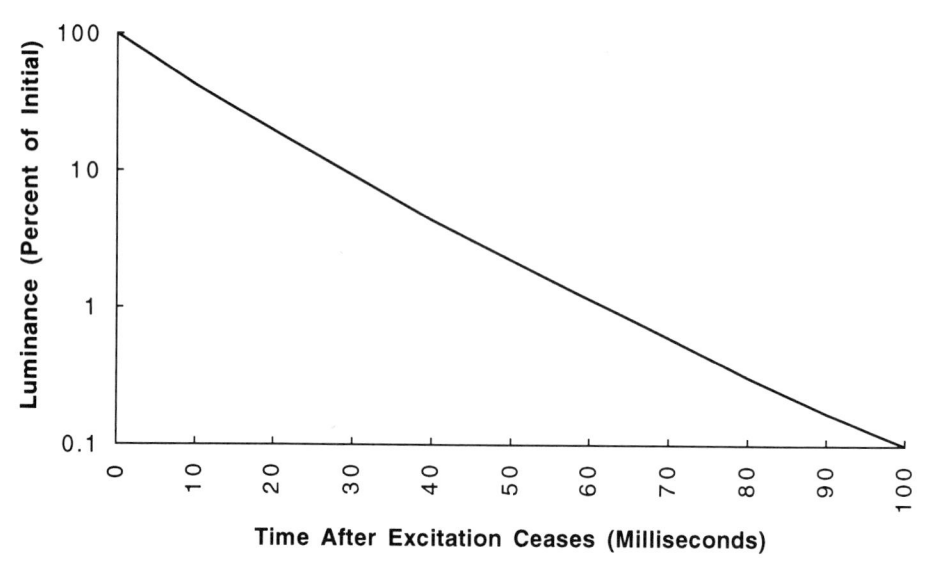

Figure 9.11. Exponential phosphor decay curve shape typical of silicate and rare-earth phosphors.

Power Law Decay

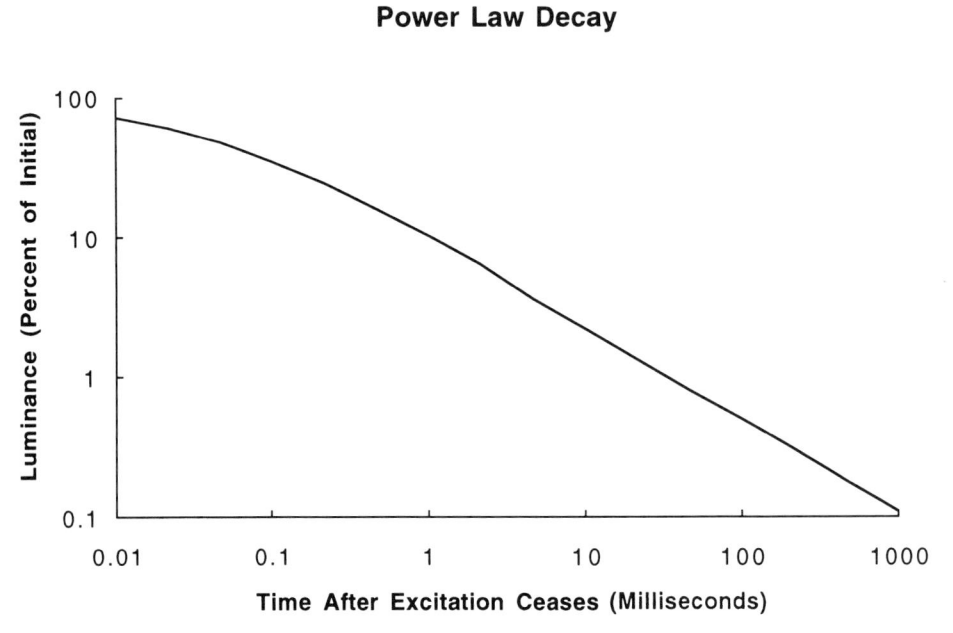

Figure 9.12. Power law phosphor decay curve shape typical of sulfide phosphors.

Pulsed Spot Method

The pulsed spot method uses a raster initially to determine the CRT DC grid 1 (G-1) drive voltage required to produce a prescribed beam current (typically 100 μA). Beam current is usually measured with a microammeter connected in series with the high-voltage anode connection of the CRT. The microammeter is mounted within a well-insulated plexiglas box for safety. It is easiest to measure the beam current in CRTs having bipotential electron guns since there are no other electrodes connected to the anode. Einzel guns have other electrodes connected to the anode. This may cause beam current measurement errors if beam-limiting aperatures intercept appreciable current and prevent it from reaching the screen. The G-1 DC drive voltage that produces the desired beam current is replaced with a 0.5-μs long pulse at a 60-Hz repetition rate that drives the grid to the same voltage level during the pulse and into beam cutoff during the remainder of the time. The raster is then disabled, thus producing a short-duration pulsed spot at the screen center. Setup in the raster mode prevents phosphor damage in the event of misapplication of excessive drive voltage.

The photodetector is placed in contact with the screen center to maximize light pickup from the spot. The output pulse is normally displayed on an oscilloscope, and the time to the 10% or other desired level is measured from the oscilloscope screen. Alternatively, a digital storage oscilloscope (DSO) or digitizer and personal computer may be used to process the signal and manipulate the data obtained. Note that the output signal from the PMT and many photodiodes will be negative in polarity, that is, more light will result in a more negative signal, which can be confusing at first. The PMT anode load resistor is chosen to provide a circuit time constant at least 10 times shorter than the phosphor decay and risetimes to be measured. This avoids distortion of the pulse characteristics being measured. The oscilloscope chosen should have good overload recovery from high-amplitude signals if the 1 and 0.1% decay times are to be measured in addition to the 10% decay.

Pulsed Line Method

The pulsed line method is used where more light to the photodetector is required. Due to the greater length of time that the photodetector receives light from the screen as the spot is scanned past it, the method is not suitable for phosphor decay or risetimes of less than 100 μs to the 10% level. The pulsed line technique is similar to the pulsed-spot method except that the raster is collapsed vertically to a line that is then reduced in length to 50 mm length using the horizontal-gain adjustment. The grid pulse used is 32 μs wide and is triggered on the horizontal-sweep sync signal. The photodetector output should be checked for saturation to avoid linearity errors, which can greatly alter the apparent shape of the measured decay curve. Photomultipliers are more likely to show saturation than silicon photodiodes because of their high gain. The easiest and probably the most accurate way to test for detector saturation is to interpose a calibrated thin-film neutral-density filter between the screen and the photodetector. Using a 50-mm-square filter with a density of 1.0, the observed pulse height on the oscilloscope should be reduced in amplitude to 10% of its previous height. It is a sure sign of saturation if it is reduced in amplitude to a lesser degree.

Pulsed Raster Method

The pulsed raster technique provides still more light to the photodetector than either of the two previously described methods. Here, the raster is left full size and operated in either the single-frame or periodic-refresh mode. If the refresh mode is used, the raster is operated long enough for the phosphor light output to stabilize. Several frames are sufficient for all but the longest decay phosphors. The raster is pulsed off during the vertical retrace interval and the decay curve displayed on the oscilloscope as described in the pulsed spot and pulsed line methods. Checking for photodetector saturation is especially important with the pulsed raster technique because of the higher peak light levels involved.

9.6 FLICKER (MODULATION)

Flicker is the term usually used to describe the visual sensation produced by a rapidly varying light intensity. The measured quantity is more correctly termed modulation. Display and fluorescent lighting modulation is easily measured using a simple photodetector such as a photoamperic silicon photodiode and an oscilloscope. The photodiode is connected to the oscilloscope vertical input which is shunted with a termination resistor (Figure 9.13). In most cases, 600 Ω will be sufficiently low to resolve the normally encountered refresh or modulation rates.

9.7 SENSOR RESPONSE TIME

For some of the previous tests, it is desirable to know the response time characteristics of the photodetector used, particularly for silicon photodiodes, and the associated circuit. Three simple means are available to determine the response time of the photodi-

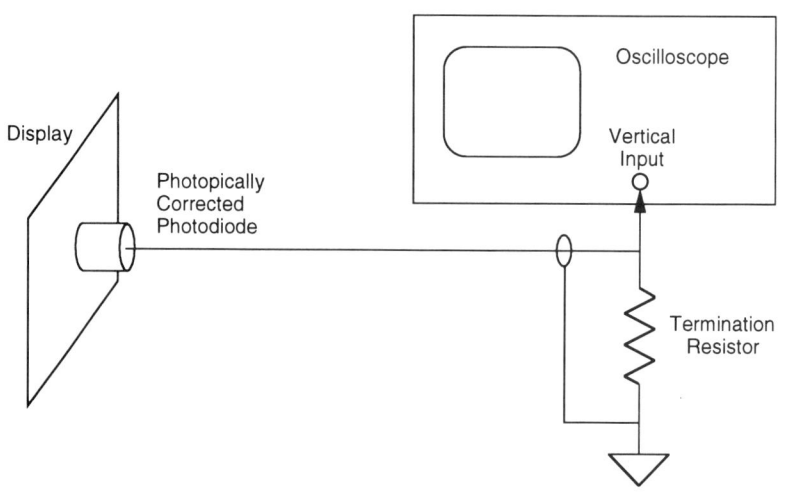

Figure 9.13. Configuration for measurement of modulation (flicker) from a display.

ode circuit. The first is to use the maximum junction capacitance published in the photodiode manufacturer's data sheet to calculate risetime. Multiply the junction capacitance by the load resistance to obtain the time constant. For all practical purposes, the oscilloscope input capacitance and cable capacitance may be ignored for all but fast-response photodiodes operated in the biased mode. Use the manufacturer's typical capacitance for the voltage level at which it is to be biased. This will be zero volts in the photoamperic mode.

A second method is to measure the junction capacitance of the actual photodiode to be used and use the same calculation described above. A digital voltmeter having a capacitance measurement function or other capacitance measurement instrument such as an impedance bridge should be used. The photodiode must be completely covered to prevent light from reaching it during the test. If no reading is obtained on the first try, reverse the connections to the photodiode and try again.

The last method involves measuring the actual response time of the circuit. A common red LED may often be driven directly by the low-impedance output of a function generator or pulse generator capable of producing fast-rise square waves or pulses (Figure 9.14). Mounting the LED in a BNC connector facilitates the test setup (Figure 9.15). Place the LED in contact with the photodiode and use the minimum generator amplitude that will provide a usable display on the oscilloscope to avoid overheating the LED. Most red LEDs have risetimes considerably better than any photodiode operated in the photoamperic mode and couple well to a 50-Ω generator output impedance. If the LED loads the generator excessively, try adding a resistor having a value the same as the generator's characteristic output impedance in series with the LED. Excessive loading may be observed using the oscilloscope connected across the generator output while connecting and disconnecting the LED. The square wave or pulse amplitude should not be reduced by much more than 50%, and the wave shape should not be distorted to a significant degree. The oscilloscope will display the time response characteristics of the photodiode, as shown in Figure 9.16.

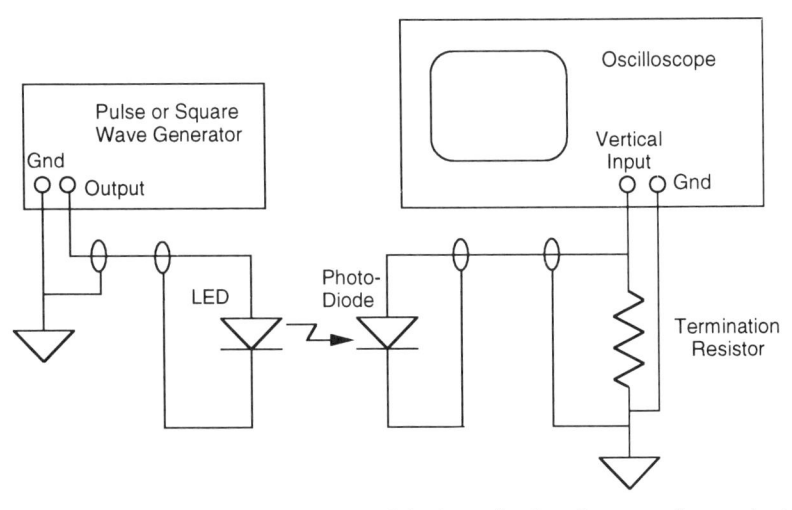

Figure 9.14. Configuration for measurement of risetime of a photodetector using a pulsed LED.

256 TIME-RELATED MEASUREMENTS

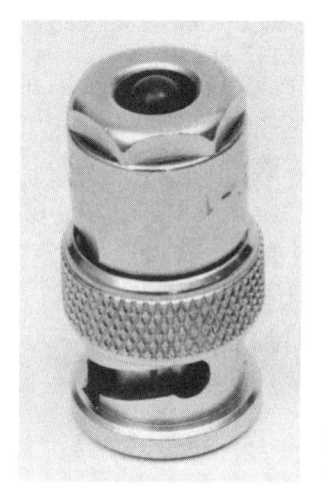

Figure 9.15. Red LED mounted in a BNC connector for ease of use.

A number of different LEDs and photodiodes tested by the author were all found to have rise- and falltimes less (in many cases substantially less) than 1 μs. Biased photodiode operation and/or use of a 50-Ω photodiode load resistor is required in order to achieve this speed of response. On the other hand, no special techniques were required for driving LEDs to submicrosecond rise- and falltimes.

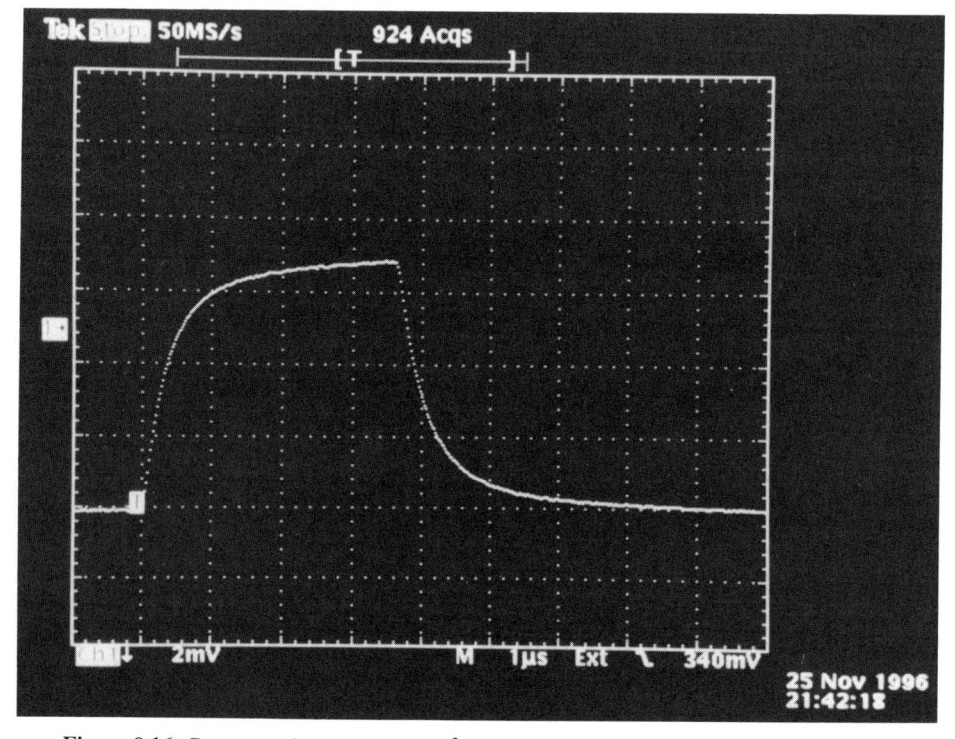

Figure 9.16. Response time of a 100-mm^2 silicon photodiode in the photoamperic mode.

REFERENCES

1. EG&G Optoelectronics, "Silicon Photodiodes," Oct. 1992.
2. R. McCluney, *Introduction to Radiometry and Photometry,* Artech House, Boston, 1994, Chap. 7.
3. W. Budde, *Optical Radiation Measurements:* Vol. 4, *Physical Detectors of Optical Radiation,* Academic, New York, 1983, pp. 232–277.
4. R. W. Engstrom, *Photomultiplier Handbook,* RCA Publication PMT-62, Lancaster, PA, 1980.
5. Burle Industries, *Photomultiplier Handbook,* Publication TP136, Burle Technologies Inc., Lancaster, PA 1980.
6. Electronic Industries Association (EIA), "Typical Characteristics of Photosensitive Surfaces," EIA TEPAC Publication No. TEP161, Dec. 1966 (reaffirmed May 1980).
7. Electronic Industries Association, (EIA), "Relative Spectral Response Data for Photosensitive Devices ('S' Curves)," EIA TEPAC Publication No. TEP150, Oct. 1964 (reaffirmed March 1981).
8. Electronic Industries Association (EIA), "Measurement of Phosphor Persistence of CRT Screens," EIA TEPAC Publication No. TEP105-14, April 1987.
9. Electronic Industries Association (EIA), "Optical Characteristics of Cathode-Ray Tube Screens," EIA TEPAC Publication No. TEP116-C, Feb. 1993.

CHAPTER 10

Calibration

10.1 GENERAL BACKGROUND

The accuracy of photometric, radiometric, and colorimetric instruments is often questioned, especially when the measured results do not match the anticipated test results. Often, the source of measurement error is in misapplication of the instrument or trying to measure beyond its capabilities. Several reasons exist for the tendency to question optical measurements more than would be the case for electrical measurements. These include the fundamental accuracy limitations of optical calibrations, the effect of nearby external influences, and the "mystery" of optical measurements to the majority of electronics engineers and technicians.

While the electronic portions of photometers, radiometers, and colorimeters may be calibrated to tenths of a percent, the optical calibration is usually limited to about 5% accuracy by the combination of spectral, geometric, thermal, and temporal effects present in every calibration. Three percent accuracy or so may be achieved with some laboratory instruments, but great care in measurement technique must be used so as not to lose this precision through inattention to details such as ambient light, reflections, and so on, when making measurements.

This chapter will concentrate on user-related aspects of calibration rather than those of primary and secondary calibration laboratories. Practical tests to verify instrument function and accuracy as well as suggestions on how to minimize measurement errors will be presented.

10.2 METROLOGY

The art and science of making measurements is termed *metrology*. Metrology includes the measurement of time, length, light, temperature, and electrical parameters, all of which are related to measurements covered in this book. Of the seven basic measures, only mass and the amount of substance (moles) are outside the sphere of importance for most light measurements.

Units

The International System of Units (SI)[1] is highly recommended for all measurements and has been required for U.S. federal agencies and departments since 1991.[2] The changeover from English units has been slow but is now generally accepted in the world of photometry and radiometry, with most new specifications and publications making use of SI units. All SI units are derived from the seven basic measurement units, the second, meter, candela, ampere, Kelvin, mole, and kilogram. They are the building blocks of all other units. Examples of derived units include the candela per square meter (nit) and the coulomb (ampere second).

Measurement Standards

Three standards are commonly employed to ensure compliance with accepted metrology practices. While aimed at calibration practices required for U.S. military contracts, MIL-STD-45662A has often been referred to for defining commercial calibration practices for lack of a similar civilian standard.[3] This situation has recently changed with the introduction of ANSI/NCSL Z540-1, a U.S. version of ISO Guide 25, and is expected to become widely used for commercial calibration laboratories.[4,5] Finally, ISO 9000, which is broader in scope and includes many quality-related guidelines for all operations of a company, including calibration practices, is rapidly becoming the international standard as many companies go through the registration process. Periodic audits ensure continued compliance.

Traceability

All calibrations in the United States should be traceable to the National Institute of Standards and Technology (NIST), formerly the National Bureau of Standards (NBS). Many other countries have their own standards organizations that do comparisons with each other to ensure consistency on a global scale. Small differences may exist between accepted standards of various countries, especially for optical measurements, and may need to be taken into consideration for products intended for international trade.

Just because an instrument is "traceable" does not automatically ensure good results. The traceability chain may include a number of calibrations, each adding its own uncertainty to the final calibration. The fewer steps in the chain, the more certainty that the accuracy meets the specified value. Of course, the care taken at each step of the chain is highly important to the final result.

The standards used in the various steps include the primary standards maintained at national standard laboratories, transfer standards used to transfer the primary calibration to other laboratories, and working standards used for everyday calibrations.

Accuracy, Uncertainty, Resolution, Stability, Precision, and Repeatability

This collection of metrology terms leads to much confusion in comparing instrument specifications and understanding their importance. In the true metrology sense, *accuracy* is a frequently misused term. Accuracy, expressed in percent, is most often misused to indicate how closely an instrument's readings agree with primary standards. This is more correctly termed *uncertainty* and may be specified in percent, parts per million (ppm), or the measurement units themselves. Accuracy, on the other hand, is more correctly the reciprocal of the uncertainty. An instrument having an uncertainty of $\pm 0.1\%$ (1000 ppm) has an accuracy of 99.90%.

Uncertainties are usually computed as the square root of the sum of the squares (RSS) of the uncertainties at each step or from each source of uncertainty in the calibration process. Since, statistically, all of the small errors at each step are unlikely to be in the same direction, they will usually partially cancel out each other. If this were not the case, it would be necessary to sum all of the errors.

Resolution is just a measure of how many steps a particular reading may be broken down in to. A good example of resolution is the use of a five-digit versus a three-digit readout on an instrument. Both instruments may have the same measurement uncertainty but the five-digit display will have 100 times the resolution of the three-digit display. A reading of 0.1 lx would be 0.1000 on the former and 0.10 on the latter. Unless the uncertainty is commensurate with the number of digits, the extra digits imply more accuracy than is available and are essentially valueless. An exception is the taking of relative readings for comparison purposes and, then, only if the instrument stability is adequate. *Stability* is a measure of the reproducibility of measurements.

In the world of metrology, *precision* is a term used to denote how closely a number of readings agree with each other without regard to how they agree with the absolute value. Precision and stability are closely related to repeatability.

Repeatability is a measure of how closely an instrument's readings agree with other readings taken at other times under the same conditions with the same instrument. It may be split into short-term and long-term repeatability since different factors come into play with time. Often, this may be the primary concern rather than absolute accuracy, particularly when making display device comparisons such as for vendor selection.

An excellent book on the subject of calibration practice is *Calibration: Philosophy in Practice* published by the Fluke Corporation.[6] Although primarily oriented toward electrical calibration, the first chapters of the book are equally applicable to photometric, radiometric, and colorimetric calibration as well.

10.3 ILLUMINANCE

Several approaches may be taken to verify the validity of photometric measurements. These range from the "sanity check" to full recalibration of the measuring instrument by an accredited calibration laboratory. The simpler approaches described below are in no way intended to replace regular calibrations on an annual basis or more often if required contractually. They are useful primarily if improper functioning of the instrument is suspected or readings are obtained that appear erroneous.

Common office illumination often can provide an easy, no-cost means of verifying proper functioning and ballpark calibration of an illuminance meter. Depending on the care taken, probably 10–20% accuracy is attainable with this approach. Many offices are designed for a minimum task illumination of about 500 lx (50 fc)[7] with a grid of dual 32–40-W fluorescent lamps on approximately 3-m centers. To compensate for lamp aging and dirt accumulation, about 20% higher readings may be found in new installations. Other light levels may be encountered that may still be used for this purpose as long as the intensity is relatively constant. These include subdued lighting in areas devoted to computer-aided design (CAD) and areas with higher light levels where very detailed work is performed. It is a good idea to mark a small area near the front edge of a chosen desk to measure and record the illuminance level when the photometer is new or freshly calibrated. The readings may then be compared in the future if incorrect readings are suspected. Be sure to use an area away from windows and stand back to avoid shadowing the cosine-corrected sensor, which has a 180° field of view. Of course, relocating nearby furniture, relamping, and painting of walls will change the readings so recheck the readings before and after any changes. Conference rooms often may be used effectively for the calibration check point since there is little furniture to interfere and there will be an unshadowed area in the center of a conference table. The only things to watch out for are dimmable or switchable lighting systems and the propensity of some organizations to periodically hang flip-chart pages all over the walls, which can have a significant effect on wall reflectance, especially in a smaller conference room.

A better means of verifying calibration is to use a dedicated "point" light source that is used only for that purpose. The device used will be governed by the accuracy needs. It may be as simple and inexpensive as a common 60-W light bulb and a constant-voltage transformer or as complex as an primary standard traceable 1000-W lamp mounted on an optical bench several meters long in an environmentally controlled room painted flat black, as in Figure 10.1.

For the simple approach, about 5–10% accuracy is possible if care is exercised. First, age a new soft-white or inside frosted 60-W bulb for about 100 h at its nominal voltage before making any measurements. The bulb end should be marked for the intended operating voltage within 5 V. For calibration, use a constant-voltage line regulation transformer of about 100 W to provide isolation from the usual AC line voltage variations that may be encountered. The light output of incandescent lamps is highly dependent on the operating voltage and will vary according to the 3.4th power of the voltage change. Thus a 1% change in voltage will result in a 3.4% change in light output. An alternative to the constant-voltage transformer would be to use a 100-W variable transformer and an AC digital voltmeter capable of accurately resolving 0.1 V to measure the voltage at the lamp socket. The transformer should be set to the same voltage and allowed to stabilize each time a measurement is made. It might even be feasible to dispense with the variable transformer for subsequent calibrations as long as it is used for the initial calibration and a graph is plotted for the light output versus lamp voltage over a range of about 115–125 V or other anticipated voltage range. Neither unregulated approach would be suitable if frequent short-term variations in line voltage are encountered.

Figure 10.1. Well-equipped calibration laboratory and optical bench.

To use these simpler methods for calibration checks, allow 20–30 min warmup of the lamp for stabilization of light output. Operating time of the lamp should be confined to that required for calibrations to minimize the falloff of light output with aging. Always measure from the same direction relative to the bulb surface since variations in light output due to filament orientation may occur if the bulb is rotated in its socket. A "free space" measurement is best to avoid reflections from nearby objects but is not absolutely required if exactly the same configuration relative to nearby objects is used each time. Choose a convenient working distance from the center of the lamp, such as 1 m. The rated initial total output of an inside frosted 60-W lamp is 870 lm. At 1 m, the 870 lm is spread over an imaginary sphere of 12.6 m^2 in area, which works out to about 70 lm/m^2 (lux).

Considerable variation in the actual readings may be expected depending on individual lamp characteristics, reflections and scattering, and operating history of the lamp, but if the same conditions are maintained for future comparisons, reasonable reproducibility may be obtained. Of these variations, reflections and scattering can be expected to be the largest contributor. The wide acceptance angle of cosine-corrected sensors and even wider angular distribution of light from an incandescent lamp aggravate the problem. Twenty percent to 40% contribution due to reflections and/or scattering are not uncommon. These may be reduced by use of a flat-black painted, apertured baffle placed on the optical axis midway between the lamp and illuminance meter or two baffles at one-third and two-thirds of the distance, as shown in Figure 10.2. The

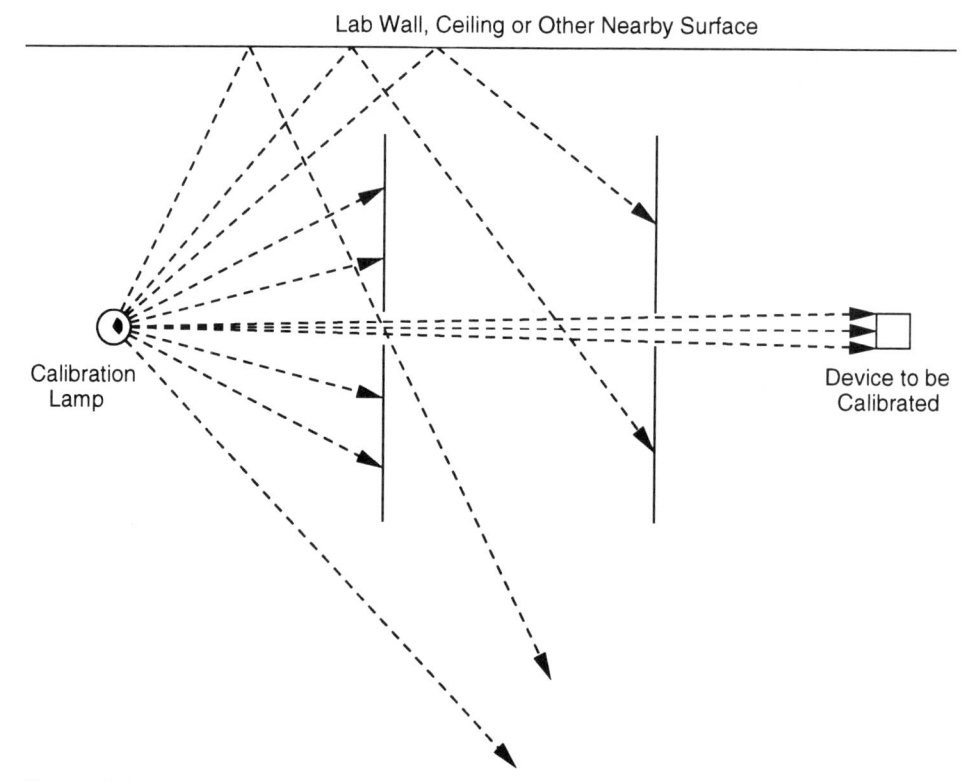

Figure 10.2. Baffling to prevent reflections from surroundings from affecting calibration accuracy.

outside size of the baffle should be as large as practicable while the aperture should be about 50% larger than the largest dimension of the active sensor area or emitting area of the lamp, whichever is the greater dimension. By having the aperture 50% larger, it becomes easier to align the system while blocking most reflections. Black photographer's cloths, black masking tape, ultra-flat-black spray paint, and black curtains are extremely useful means to reduce scattering and reflections at low cost. These items are not limited to the simple tests. They will be found in the best of calibration facilities as well. Keep in mind that even black surfaces and objects reflect a small amount of light. Five percent reflectance from black surfaces is common.

Note that the previously described tests are intended only as "sanity checks" when measurements do not look right. Again, these are not intended to replace the regular calibration cycle for the instrument as recommended by the manufacturer or as specified by governmental or contractual requirements. Just because you suddenly start getting different readings on your "calibrated desktop" is not necessarily cause to complain to the photometer manufacturer that the instrument is faulty; the maintenance crew may have just replaced the fluorescent lamps with another brand or type during the night, one or more lamps may have failed, the desk plant may have grown enough to absorb or block some light, and so on.

The preferred method for verifying illuminance calibration involves the use of an optical bench in an optically controlled environment. This is actually just a highly refined version of the techniques previously described using a better characterized and more stable lamp, regulated power supply, and baffles in a properly designed room. The accuracy under such conditions should be considerably better than 5%, which will be sufficient for the actual calibration of most photometers. Such a facility needs the attention to detail of design and execution by an experienced optical metrologist. The same system may be readily adapted to luminance and other photometric, radiometric, and colorimetric measurements and calibrations.

The lamp should be from a reputable calibration facility with traceability to NIST or other national standards. Tungsten–halogen lamps are preferred as they operate at a higher color temperature and hence are richer in output in the blue portion of the spectrum. Side benefits of tungsten–halogen lamps include greater efficacy, reduced current loss with age, and freedom from darkening of the bulb since the tungsten is redeposited on the filament by the tungsten–halogen cycle rather than the lamp walls. A constant-current lamp power supply is usually used for best stability with time. The optical bench, which may be as long as several meters in overall length, has ruled markings for distance from the lamp-emitting plane to aid in reproducibility of setups and the use of other distances to make use of the effects of the inverse square law for reduced or increased light levels. Adjustable carriers hold the sensor, baffles, and accessories such as filters or diffusers in precise alignment at their desired distances. Free space around the optical bench, temperature and humidity control of the laboratory, and liberal use of flat-black paint on almost everything including the walls and ceiling complete the facility.

One variation on this is the use of calibrated detectors rather than calibrated lamps. The calibrated detector may then be used to calibrate any high-quality lamp. Since the detector stability and life greatly exceed those of even the best lamps, this technique is now coming into widespread usage.

10.4 LUMINANCE

Similar tests to those for illuminance may be used for luminance photometers. Here a few sheets of high-quality white paper or a Kodak R27 reflectance card may be used to good advantage. A stable computer monitor and a photographic light box are other examples of devices that are useful in performing a "sanity check" when all is not going well in the measurement department. As with any of the tests described in this chapter, an hour or two spent devising a simple test in advance can pay rich dividends in peace of mind for that coming day when "you can't get any of the numbers to come out right" and the inspector is coming for yet another quality audit.

The same corner of the desk as calibrated in the previously described illuminance test may be used for luminance depending on the instrument to be tested. Some cannot be used in a suitable position to view the surface without shadowing it. In this situation, a shoulder-high section of wall that is relatively open and without nearby furniture or other reflective objects and is uniformly illuminated by overhead diffuse lighting is well suited for the purpose. Staple several sheets of white photocopy paper together and place in the chosen location. Measure the luminance of the center of the surface with the photometer when first purchased or freshly calibrated, record the reading, and save it along with the paper stack used. For spot photometers, a small area near the center of the paper should be marked off so that the border is visible within the viewfinder but outside the measurement area. Since the paper will probably have a reflectance close to 85% (a reflectance factor of 0.85), measurements may be cross checked with illuminance readings taken at the same location for further confidence. The reading in lux divided by π and multiplied by the reflectance factor of the paper (0.85) should result in a number close to that of the measured luminance in candelas per square meter (nits). Better alternatives include the Kodak R-27 Gray Card, which is 18% reflectance gray on one side and 90% white on the other, or calibrated opal glass reflectance standards. The latter is more durable but more limited in the maximum size available.

A photographic light box such as used for viewing slides or X-ray film makes a good portable reference light source for any type of luminance meter, particularly if operated from a constant-voltage transformer. The light box's fluorescent lamp phosphor spectrum, usually 5000 K, is well suited to simulate the color temperature of CRT phosphors and cold-cathode fluorescent lamps as used for LCD backlighting. The luminance level is usually around 1400 ± 300 cd/m^2 (nits) according to industry standards. Calibrated thin-film neutral-density filters may be used to reduce the light level to be commensurate with that of typical displays. Purchase a light box of suitable size for the photometer(s) to be tested and set it aside from general use so as to avoid aging the lamp(s). Preage the lamps 100–200 h for stabilization before calibrating the light box since the fluorescent lamps may change rapidly in intensity, color, and stability during the first hours of operation. Always measure the same spot in the center of the viewing area for consistency and provide a warmup period of about 20–30 min before making any measurements. Avoid ambient light for this test. Confidence in the stability of the light box (and to a lesser degree the photometer) may be gained by routinely using it before any display measurements are made. Repeatability should be on the order of 10% or so.

A computer CRT display monitor can also be used in a pinch, although some units are subject to short- and long-term drift. The author has successfully used both monochrome (Macintosh SE) and color (Trinitron) monitors for this purpose when no other light source was available. Adequate warmup is a key to success. Use the preset detent on the brightness control if available or use the maximum brightness setting if no blooming is observed and if it does not appear excessively bright (100 cd/m^2 or so). Just display a blank white page using word processing software and make your measurement in the center of the screen. Periodically remeasure it to verify the stability of the display. Usually, this test is performed with a monitor used for everyday office use so a few percent decrease in light output over a long period of time is not unreasonable due to CRT cathode and phosphor aging as well as electron and internal X-ray browning of the CRT's glass faceplate. Screensaver software that produces randomly moving patterns with large dark areas of the screen when not in actual use can help reduce these changes. The author has observed some decorative "screensavers" consisting of fixed images of normal intensity that are little better than just leaving menus or a document displayed.

Stand-alone luminance standards are commercially available. These may take the form of a diffuser and incandescent lamp in a portable enclosure (Figure 10.3) or be less portable and consist of an incandescent lamp in an integrating sphere (Figure 10.4). Color temperatures may range from 1800 to 3100 K for adjustable units while fixed intensity units may be about 2854 or 3000 K depending on whether a conventional incandescent or a tungsten–halogen lamp is employed respectively. The specified luminance accuracy of commercially available units is typically 2%. Figure 10.5 shows a photometer being verified using a portable luminance standard and constant-voltage transformer.

Other systems for calibration of luminance by calibration laboratories are similar to the optical benches previously described for illuminance calibration. The primary difference is the addition of a calibrated diffuser. Due to limitations of the diffuser calibration and geometry, the accuracy is probably not much better that that of the stand-alone units (about 2%).

Figure 10.3. Portable fixed-luminance reference source suitable for checking photometer performance. It consists of an incandescent lamp and diffuser arrangement.

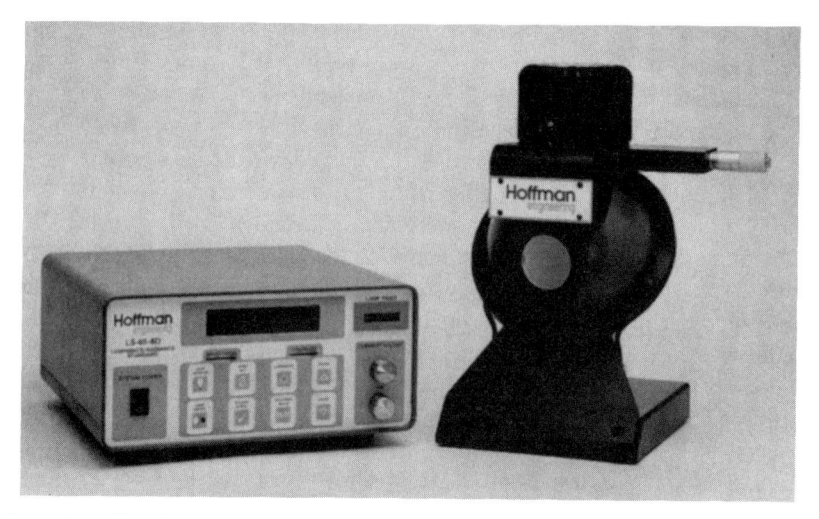

Figure 10.4. An adjustable luminance reference standard employing an integrating sphere. (Courtesy of Hoffman Engineering.)

10.5 CHROMATICITY

The requirements for verifying or calibrating colorimeters and spectroradiometers are quite similar to those for luminance with one or two exceptions. While incandescent light sources are often used for the purpose, the trend is toward the use of phosphor light sources having a color temperature close to 6500 K to better simulate a typical display. The incandescent lamp spectrum with its large infrared output is quite different spectrally from that of displays. Significant errors can result when comparing chromaticity measurements of incandescent and phosphor light sources if there is any unwanted infrared leak-

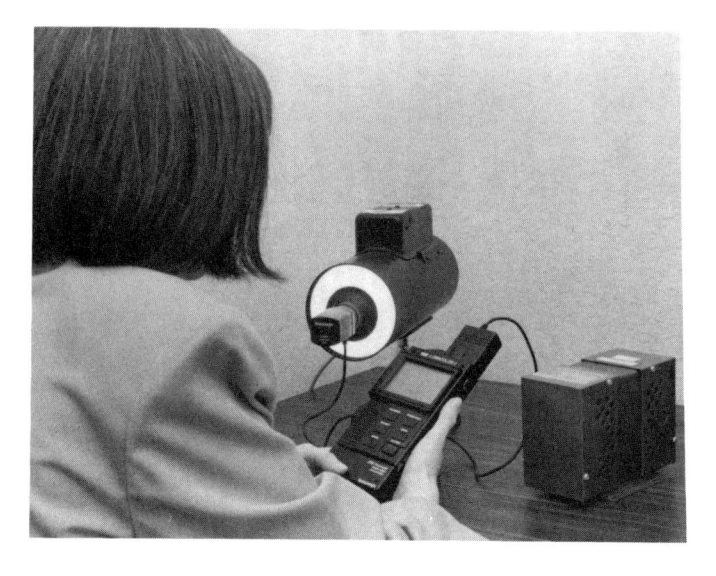

Figure 10.5. Verifying the operation of a portable luminance meter using a reference luminance source.

age of the detector. The high sensitivity of the commonly used silicon photodiode or CCD array to infrared increases this likelihood. For this reason, it is recommended to avoid the use of incandescent lamps and "phosphor simulation" filters. The filters often offer little or no infrared attenuation and thus are far from matching a phosphor spectrum.

The 5000 K photographic light box operated from a constant-voltage transformer is fairly stable, portable, and easy to use as a check source. Drawbacks include too high a light output and common use of a mix of only two phosphors, yellow and blue, to produce white. The latter, however, is only a negative for CRT and other self-luminous color displays, which usually use red, green, and blue phosphors to produce white. Monochrome CRTs and LCD backlights have emission spectra quite similar to that of the light box. Follow the recommendations in the section on luminance. Just record the chromaticity values in addition to luminance. Sufficient warmup for stabilization and operation at a constant ambient temperature is even more important for the light box used for checking chromaticity than for luminance. The proportion of light contributed by each phosphor is often different with temperature since different phosphor types have their own unique characteristic curve of light output versus temperature.

A color CRT display is often the best of the simple approaches. A stable monitor operated at a preset luminance may be used, providing it is not apt to be readjusted between uses, for verifying chromaticity and luminance measurements. Frequent measurements to verify its stability are important to gain confidence in its reproducibility. For laboratories having the equipment, a better technique is to use an accurately calibrated colorimeter or spectroradiometer to measure chromaticity of the monitor and to then compare filter colorimeters with it directly.

Of course, the incandescent light source may be used in a manner similar to that described for luminance. If it is always being used for periodically verifying the same instrument, a little infrared leakage in the detector may be tolerable as long as it is always the same amount and the same detector. The optical bench for precision calibrations may also rely on the incandescent lamp, preferably tungsten-halogen, which has a better balance across the visible spectrum.

Infrared leakage of any sensor for illuminance, luminance, or chromaticity can be checked by using a tungsten–halogen lamp by placing a 3-mm-thick Schott RG-780 glass color filter in front of the sensor and looking at the output as described in CIE Publication Number 69.[8] The light level used should be at least 1000 times greater than the minimum resolvable signal of the instrument under test. The percentage of leakage with the filter compared to the reading without the filter can then be computed. For filter colorimeters, use the raw X, Y, and Z values if available and measure the infrared leakage for each. Divide the reading with the filter by the reading without it to compute the infrared leakage. Above 1–2% infrared leakage, accuracy will suffer.

10.6 LIGHT-EMITTING DIODES

Light-emitting diodes can be difficult display devices to measure accurately. This is because of the relatively narrow beams emitted by many; the narrower the beam, the harder it is to get consistent results and comparisons. The narrow spectral bandwidths of LEDs further add difficulties if the photometric equipment is not closely matched to the CIE Standard Observer (photopic response curve). No adequate LED measurement standards exist to aid comparisons. This leads to frequent disputes between manufacturers and users.

Probably the best solution to measurement discrepancies is the use of one or more calibrated "golden" standards that may be shared by manufacturer and user. The LEDs of the type in dispute can be sent to independent calibration laboratories for measurement and used to verify any measurements made or at least to determine the correction factor required to make data agree between the parties involved. The LEDs set aside for this purpose should be aged at the current to be used for 100 h or until the light output stabilizes. Choose an operating current well below the maximum rating to reduce aging during tests and to minimize the chip temperature rise, which can degrade its stability when elevated. Naturally, the current must be set and maintained accurately for all testing. As with all light sources used as reference standards, once the initial aging and calibration are completed, the operating time should be kept to a minimum consistent with proper stabilization and testing duration.

The angular effects of narrow-angle LEDs are particularly difficult to deal with. The beam may be aligned slightly off-axis so differences in fixturing and alignment technique can give rise to measurement differences. Also, for example, if one photometer intercepts a 3° portion of the beam and another intercepts 5°, the narrower one will read higher but be more critical in alignment and will be subject to greater variation in readings. The photometer that samples the larger angle is averaging the high peak luminous intensity of the beam center with the lower intensity off-axis. Both are probably working correctly and may read identically on the diffuse light sources used for calibration or wider angle LEDs but may produce substantial differences in readings. Yet, each is indicating exactly what it sees. Again, a calibrated reference LED of the same type will help resolve these differences. If a relatively wide angle diffused calibrated LED of the same color is also available and good agreement is obtained with it when measured with both instruments, any differences observed with narrow-angle LEDs may be safely assumed to be related to alignment and/or different sampled angles of the beam.

REFERENCES

1. American National Standards Institute (ANSI)/Institute of Electrical and Electronics Engineers (IEEE), "American National Standard for Metric Practice," ANSI/IEEE Std. 268-1992, Oct. 28, 1992.
2. National Institute of Standards and Technology (NIST), "Guide for the Use of the International System of Units," NIST Special Publication 811, April 1995.
3. Department of Defense (DOD), "Calibration System Requirements," MIL-STD-45662A, Aug. 1, 1988.
4. American National Standards Institute (ANSI), "American National Standard for Calibration—Calibration Laboratories and Measuring Test Equipment—General Requirements," ANSI/NCSL Z540-1-1994, July 27, 1994.
5. J. Lloyd, "ANSI/NCSL Z540-1: A New National Standard for Calibration Laboratories and Equipment," Cal Lab., Mar./Apr. 1995, pp. 7–9.
6. Fluke Corporation, *Calibration: Philosophy in Practice*, 2nd ed., 1994.
7. Illuminating Engineering Society of North America (IESNA), *Lighting Handbook,* 8th ed., IESNA, New York, Chapter 15, 1993.
8. Commission Internationale de l'Eclairage (CIE), "Methods of Characterizing Illuminance Meters and Luminance Meters," CIE Publication No. 69, 1987, p. 11.

CHAPTER 11

Display and Related Standards

Many standards exist for definitions of terminology and photometric, radiometric, colorimetric, and display measurement procedures. These include documents from ANSI, ASTM, CIE, DIN, EBU, EIA, EIAJ, HFES, IEC, IEEE, IESNA, ISO, ITSB, MIL, NAPM, NIDL, SAE, SMPTE, VESA, and others. The following sections list some of the more pertinent published display standards, both current and in preparation. Other display related organizations are included as well since they often publish useful papers, articles, and books on the subject. A listing of addresses of standards organizations is included in Appendix M. Periodic reviews of display standards efforts are published sporadically as papers and articles in display-related journals and magazines.[1-8]

Several display standards are always in preparation as the state of the art rapidly advances, especially in the area of flat-panel displays, although there is much commonality between flat-panel and CRT display color measurements. After all, the correct and reproducible appearance of the display to the human eye is the ultimate goal. That said, we will now delve into the alphabet soup of display standards.

11.1 ANSI

The American National Standards Institute (ANSI) is the central organization responsible for coordinating consistent voluntary standards termed, appropriately enough, American National Standards.[9] ANSI is the U.S. member of international standards organizations, including the International Electrotechnical Commission (IEC), International Standards Organization (ISO), and Pacific Area Standards Congress (PASC).

ANSI standards are often published jointly with other standards formulating organizations. These are identified, for instance, as ANSI/IEEE Std. 268-1992,[10] ANSI/HFS 100-1988,[11] and so on, with the last four digits indicating the year of publication. ANSI itself also has published many standards.

As part of ANSI coordinating efforts, the Image Technology Standards Board was formed to identify potential conflicts between other organizations formulating imaging standards. Committees have been formed within ITSB for several categories of imaging technology. It is the task of each committee to follow the standards work of other organizations pertinent to their category.

11.2 ASTM

Formerly the American Society for Testing and Materials, this standards organization now uses just the abbreviation ASTM as its official name. As indicated by its original name, its focus is on materials and testing pertinent to them. ASTM was founded in 1898 and has a primary function of ensuring the quality and safety of materials, products, systems, and services through voluntary standards.

ASTM has developed many standards pertaining to the appearance of materials. These include color, texture, reflectance, and many related tests that are a result of the work of the ASTM Committee E12 on Appearance of Materials.[12] A number of these standards are relevant to the measurement of display reflectance, materials, and processes. Of special interest to display designers and users is the work of Committee E12.06 on the Appearance of Displays.[13] Also of interest to those making display color measurement by spectroradiometric techniques is ASTM standard E1336-91.[14]

11.3 CIE

The Commission Internationale de l'Eclairage (CIE) is the foremost organization in the field of light and color. CIE is a truly international organization headquartered in Vienna and has been active in photometric and colorimetric standards since its formation in 1913. CIE's activities are predated by the Commission Internationale de Photometrie (CIP), which existed from 1900 to 1913. As a historical sidelight, CIP grew out of a paper presented at the 1900 International Gas Conference in Paris on the need for photometry techniques for the incandescent *gas* lamp.[15] CIE has since been expanded to include the following divisions[16]:

Division 1	Vision and Colour
Division 2	Physical Measurement of Light and Radiation
Division 3	Interior Environment and Lighting Design
Division 4	Lighting and Signalling for Transport
Division 5	Exterior and Other Lighting Applications
Division 6	Photobiology and Photochemistry

Within each division are a number of committees charged with specific tasks usually culminating in official CIE publications. Working meetings of the individual divisions are held periodically while quadrennial sessions of the entire CIE are held in major international cities. While slow in being officially approved since so many countries must come to agreement, standards formulated by CIE have had considerable acceptance and longevity as witness the 1924 CIE Standard Observer and 1931 Chromaticity Diagram. Much of the foundations of photometry and colorimetry are the result of CIE efforts.

A number of committees and CIE publications are of importance to the display industry. These publications include the "International Lighting Vocabulary,"[17] "Colorimetry of Self-Luminous Displays: A Bibliography,"[18] "Colorimetry,"[19] and "Methods of Characterizing Illuminance Meters and Luminance Meters."[20]

11.4 CORM

The Council for Optical Radiation Measurements (CORM) is a liaison between industry and the National Institute of Standards and Technology (NIST) in the area of light measurement. CORM publishes no standards documents but serves to arrive at a consensus of industry's most pressing needs in furthering the development of optical radiation metrology. Currently, the need identified that is of most interest to the display community is the need for better physical standards and published standardized procedures for colorimetric measurement of displays.[21] An annual CORM conference is held at NIST in Gaithersburg, Maryland each May.

11.5 DIN

Deutsche Industrie Norms (DIN) are German standards documents. Although published only in German, the one DIN publication of most importance for display measurement is DIN 5032 Part 7, "Lichtmessung."[22] This document outlines the classes (A, B, C, and L) of photometric instruments as determined by the f_1–f_{12} and other parameters defined in DIN 5032 Part 6 on the characterization of photometers. Part 6 is similar to CIE Publication No. 69. One other DIN standard, DIN 66234 on the ergonomics of workstations, appears to have evolved into ISO 9241.[23–25]

11.6 EBU

The European Broadcast Union (EBU) of Brussels has published several standards related to colorimetry for television cameras and studio monitors. These include measurement of colorimetric fidelity of cameras,[26,27] phosphor chromaticity,[28] adjustment of color monitors,[29] and color studio monitor chromaticity tolerances.[30]

11.7 EIA

The Electronic Industries Association (EIA) evolved from the earlier Radio Manufacturers Association (RMA). Formed in 1924, RMA went through a progression of name changes reflecting the rapid change in electronics technology during the late 1940s and early 1950s. It became the Radio-Television Manufacturers Association (RTMA) in 1950, the Radio-Electronics-Television Manufacturers Association in 1953, and finally, the Electronics Industries Association in 1957.[31] From the 1940s on, it developed many standards pertaining to the display industry. These began with standards for television and registration of CRT and phosphor characteristics on an industrywide basis for interchangeability between manufacturers. The division responsible for device standards began as the Joint Electron Tube Engineering Council (JETEC), evolved to Joint Electron Device Engineering Council (JEDEC), and finally became the Tube Engineering Panel Advisory Council (TEPAC). References to JEDEC are still commonly found in display publications.

Current display publications from EIA include CRT terms and definitions,[32,33] CRT test methods,[34–45] glass bulb dimensional[46–50] and glass quality[51–54] standards, CRT deflection yokes,[55,56] phosphor[57] and CRT type registrations under the Worldwide Type Designation (WTDS),[58] display X-radiation measurement,[59–66] and technical bulletins on a variety of display topics.[67–72]

11.8 EIAJ

The Electronic Industries Association of Japan is the Japanese counterpart to EIA. A number of standards have been produced by EIAJ that are related to both flat-panel displays and CRTs. No comprehensive list was available at the time of publication, but the known examples include LCD terms and definitions,[73] liquid crystal measurement methods,[74] CRT registration, CRT measurements[75-78] reportedly similar to the EIA TEP-105 series, CRT X-Radiation,[79-86] and CRT glass quality.[87-91]

11.9 EN

The formation of the European Community has led to the consolidation of the many individual standards of the participating countries into common standards designated European Norms (EN). This has been good from the standpoint of reducing the need for new products to meet the standards of each country that they will be sold in. It has caused some confusion during the interim period since both the old and new standards have been quoted in device specifications, making comparisons difficult.

The primary ENs related to displays at this time deal with limits and measurement methods for electromagnetic compatibility and frequently include references to IEC standards. A safety standard, EN 61010-1, is in preparation and should combine IEC 1010-1 (which replaced IEC 348), CSA 1010-1, and UL 3111 into one international standard for test and measurement instruments.

11.10 HFES

The Human Factors and Ergonomics Society (HFES), formerly the Human Factors Society (HFS), issued a joint ANSI/HFS document on human factors for displays as ANSI/HFS 100 in 1987 that is particularly useful for display ergonomics.[92] Some information is included in the standard pertaining to appearance and viewability of displays. An updated version is reported to be in preparation.

HFES has a Web homepage at http://www.hfes.vt.edu/HFES/.

11.11 IEC

The International Electrotechnical Commission (IEC or CEI) has many published standards for safety, electromagnetic compatibility, and the performance of electronic devices. These have been assuming greater importance in recent years with the emergence of the European Community and have formed the basis for the European Norms (EN) with a corresponding deemphasis of individual country standards.

IEC Technical Committee Number 47 is formulating LCD standards including specification sheet format and measuring methods. Other IEC publications relate to CRT specifications[93] and CRT measurements.[94-97]

11.12 IEEE

The Institute of Electrical and Electronics Engineers (IEEE), formerly the Institute of Radio Engineers (IRE), surprisingly has no direct involvement in formulating display standards. Probably the greatest contributions of the IEEE to displays have been the standards of units and quantities,[98,99] definitions,[100] and, previously, television. The latter is evidenced in the continued use of "IRE units" for video signal levels.

11.13 IESNA

The Illuminating Engineering Society of North America (IESNA) is the primary U.S. developer of lighting standards. The IESNA was previously known as the IES. Of interest to the display user is the IESNA Recommended Practice RP-24, "IES Recommended Practice for Lighting Offices Containing Computer Visual Display Terminals." In addition, the IESNA's 989-page *Lighting Handbook* contains a wealth of information on light sources, photometry, color, and lighting practice.[101] Periodicals published by the IESNA include *Lighting Design and Application* and *Journal of the IESNA*, both of which are included with IESNA membership. The former contains primarily nontechnical articles about lighting installations and new lighting products while the latter contains technical papers on the development of improved light sources, installation, and measurements.

11.14 ISCC

The Intersociety Color Council (ISCC), formed in 1931, performs the function of providing interaction among the many diverse technical societies having an interest in color measurement and specification. The aims and purposes are summed up in the ISCC Constitution as follows:

> A. To stimulate and coordinate the work being carried out by the various members leading to the uniformity of description and specification by these members.
>
> B. To promote the practical application of this work to color problems arising in science, art, and industry, for the benefit of the public at large.
>
> C. To promote communication between technically oriented specialists in color and creative workers in art, design, and education, so as to facilitate more effective use of color by the public through dissemination of information about color in both scientific and artistic applications.
>
> D. To promote educational activities and the interchange of ideas on the subjects of color and appearance among its members and the public generally.
>
> E. To cooperate with other organizations, both public and private, to accomplish these objectives for the direct and indirect enjoyment and benefit of the public at large.

To illustrate the diversity of interests represented within the ISCC, the member organizations include the following:

American Association of Textile Chemists and Colorists (AATCC)
American Society for Testing and Materials (ASTM)*
American Society for Photogrammetry and Remote Sensing (ASPRS)
Color Association of the United States, Inc. (CAUS)
Color Marketing Group (CMG)
Color Pigments Manufacturer's Association (CPMA)
Detroit Color Council (DCC)
Federation of Societies for Coatings Technology (FSCT)
Gemological Institute of America (GIA)
Graphic Arts Technical Foundation (GATF)
Human Factors and Ergonomics Society (HFES)*
Illuminating Engineering Society of North America (IESNA)*
Individual Member Group (IMG)
National Association of Printing Ink Manufacturers (NAPIM)
Optical Society of America (OSA)*
Society for Imaging Science and Technology (SPSE)
Society for Information Display (SID)*
Society of Plastic Engineers, Color and Appearance Division (SPE)
Technical Association of the Graphic Arts (TAGA)

(Asterisks denote those member organizations involved in standards development or publishing material pertinent to displays.) The ISCC has not developed standards itself, but members participate in the color standards work of the member societies. The ISCC project committee 32, Image Technology, is investigating the needs for video display color calibration as well as defining color mapping for transfer of information from video display color space to hard-copy color space.[102]

11.15 ISO

The International Organization for Standardization (ISO) is known today primarily for ISO 9000, which is becoming the international standard for quality. Under ISO 9000, companies that comply are registered and undergo periodic audits to assure compliance. ISO 9000 and its subsequent documents (ISO 9001, ISO 9002, etc.) define almost all aspects of the participating company's internal quality-related processes.

The most important ISO publication for display measurement is ISO 9241.[103-105] A similar standard for flat-panel displays[106] is evolving in the proposed ISO Standard 13406.

11.16 IS&T

The Society for Imaging Science and Technology (IS&T) is an international society devoted toward advancing the understanding and application of imaging science and technology in the areas of graphic arts, electronic imaging, hybrid imaging systems, photofinishing, and other related photographic and electronic imaging technologies. IS&T

publishes no display standards per se but publishes the *Journal of Imaging Science and Technology,* copublishes the *Journal of Electronic Imaging,* and holds several imaging conferences each year. The membership is approximately 2000.

IS&T has a Web homepage at HTTP://www.imaging.org/.

11.17 ITSB

See ANSI.

11.18 MIL

Military Standards (MIL) have long been a part of the U.S. armed forces procurement system and cover everything from paper clips to thermonuclear devices. Environmental testing, calibration, and performance characteristics are all specified in great and unrelenting detail. The end of the cold war with the reduction in military spending has reduced the importance of MIL standards to some degree at a time when civilian safety, environmental, and product quality standards requirements are mushrooming.

MIL standards applicable to displays cover general specifications[107] and test methods[108] for electron tubes, including CRTs, and airborne display specifications.[109] MIL-STD-45662,[110] has been widely used for specifying calibration requirements for commercial calibration laboratories for lack of a better civilian standard. This is just now being supplanted by ANSI/NCSL Z540, which was developed by the National Conference of Standards Laboratories.[111] This can be expected to include photometric, radiometric, and colorimetric calibrations in the future.

11.19 NAPM

The National Association of Photographic Manufacturers (NAPM) and ANSI have jointly published at least two standards of particular interest to the display community. These pertain to the measurement of LCDs for overhead projectors[112] and data projectors/large-screen displays.[113]

11.20 NIDL

From the *National Information Display Laboratory (NIDL) Introduction*:*

> The National Information Display Laboratory (NIDL) brings together commercial and academic leaders in advanced display hardware, softcopy information processing tools, and information collaboration and communications techniques to help government users better accomplish their jobs. One element of this is high resolution imaging and displays which are critical technologies identified in the White House critical technologies list. NIDL serves a large number of Intelligence Community and Department of Defense users and, under its

*Copyright 1995 David Sarnoff Research Center, Inc. All rights reserved. Reprinted by permission

dual use philosophy, an increasing number of other government users such as the U.S. Geological Survey, IRS, FBI, Department of Health and Human Services, and others.

In establishing NIDL in 1990, the government sought to leverage the resources of the world's commercial and university leaders in crucial technologies. Recognizing the dynamic developments in the commercial marketplace, the concern was to take advantage of the commercial markets and replace the old acquisition paradigm that was slow, commercially incompatible, and expensive.

The NIDL is hosted by the David Sarnoff Research Center in Princeton, New Jersey, a world research leader in high definition digital TV, advanced displays, computing and softcopy tools. NIDL is a distributed lab, encompassing many industrial and academic partners who are leaders in their respective fields. The goal is to obtain the best solution for government needs regardless of location or company. No one organization can satisfy the range of requirements of government program offices for information related solutions and thus the NIDL seeks the best solution wherever available. The NIDL often serves as an agent for advanced research within the academic community for the government.

The NIDL focuses on government user's needs which are often several years in advance of those of the commercial marketplace. One of the goals of the NIDL is to foster research in advanced capabilities in a manner that provides incentives for commercialization. When successful, this benefits government users in future years with commercially available technology and low-cost products driven by the commercial marketplace.

While not a standards organization per se, NIDL has developed proposed standards for monochrome CRT displays[114] and color CRT[115] display measurements. Both have been submitted to the EIA/SID JT-20 committee for consideration.

11.21 NIST

The National Institute of Standards and Technology (NIST), formerly the National Bureau of Standards (NBS), is the focal point in the United States for physical standards rather than standards documents themselves. Of particular interest is the work performed by NIST on photometric, radiometric, and colorimetric calibrations. Traceability of most U.S. calibrations is to NIST reference standards. A number of international color standards are based on NBS research by Judd,[116] Kelly,[117] and others.

NIST publishes a guide to its calibration services that is of primary value to calibration laboratories for display measurement equipment rather than display designers and users.[118]

The NIST Flat Panel Display Laboratory has developed a proposed flat-panel display measurement standard for VESA. This badly needed document is still in the early stages of formulation at this time.[119]

11.22 SAE

The Society of Automotive Engineers (SAE) involvement in display standards stems primarily from aircraft cockpit displays as driven by commercial and military aviation needs. As cockpit displays have progressed from the traditional mechanical gauge to the CRT and now flat-panel displays, the need to be able to specify electronic display performance in a consistent manner has increased. The SAE A-4 Aircraft Instruments Committee has published several documents for this purpose. Probably the most important from the display standpoint is ARP-1782, "Photometric and Colorimetric Measurement Procedures for Airborne Direct View CRT Displays,"[120] published in January of 1989. A similar document for matrix-addressed flat-panel displays is in preparation and will carry the designation ARP-4260.[121] The ARP designations denote a recommended practice while AS denotes a standard. Related publications include AS-8034, "Minimum Performance Standard for Airborne Multipurpose Electronic Displays;"[122] ARP-1874, "Design Objectives for CRT Displays for Part 25 (Transport) Aircraft;"[123] ARP-4256, "Design Objectives for Liquid Crystal Displays for Part 25 (Transport) Aircraft;"[124] and ARP-4067, "Design Objectives for CRT Displays for Part 23 Aircraft."[125]

SAE has a Web homepage at http://www.sae.org.

11.23 SID

The Society for Information Display (SID) is an international organization devoted to all aspects of information display from research to final product marketing. Anyone with an interest in displays should consider a membership in the SID. This organization is devoted exclusively to display design, application, and measurement. The annual SID International Symposium (always held in May in the United States but with a strong international attendance) and the International Display Research Conference (usually held abroad in October) are the events at which to learn and see the latest in display technology. Periodicals published by SID include *Information Display* and *Journal of the SID,* both of which are included with SID membership. The former contains primarily general interest articles about display devices and systems and about new display products while the latter contains technical papers on the development of improved displays and their measurement.

SID does not produce standards. However, through its standards subcommittee and its membership, it maintains active participation in standards activities on a world-wide basis.

SID has a Web homepage at http://www.sid.org.

11.24 SMPTE

The Society of Motion Picture and Television Engineers (SMPTE) is a leader in the advancement of the fields of film, imaging, and television technology. Many SMPTE standards have been promulgated over the years. These include television colorimetry,[126–130] video signals,[131,132] test patterns,[133–136] projection screens,[137,138] and, of course, many on film and film recording.

SMPTE has a Web homepage at http://www.smpte.org.

11.25 SPIE

The International Society for Optical Engineering (SPIE) is an organization oriented toward research, engineering, and application in the fields of optics, imaging, and electronics. No display standards are published by SPIE but members are active in imaging technology, which of course includes displays.

SPIE has a Web homepage at http://www.spie.org/.

11.26 VESA

The Video Electronics Standards Association (VESA) is a recent entry into display standards formulation. VESA was organized in 1989 to bring more order to display interfaces for the ease of system integration.[139]

VESA has published one standard for CRT display measurements. Oddly enough, this standard bears no identifying number, just its title "Display Specifications and Test Procedure, Version 1.0, Rev. 1.0."[140] As a first attempt at measurement standards, it does have some rough edges and a few errors. Especially noticeable are the use of the term "brightness" where "luminance" was meant and the use of the ANSI/EIA deprecated unit the centimeter instead of the millimeter. It is, however, a step in the right direction and it is hoped that the next VESA measurement standard on flat-panel display measurements,[141] now in draft form, will be more polished. VESA does use a "fast-track" method of preparing standards to be able to publish them in a timely manner in a fast-moving field. The down side is that the accuracy and refinement of standards developed by the more ponderous traditional standards organizations are lacking. National standards often take several years to reach consensus and international standards can be measured in decades. Later versions should correct the shortcomings in the VESA standard, and having something to work with early on is certainly better than having nothing.

VESA has a Web homepage at http://www.vesa.org.

11.27 OTHERS

Several organizations are responsible for safety, electromagnetic radiation/susceptibility, and quality standards and regulations applicable to displays. These include Bureau of Radiological Health (BRH), British Standards Association (BSA), Canadian Standards Association (CSA), European Norm (EN), Federal Communications Commission (FCC), International Electrotechnical Commission (IEC), International Organization for Standardization (ISO), National Board for Measurement and Testing (MPR) in Sweden, Underwriters Laboratories (UL), Verband Deutscher Electrotechniken (VDE) in Germany, and others. Testing to these standards is outside of the scope of this book. The reader is directed to the standards themselves, a number of which are included in the references for this chapter. Many of these are mandatory to meet legal and export requirements.

REFERENCES

1. R. J. Zavada, "Documenting Technology for Creativity," *SMPTE J.*, pp. 212–215, Mar. 1984.
2. P. A. Keller and R. J. Zavada, "A Survey of Display Standards Activities," *Information Display*, pp. 21–26, Dec. 1989.
3. J. C. Greeson, "International Standards Challenge Flat-Panel Displays," *Information Display*, pp. 14–20, July/Aug. 1990.
4. J. C. Greeson, "Display Standards," *Information Display*, pp. 26–28, Dec. 1990.
5. J. C. Greeson, "Display Standards Update," *Information Display*, pp. 22–27, Dec. 1991.
6. J. C. Greeson, "Display Standards in Trouble," *Information Display*, pp. 24–27, Dec. 1994.
7. E. F. Kelley, "A Survey of the Components of Display Measurement Standards," *J. SID*, Vol. 3/4, pp. 219–222, 1995.

8. J. E. Schuessler, "Flat Panel Standardization Activities in the United States," Asia Display '95, 1995, pp. 955–956.
9. American National Standards Institute (ANSI), "Procedures for the Development and Coordination of American National Standards," ANSI, Sept. 9, 1987.
10. American National Standards Institute (ANSI)/Institute of Electrical and Electronics Engineers (IEEE), "American National Standard for Metric Practice," ANSI/IEEE Std. 268-1992, Oct. 28, 1992.
11. American National Standards Institute(ANSI)/Human Factors Society (HFS), "American National Standard for Human Factors Engineering of Visual Display Terminal Workstations," ANSI/HFS 100-1988, Feb. 4, 1988.
12. American Society for Testing and Materials (ASTM), *ASTM Standards on Color and Appearance Measurement,* 4th ed., ASTM, Philadelphia, PA, 1994.
13. P. J. Alessi, "Overview of ISCC and ASTM Committee Work on Video Displays," *Color Res. Applicat.,* Vol. 11 (Suppl.), p. S29, 1986.
14. American Society for Testing and Materials (ASTM), "Standard Test Method for Obtaining Colorimetric Data from a Video Display Unit by Spectroradiometry," ASTM E1336-91, Apr., 1991.
15. Commission Internationale de l'Eclairage (CIE), "History of the International Commission on Illumination," CIE Publication No. 9, 1963.
16. Commission Internationale de l'Eclairage (CIE), *CIE News*, No. 14, pp. 2–4, June 1990.
17. Commission Internationale de l'Eclairage (CIE)/International Electrotechnical Commission (IEC), "International Lighting Vocabulary," CIE Publication No. 17.4/IEC Publication No. 50, 1987.
18. Commission Internationale de l'Eclairage (CIE), "Colorimetry of Self-Luminous Displays— A Bibliography," CIE Publication No. 87, 1990.
19. Commission Internationale de l'Eclairage (CIE), "Colorimetry," CIE Publication No. 15.2, 1986.
20. Commission Internationale de l'Eclairage (CIE), "Methods of Characterizing Illuminance Meters and Luminance Meters," CIE Publication No. 69, 1987.
21. Council for Optical Radiation Measurements (CORM), "CORM Sixth Report," Dec. 1995.
22. Deutsche Industrie Norms (DIN), "Lichtmessung," DIN 5032, Teil 7 (Part 7), Dec. 1985.
23. Deutsche Industrie Norms (DIN), "VDU Workstations," DIN 66234, various publication dates from the 1980s for individual parts.
24. B. A. Rupp, "Visual Display Standards: A Review of Issues," *Proc. SID,* Vol. 22, No. 1, pp. 63–72, 1981.
25. H. Koch and R. Schafer, "West German Standards for Display Workstations: A New Report," *Displays,* pp. 73–77, Apr. 1986.
26. European Broadcast Union (EBU), "Methods of Measurement of the Colorimetric Fidelity of Television Cameras," EBU Tech. 3237-E, May 1983.
27. European Broadcast Union (EBU), "Methods of Measurement of the Colorimetric Fidelity of Television Cameras," EBU Supplement 1 to Tech. 3237-E, Nov. 1989.
28. European Broadcast Union (EBU), "The Chromaticity of the Luminophors of Television Receivers," EBU Technical Statement D28, 1980.
29. European Broadcast Union (EBU), "Procedure for the Operational Alignment of Grade-1 Colour Monitors," EBU Technical Recommendation R23, 1980.
30. European Broadcast Union (EBU), "EBU Standard for Chromaticity Tolerances for Studio Monitors," EBU Tech. 3213-E, Aug. 1975 (reaffirmed 1981).
31. J. D. Secrest, *Electronics Industries Association: The First Fifty Years,* Electronics Industries Association, Washington, D.C., 1974.

32. Electronics Industries Association, "Glossary of CRT Terms and Definitions," EIA Publication No. TEP192, Sept. 1984.
33. Electronics Industries Association (EIA), "Display Storage Tube Nomenclature," EIA Publication No. EIA-305, Feb. 1965 (reaffirmed Dec. 1987).
34. Electronics Industries Association (EIA), "Industrial Cathode-Ray Tube Test Methods," EIA Publication No. TEP105, Feb. 1981.
35. Electronics Industries Association (EIA), "Line Profile Measurements in Monochrome Cathode-Ray Tubes," EIA Publication No. TEP105-7-A, Jan. 1987.
36. Electronics Industries Association (EIA), "Raster Response Measurement of Monochrome Cathode-Ray Tubes," EIA Publication No. TEP105-8, Jan. 1987.
37. Electronics Industries Association (EIA), "Line Profile Measurements in Shadow-Mask and Other Structured Screen Cathode-Ray Tubes," EIA Publication No. TEP105-9, Jan. 1987.
38. Electronics Industries Association (EIA), "Contrast Measurement of Cathode-Ray Tubes," EIA Publication No. TEP105-10, Apr. 1987.
39. Electronics Industries Association (EIA), "Measurement of the Color of CRT Screens," EIA Publication No. TEP105-11-A, Dec. 1988.
40. Electronics Industries Association (EIA), "Test Method for Tube Face Reflectivity," EIA Publication No. TEP105-12, Apr. 1987.
41. Electronics Industries Association (EIA), "Test Method for Specular Gloss," EIA Publication No. TEP105-13, Apr. 1987.
42. Electronics Industries Association (EIA), "Measurement of Phosphor Persistence of Cathode-Ray Tube Screens," EIA Publication No. TEP105-14, Apr. 1987.
43. Electronics Industries Association (EIA), "CRT Screen and Glass Aging Procedures," EIA Publication No. TEP105-15, Oct. 1988.
44. Electronics Industries Association (EIA), "Phosphor Linearity Tests Using Illuminance and Microphotometer Detectors," EIA Publication No. TEP105-16, Aug. 1990.
45. Electronics Industries Association (EIA), "MTF Test Method for Monochrome CRT Display Systems," EIA Publication No. TEP105-17, July 1990.
46. Electronics Industries Association (EAI), "Method for Calculating Refractive Gain for Color TV Picture Tube and Color Monitor Tube Screens," EIA Publication No. TEP195, Dec. 1987.
47. Electronics Industries Association (EIA), "Registered Screen Dimensions for Monochrome Picture Tubes," EIA Publication No. EIA-266-A, Sept. 1979.
48. Electronics Industries Association (EIA), "Registered Screen Dimensions for Color Picture Tubes," EIA Publication No. ANSI/EIA-324-A, April 1982.
49. Electronics Industries Association (EIA), "Screen Definition for Color Picture Tubes," EIA Publication No. ANSI/EIA-527, June 1986 (reaffirmed June 1993).
50. Electronics Industries Association (EIA), "Typical EIA/TEPAC Picture Tube Screen Dimensions," EIA Publication No. TEP-162, May 1980.
51. Electronics Industries Association (EIA), "Monochrome Cathode-Ray Tube Bulb Criteria," EIA Publication No. TEP110-A, Oct. 1985.
52. Electronics Industries Association (EIA), "Glossary of Terms Used in the Description of Glass Components and of Their Defects," EIA Publication No. TEP123-A, March 1962 (reaffirmed May 1980).
53. Electronics Industries Association (EIA), "Criteria of Bulbs and Implosion Panel for Television Picture Tubes," EIA Publication No. TEP131, March 1966 (reaffirmed Oct. 1982).
54. Electronics Industries Association (EIA), "Criteria for Bulb Parts for Color Television Picture Tubes," EIA Publication No. TEP131 (Color), July 1973 (reaffirmed Oct. 1981).
55. Electronics Industries Association (EIA), "Deflecting Yokes for Cathode-Ray Tubes," EIA Publication No. EIA-256-A, Jan. 1965 (reaffirmed June 1980).

56. Electronics Industries Association (EIA), "Magnetic Deflection Yokes," EIA TEPAC Engineering Bulletin TEB-22, Nov. 1979.
57. Electronics Industries Association (EIA), "Optical Characteristics of Cathode-Ray Tube Screens," EIA Publication No. TEP116-C, Feb. 1993.
58. Electronics Industries Association (EIA), "Worldwide Type Designation System for TV Picture Tubes and Monitor Tubes," EIA Publication No. TEP-106-B, June 1988.
59. Electronics Industries Association (EIA), "Recommended Practice for Measurement of X-Radiation from Projection Cathode-Ray Tubes," EIA Publication No. ANSI/EIA-500-A, Feb. 1989.
60. Electronics Industries Association (EIA), "Recommended Practice for Measurement of X-Radiation from Raster-Scanned Direct-View Data Display Cathode-Ray Tubes," EIA Publication No. ANSI/EIA-501-A, May 1990.
61. Electronics Industries Association (EIA), "Recommended Practice for Measurement of X-Radiation from Non-Raster-Scanned Direct-View Data Display Cathode-Ray Tubes," EIA Publication No. ANSI/EIA-502-A, March 1989.
62. Electronics Industries Association (EIA), "Recommended Practice for Measurement of X-Radiation from Direct-View Television Picture Tubes," EIA Publication No. ANSI/EIA-503-A, May 1990.
63. Electronics Industries Association (EIA), "Considerations Used in Establishing the X-Radiation Ratings of Monochrome and Color Direct-View Television Picture and Data Display Tubes," EIA Publication No. TEP-194, July 1986.
64. Electronics Industries Association (EIA), Amendment No. 2 to TEP-194, July 1986.
65. Electronics Industries Association (EIA), "Cathode-Ray Tube X-Radiation Round Robin Procedures," EIA Publication No. TEP-196, Oct. 1989.
66. Electronics Industries Association (EIA), "Preparation of X-Radiation Characteristic Curves for Cathode-Ray Tubes," EIA Publication No. TEP-197, July 1989.
67. Electronics Industries Association (EIA), "CRT Considerations for Raster Dot Alpha-Numeric Presentations," EIA TEPAC Engineering Bulletin TEB-21, Apr. 1979.
68. Electronics Industries Association (EIA), "Spot Pulse Width and Repetition Rate in Raster Dot Alpha-Numeric CRT Presentations for Rectangular Pulses," EIA TEPAC Engineering Bulletin TEB-23-A, Nov. 1981.
69. Electronics Industries Association (EIA), "The Effect of Pulse Shape in Raster Dot Alpha-CRT Numeric Presentations on Spot Luminance and Luminance Distribution," EIA TEPAC Engineering Bulletin TEB-24, Nov. 1981.
70. Electronics Industries Association (EIA), "A Survey of Data-Display CRT Resolution Measurement Techniques," EIA TEPAC Engineering Bulletin TEB-25, June 1985.
71. Electronics Industries Association (EIA), "1976 CIE-UCS Chromaticity Diagram with Color Boundaries," EIA TEPAC Engineering Bulletin TEB-26, July 1988.
72. Electronics Industries Association (EIA), "Relating CRT Resolution and Addressability," EIA TEPAC Engineering Bulletin TEB-27, Sept. 1988.
73. Electronic Industries Association of Japan (EIAJ) "Terms and Definitions for Liquid Crystal Display Devices," EIAJ Publication No. LD-101, 1980.
74. Electronic Industries Association of Japan (EIAJ), "Measuring Methods for Liquid Crystal Display Panels and Constructive Materials," EIAJ Publication No. LD-201, 1984.
75. Electronic Industries Association of Japan (EIAJ), "Testing Methods for Cathode-Ray Tubes," EIAJ Publication No. ED-2131, March 1991.
76. Electronic Industries Association of Japan (EIAJ), "Testing Methods for Cathode-Ray Tubes with Deflection Yokes," EIAJ Publication No. ED-2101A, May 1989.
77. Electronic Industries Association of Japan (EIAJ), "Measuring Method of Phosphor Persistence for CRT Screen," EIAJ Publication No. ED-2102, Sept. 1988.

78. Electronic Industries Association of Japan (EIAJ), "Guidance for Conforming to the F.R.G.'s 'Self-Protected CRT,'" EIAJ Publication No. ED-2161A, Dec. 1991.
79. Electronic Industries Association of Japan (EIAJ), "The Designation System for the X-Radiation Limit Curve for Cathode-Ray Tubes and the Unification System of X-Radiation Characteristics Standards," EIAJ Publication No. ED-2111, July 1991.
80. Electronic Industries Association of Japan (EIAJ), "Measuring Method of X-Radiation from TV Picture Tubes," EIAJ Publication No. ED-2112, July 1991.
81. Electronic Industries Association of Japan (EIAJ), "Practice for the Preparation of X-Radiation Limit Curves of Cathode-Ray Tubes," EIAJ Publication No. ED-2113, July 1991.
82. Electronic Industries Association of Japan (EIAJ), "The Measuring Method of X-Radiation from Raster Scanned Direct View Monitor Display Cathode-Ray Tubes," EIAJ Publication No. ED-2115, July 1991.
83. Electronic Industries Association of Japan (EIAJ), "The Measuring Method of X-Radiation from Projection Cathode-Ray Tubes," EIAJ Publication No. ED-2116, July 1991.
84. Electronic Industries Association of Japan (EIAJ), "The Measuring Method of X-Radiation from Non-Raster-Scanned Direct-View Display Cathode-Ray Tubes," EIAJ Publication No. ED-2117, July 1991.
85. Electronic Industries Association of Japan (EIAJ), "Practice for the Preparation of X-Radiation Limit Curves of Non-Raster-Scanned Direct-View Cathode-Ray Tubes," EIAJ Publication No. ED-2118, Mar. 1990.
86. Electronic Industries Association of Japan (EIAJ), "The Designation System for the X-Radiation Limit Curves for Non-Raster-Scanned Direct-View Cathode-Ray Tubes, and the Unification System of X-Radiation Characteristics Standards," EIAJ Publication No. ED-2119, July 1991.
87. Electronic Industries Association of Japan (EIAJ), "Inspection Standards of Glass Bulbs for Cathode-Ray Tubes," EIAJ Publication No. ED-2135, Mar. 1990.
88. Electronic Industries Association of Japan (EIAJ), "Type Designation System of Glass Bulbs for Cathode-Ray Tubes," EIAJ Publication No. ED-2134A, June 1992.
89. Electronic Industries Association of Japan (EIAJ), "Useful Screen Dimensions and Area of Glass Bulbs for Cathode-Ray Tubes," EIAJ Publication No. ED-2136, Dec. 1989.
90. Electronic Industries Association of Japan (EIAJ), "Glass Bulbs for Cathode-Ray Tubes, Characteristics Standards for," EIAJ Publication No. ED-2138, Feb. 1992.
91. Electronic Industries Association of Japan (EIAJ), "Fault and Defect Criteria in Useful Screen Area for Cathode-Ray Tubes," EIAJ Publication No. ED-2136, Dec. 1989.
92. American National Standards Institute (ANSI)/Human Factors Society (HFS), "American National Standard for Human Factors Engineering of Visual Display Terminal Workstations," ANSI/HFS 100-1988, Feb. 4, 1988.
93. International Electrotechnical Commission (IEC), "Detail Specification: Monochrome Cathode-Ray Tube for Alphanumeric/Video Display," IEC Publication No. PQC39/US0001, May 1990.
94. International Electrotechnical Commission (IEC), "Measurements of the Electrical Properties of Electronic Tubes, Part 14: Methods of Measurement of Radar and Oscilloscope Cathode-Ray Tubes," IEC Publication No. 151-16, 1968.
95. International Electrotechnical Commission (IEC), "Measurements of the Electrical Properties of Electronic Tubes, Part 16: Methods of Measurement for Television Picture Tubes," IEC Publication No. 151-16, 1968.
96. International Electrotechnical Commission (IEC), "Measurements of the Electrical Properties of Electronic Tubes, Part 28: Methods of Measurement of Colour Television Picture Tubes," IEC Publication No. 151-28, 1978.

97. International Electrotechnical Commission (IEC), "Photometric and Colorimetric Methods of Measurement of the Light Emitted by a Cathode-Ray Tube Screen," IEC Publication No. 441, 1974.

98. American National Standards Institute (ANSI)/Institute of Electrical and Electronics Engineers (IEEE), "Letter Symbols for Units of Measurement," ANSI/IEEE Std. 260-1978, 1978 (reaffirmed 1985).

99. American National Standards Institute (ANSI)/Institute of Electrical and Electronics Engineers (IEEE), "American National Standard for Metric Practice," ANSI/IEEE Std. 268-1992, Oct. 28, 1992.

100. American National Standards Institute (ANSI)/Institute of Electrical and Electronics Engineers (IEEE), "IEEE Dictionary of Electrical and Electronics Terms," ANSI/IEEE Std. 100-1988, 1988.

101. Illuminating Engineering Society of North America (IESNA), *Lighting Handbook,* 8th ed., IESNA, New York, 1993.

102. P. J. Alessi, "Overview of ISCC and ASTM Committee Work on Video Displays," *Color Res. Applicat.,* Vol. 11 (Suppl.), p. S29, 1986.

103. International Organization for Standardization (ISO), "Ergonomic Requirements for Office Work with Visual Display Terminals (VDTs), Part 3: Visual Display Requirements," ISO Publication No. 9241-3, 1992.

104. International Organization for Standardization (ISO), "Ergonomic Requirements for Office Work with Visual Display Terminals (VDTs), Part 7: Display Requirements with Reflections," ISO Publication No. 9241-7.2, 1991 (Draft).

105. International Organization for Standardization (ISO), "Ergonomic Requirements for Office Work with Visual Display Terminals (VDTs), Part 8: Requirements for Displayed Colours," ISO Publication No. 9241-8, 1992 (Draft).

106. International Organization for Standardization (ISO), "Ergonomic Requirements for Office Work with Visual Display Terminals (VDTs), Part 3: Flat Panel Display Addendum," ISO Publication No. 9241-3, Sept. 27, 1991 (Draft to become ISO 13406).

107. Department of Defense (DOD), "Performance Specification—Electron Tubes, General Specification for," MIL-PRF-1L, Dec. 30, 1994.

108. Department of Defense (DOD), "Test Methods for Electron Tubes," MIL-STD-1311B, Mar. 28, 1975.

109. U.S. Air Force (USAF), "Displays, Airborne, Electronically/Optically Generated," AFGS-87213A, Nov. 30, 1987.

110. Department of Defense (DOD), "Calibration Systems Requirements," MIL-STD-45662A, Aug. 1, 1988.

111. American National Standards Institute (ANSI)/ National Conference of Standards Laboratories (NCSL), "American National Standard for Calibration—Calibration Laboratories and Measuring Test Equipment—General Requirements," ANSI/NCSL Z540-1-1994, July 27, 1994.

112. American National Standards Institute (ANSI)/National Association of Photographic Manufacturers (NAPM), "Liquid-Crystal Imaging Devices for Use with Overhead Projectors—Method for Measuring and Reporting Performance Characteristics and Features," ANSI IT7.228-1990, Nov. 8, 1990.

113. American National Standards Institute(ANSI)/National Association of Photographic Manufacturers (NAPM), "Data Projection Equipment and Large Screen Data Displays—Test Methods and Performance Characteristics," ANSI IT7.215-1992, Aug. 6, 1992.

114. National Information Display Laboratory (NIDL), "Display Monitor Measurement Methods, Part 1: Monochrome CRT Monitor Performance," NIDL Publication No. 171795-036, Draft Version 2.0, July 12, 1995, p. 78.

115. National Information Display Laboratory (NIDL), "Display Monitor Measurement Methods, Part 2: Color CRT Monitor Performance," NIDL Publication No. 171795-037, Draft Version 2.0, July 12, 1995, p. 37.
116. D. B. Judd, "Contributions to Color Science," NBS Special Publication 545, 1979.
117. K. L. Kelly and D. B. Judd, "Color: Universal Language and Dictionary of Color Names," NBS Special Publication 440, Dec. 1976.
118. J. D. Simmons, "NIST Calibration Services Guide," NIST Special Publication 250, 1991.
119. National Institute of Standards and Technology (NIST), "Flat Panel Display Measurements," draft submitted to VESA Sept. 1, 1995, revised Oct. 6, 1995.
120. Society of Automotive Engineers (SAE), "Photometric and Colorimetric Procedures for Direct View CRT Displays," SAE Recommended Practice ARP1782, Jan. 1989.
121. Society of Automotive Engineers (SAE), "Photometric and Colorimetric Measurement Procedures for Airborne Flat Panel Displays," Draft, SAE Recommended Practice ARP4260, Dec. 12, 1996.
122. Society of Automotive Engineers (SAE), "Minimum Performance Standard for Airborne Multipurpose Electronic Displays," SAE Aerospace Std. AS-8034, Dec. 30, 1982 (reaffirmed May 1993).
123. Society of Automotive Engineers (SAE), "Design Objectives for CRT Displays for Part 25 (Transport) Aircraft," SAE Recommended Practice ARP1874, May 1988 (reaffirmed May 1993).
124. Society of Automotive Engineers (SAE), "Design Objectives for Liquid Crystal Displays for Part 25 (Transport) Aircraft," Draft, SAE Recommended Practice ARP4256, 1995.
125. Society of Automotive Engineers (SAE), "Design Objectives for CRT Displays for Part 25 Aircraft," SAE Recommended Practice ARP4067, Nov. 1989.
126. Society of Motion Picture and Television Engineers (SMPTE), "Setting Chromaticity and Luminance of White for Color Television Monitors using Shadow-Mask Picture Tubes," SMPTE Recommended Practice RP 71-1977, Jan. 27, 1977.
127. Society of Motion Picture and Television Engineers (SMPTE), "SMPTE C Color Monitor Colorimetry," SMPTE Recommended Practice RP 145-1994, June 1, 1994.
128. Society of Motion Picture and Television Engineers (SMPTE), "Alignment Color Bar Test Signal for Television Picture Monitors," SMPTE Engineering Guideline EG 1-1990, Mar. 28, 1990.
129. Society of Motion Picture and Television Engineers (SMPTE), "Critical Viewing Conditions for Evaluation of Color Television Pictures," SMPTE Recommended Practice RP 166, undated (proposed).
130. Society of Motion Picture and Television Engineers (SMPTE), "Alignment of Professional Television Color Monitors," SMPTE Recommended Practice RP 167, Sept. 1992 (proposed).
131. Society of Motion Picture and Television Engineers (SMPTE), "Composite Analog Video Signal—NTSC for Studio Applications," SMPTE Recommended Practice RP 170M-1994, Oct. 19, 1994.
132. Society of Motion Picture and Television Engineers (SMPTE), "Signal Parameters—1125-Line High-Definition Production Systems," SMPTE Recommended Practice RP 240M-1994, June 1, 1994.
133. Society of Motion Picture and Television Engineers (SMPTE), "Specifications for Operational Alignment Test Pattern for Television," SMPTE Recommended Practice RP 27.1-1989, 1989.
134. Society of Motion Picture and Television Engineers (SMPTE), "Specifications for Deflection Linearity Test Pattern for Television," SMPTE Recommended Practice RP 38.1-1989, 1989.
135. Society of Motion Picture and Television Engineers (SMPTE), "Specifications for Medical Diagnostic Imaging Test Pattern for Television Monitors and Hard-Copy Recording Cameras," SMPTE Recommended Practice RP 133-1986, Jan. 8, 1986.

136. J. E. Gray, K. G. Lisk, D. H. Haddick, J. H. Harshbarger, A. Oosterhof, and R. Schwenker, "Test Pattern for Video Displays and Hard-Copy Cameras," *Radiology,* Vol. 152, No. 2, pp. 519–527, 1985.
137. Society of Motion Picture and Television Engineers (SMPTE), "Measurement of Screen Luminance in Theaters," SMPTE Recommended Practice RP 98-1990, Feb. 10, 1990.
138. Society of Motion Picture and Television Engineers (SMPTE), "Installation of Gain Screens," SMPTE Recommended Practice RP 95-1994, Jan. 12, 1994.
139. J. E. Schuessler, "Flat Panel Standardization Activities in the United States," Asia Display '95, 1995, pp. 955–956.
140. "Video Electronics Standards Association (VESA), "Display Specifications and Test Procedure," Version 1.0, Rev. 1.0, Oct. 3, 1994.
141. National Institute of Standards and Technology (NIST), "Flat Panel Display Measurements," draft submitted to VESA Sept. 1, 1995, revised Oct. 6, 1995.

APPENDIX A

Unit Abbreviations

Candela	cd
Candle power*	cp
Centimeter†	cm
Footcandle	fc
Footlambert	fL
Kelvin	K
Lumen	lm
Lux	lx
Mean spherical candlepower‡	mscp
Meter	m
Nit	nt
Second	s
Steradian	sr
Watt	W

*Replaced by the candela.
†Deprecated.
‡Replaced by mean spherical luminous intensity.

APPENDIX B

SI Prefixes

Prefix	Exponent	Symbol
Yotta	10^{24}	Y
Zetta	10^{21}	Z
Exa	10^{18}	E
Peta	10^{15}	P
Tera	10^{12}	T
Giga	10^{9}	G
Mega	10^{6}	M
Kilo	10^{3}	k
Hecto*	10^{2}	h
Deka*	10^{1}	da
Deci*	10^{-1}	d
Centi*	10^{-2}	c
Milli	10^{-3}	m
Micro	10^{-6}	μ
Nano	10^{-9}	n
Pico	10^{-12}	p
Femto	10^{-15}	f
Atto	10^{-18}	a
Zepto	10^{-21}	z
Yocto	10^{-24}	y

*Deprecated

Source: Derived from *American National Standard for Metric Practice*, ANSI/IEEE Std. 268-1992, 1992.

APPENDIX C

Laws, Formulas, and Constants

Boltzmann's Constant

$$1.3807 \times 10^{-23} \text{ J/K}$$

Gamma

$$B = KE^\gamma$$

where

B = luminance
E = CRT control grid drive above cutoff
γ = exponent representing gamma
K = a constant

Grassman's Laws

1. Lights of the same color produce identical effects in mixtures regardless of their spectral composition.
2. The luminance produced by the additive mixture of a number of lights is the sum of the luminances produced separately by each of the lights.

Inverse Square Law   Used to compute light intensity of a point source versus distance:

$$\frac{E_1}{E_2} = \left(\frac{d_2}{d_1}\right)^2$$

where

E_1 = initial illuminance or irradiance
E_2 = illuminance or irradiance at d_2
d_1 = initial distance
d_2 = new distance

Kirchhoff's Law Relates the radiant exitance of a surface (M) in watts per square meter to that of a true blackbody emitter (M_b) as a function of the surface's absorptance (α), which is also equal to its emissivity (ϵ):

$$\frac{M}{\alpha} = M_b$$

Lambert's Law An infinitely large and perfectly diffusing (Lambertian) surface will have equal luminance when measured from any direction due to the increasing area being measured as the angle from normal increases. This exactly compensates for the cosine decrease of light from a given point on the surface.

Planck's Constant (h)

$$h = 6.626 \times 10^{-34} \text{ Js}$$

Planck's Radiation Law The basic law of thermal radiation that describes the radiant exitance (M) of a blackbody (Planckian) radiator as a function of temperature in Kelvin and wavelength (λ):

$$M_{e,\lambda}(\lambda, T) = c_1 \lambda^{-5} (e^{c_2/\lambda T} - 1) \qquad \text{W} \cdot \text{m}^{-3}$$

where

$$c_1 = 3.74150 \times 10^{-16} \text{ W} \cdot \text{m}_2 \qquad c_2 = 1.4388 \times 10^{-2} \text{ m} \cdot \text{K}$$

Stefan–Boltzmann Constant

$$\sigma = 5.67 \times 10^{-8} \text{ W m}^{-2} \text{ K}^{-4}$$

Stefan–Boltzmann Law Computes the radiant exitance (M) in watts per square meter of a blackbody source with temperature in Kelvin:

$$M = \sigma K^4$$

where

σ = Stefan–Boltzmann constant
K^4 = temperature in Kelvin

Talbott's Law If a point on the retina is excited by light undergoing periodic intensity variations at a frequency exceeding the fusion frequency, the sensation produced is identical with that of a steady light whose intensity equals the mean of the variable intensity taken over one period.

Velocity of Light in Vacuum

$$c = 299.8 \times 10^6 \text{ m/s}$$

290 APPENDIX C

Wien Displacement Law Relates the peak wavelength (λ_{max}) of a blackbody radiator to its temperature in Kelvin:

$$\lambda_{max} = \frac{2898}{K}$$

Incandescent Lamp Formulas (Unless otherwise noted the following is from the General Electric Company, "Incandescent Lamps," GE Publication TP-110R, 1977, p. 29.)

In the following formulas, the subscript 1 is the normal rated value from the manufacturer's data and the subscript 2 is the changed value. Exponents shown in the equations are typical for gas-filled lamps that are normally lamps of about 40 W and higher. Smaller lamps are usually vacuum lamps and their typical exponents are given following the equations.

Lumens versus voltage:

$$\frac{lm_2}{lm_1} = \left(\frac{V_2}{V_1}\right)^{3.38}$$

(Use exponent of 3.51 for vacuum lamps.)

Efficacy versus voltage:

$$\frac{(lm/W)_2}{(lm/W)_1} = \left(\frac{V_2}{V_1}\right)^{1.84}$$

(Use exponent of 1.93 for vacuum lamps.)

Watts versus voltage:

$$\frac{W_2}{W_1} = \left(\frac{V_2}{V_1}\right)^{1.54}$$

(Use exponent of 1.58 for vacuum lamps.)

Amperes versus voltage:

$$\frac{A_2}{A_1} = \left(\frac{V_2}{V_1}\right)^{0.541}$$

(Use exponent of 0.580 for vacuum lamps.)

Life versus voltage:

$$\frac{A_2}{A_1} = \left(\frac{V_1}{V_2}\right)^{13.1}$$

(Use exponent of 13.5 for vacuum lamps.)

Color Temperature in kelvin versus voltage:

$$\frac{K_2}{K_1} = \left(\frac{V_1}{V_2}\right)^{0.42}$$

(*Source:* Illuminating Engineering Society of North America, *Lighting Handbook,* 8th ed., 1993, p. 186.)

Transmittance Versus Filter Thickness (absorbing filters such as glass and plastic)

$$\tau_2 = \tau_1^{d_2/d_1}$$

where

d = filter thickness
1 = initial values
2 = changed values
τ = filter internal transmission factor (as a ratio, not percent; excludes surface reflectance losses)

Optical Density (commonly used to specify neutral-density filters)

$$D = \log\left(\frac{1}{\tau}\right)$$

where

τ = filter internal transmission factor (as a ratio, not percent; usually excludes surface reflectance losses)

Transmittance of Filters in Series

$$\tau_\lambda = \tau_{1\lambda} \cdot \tau_{2\lambda} \cdot \tau_{3\lambda} \cdots$$

where

τ = transmittance factor
λ = wavelength

Surface Reflection

$$r = \left(\frac{n-1}{n+1}\right)^2$$

where

n = index of refraction of the material

Formulas for Calculating Scan Frequencies, Video Bandwidth, and Digital Clock Frequencies

$$f_h = f_v \cdot \frac{l_v}{df_v}$$

$$f_{clk} = f_h \times 10^{-6} \left(\frac{p_h}{df_h}\right)$$

$$\text{bw} = f_{clk} \cdot k$$

where

- f = frequency, Hz
- h = horizontal
- v = vertical
- l = number of scan lines
- df = duty factor, typically 0.98 for vertical and 0.8 for horizontal (active scan/scan period)
- f_{clk} = pixel clock frequency, MHz
- p_h = number of horizontal pixels
- bw = video bandwidth, MHz
- k = constant of about 0.6

APPENDIX D

Radiometric and Photometric Conversions

Footcandles to lux	lx = 10.764 · fc
Lux to footcandles	fc = lx/10.764
Footlamberts to cd/m² (nits)	cd/m² = 3.426 · fL
cd/m² (nits) to footlamberts	fL = (cd/m²)/3.426
Lux to candelas	c = l × d^2 (where d = distance in meters)
Footcandles to candelas	cd = fc × d^2 (where d = distance in feet)
Wavelength to frequency	THz = 300,000/nm
Frequency to wavelength	nm = 300,000/THz
Wavelength to electron volts	eV = 1239.5/nm
Electron volts to wavelength	nm = 1239.5/eV
Footcandles to watts per square meter at 555 nm	1 fc = 15.8 mW/m²
Lux to watts per square meter at 555 nm	1 lx = 1.47 mW/m²
Footcandles to watts per square meter at 632.8 nm	1 fc = 62.4 mW/m²
Lux to watts per square meter at 632.8 nm	1 lx = 5.8 mW/m²
Angstroms to nanometers	nm = Å/10
Nanometers to angstroms	Å = nm · 10
Joules to ergs	1 J = 1 × 10^7 ergs
Ergs to joules	1 erg = 1 × 10^{-7} J
Watts per square centimeter to watts per square meter	W/m² = W/cm² · 10,000
Watts per square meter to watts per square centimeter	W/cm² = W/m²/10,000
Micrometers to nanometers	nm = 1000 · μm
Nanometers to micrometers	μm = nm/1000

APPENDIX E

1 nm CIE Tristimulus Values

nm	X	Y	Z	X + Y + Z	nm
380	0.00136	0.00003	0.00645	0.00784	380
381	0.00150	0.00004	0.00708	0.00862	381
382	0.00164	0.00004	0.00774	0.00942	382
383	0.00180	0.00005	0.00850	0.01035	383
384	0.00199	0.00005	0.00941	0.01145	384
385	0.00223	0.00006	0.01055	0.01284	385
386	0.00253	0.00007	0.01196	0.01456	386
387	0.00289	0.00008	0.01365	0.01662	387
388	0.00330	0.00009	0.01558	0.01897	388
389	0.00375	0.00010	0.01773	0.02158	389
390	0.00423	0.00012	0.02005	0.02440	390
391	0.00476	0.00013	0.02251	0.02740	391
392	0.00533	0.00015	0.02520	0.03068	392
393	0.00597	0.00017	0.02828	0.03442	393
394	0.00674	0.00019	0.03189	0.03882	394
395	0.00765	0.00021	0.03621	0.04407	395
396	0.00875	0.00024	0.04143	0.05042	396
397	0.01002	0.00028	0.04750	0.05780	397
398	0.01142	0.00031	0.05412	0.06585	398
399	0.01286	0.00035	0.06099	0.07420	399
400	0.01431	0.00039	0.06785	0.08255	400
401	0.01570	0.00043	0.07448	0.09061	401
402	0.01714	0.00047	0.08136	0.09897	402
403	0.01878	0.00051	0.08915	0.10844	403
404	0.02074	0.00057	0.09854	0.11985	404
405	0.02319	0.00064	0.11020	0.13403	405
406	0.02620	0.00072	0.12461	0.15153	406
407	0.02978	0.00082	0.14170	0.17230	407
408	0.03388	0.00094	0.16130	0.19612	408
409	0.03846	0.00107	0.18325	0.22278	409
410	0.04351	0.00121	0.20740	0.25212	410
411	0.04899	0.00136	0.23369	0.28404	411
412	0.05502	0.00153	0.26261	0.31916	412
413	0.06171	0.00172	0.29477	0.35820	413
414	0.06921	0.00193	0.33079	0.40193	414
415	0.07763	0.00218	0.37130	0.45111	415
416	0.08695	0.00245	0.41620	0.50560	416

nm	X	Y	Z	X + Y + Z	nm
417	0.09717	0.00276	0.46546	0.56539	417
418	0.10840	0.00311	0.51969	0.63120	418
419	0.12076	0.00352	0.57953	0.70381	419
420	0.13438	0.00400	0.64560	0.78398	420
421	0.14935	0.00454	0.71848	0.87237	421
422	0.16539	0.00515	0.79671	0.96725	422
423	0.18198	0.00582	0.87784	1.06564	423
424	0.19861	0.00654	0.95943	1.16458	424
425	0.21477	0.00730	1.03905	1.26112	425
426	0.23018	0.00808	1.11536	1.35362	426
427	0.24488	0.00890	1.18849	1.44227	427
428	0.25877	0.00976	1.25812	1.52665	428
429	0.27180	0.01066	1.32393	1.60639	429
430	0.28390	0.01160	1.38560	1.68110	430
431	0.29494	0.01257	1.44263	1.75014	431
432	0.30489	0.01358	1.49480	1.81327	432
433	0.31378	0.01463	1.54219	1.87060	433
434	0.32164	0.01571	1.58488	1.92223	434
435	0.32850	0.01684	1.62296	1.96830	435
436	0.33435	0.01800	1.65640	2.00875	436
437	0.33921	0.01921	1.68529	2.04371	437
438	0.34312	0.02045	1.70987	2.07344	438
439	0.34613	0.02171	1.73038	2.09822	439
440	0.34828	0.02300	1.74706	2.11834	440
441	0.34960	0.02429	1.76004	2.13393	441
442	0.35014	0.02561	1.76962	2.14537	442
443	0.35001	0.02695	1.77626	2.15322	443
444	0.34928	0.02835	1.78043	2.15806	444
445	0.34806	0.02980	1.78260	2.16046	445
446	0.34637	0.03131	1.78296	2.16064	446
447	0.34426	0.03288	1.78170	2.15884	447
448	0.34180	0.03452	1.77919	2.15551	448
449	0.33909	0.03622	1.77586	2.15117	449
450	0.33620	0.03800	1.77211	2.14631	450
451	0.33319	0.03984	1.76825	2.14128	451
452	0.33004	0.04176	1.76403	2.13583	452
453	0.32663	0.04376	1.75894	2.12933	453
454	0.32288	0.04584	1.75246	2.12118	454
455	0.31870	0.04800	1.74410	2.11080	455
456	0.31402	0.05024	1.73355	2.09781	456
457	0.30888	0.05257	1.72085	2.08230	457
458	0.30329	0.05498	1.70593	2.06420	458
459	0.29725	0.05745	1.68873	2.04343	459
460	0.29080	0.06000	1.66920	2.02000	460
461	0.28397	0.06260	1.64752	1.99409	461
462	0.27672	0.06527	1.62341	1.96540	462
463	0.26891	0.06804	1.59602	1.93297	463
464	0.26042	0.07091	1.56452	1.89585	464
465	0.25110	0.07390	1.52810	1.85310	465
466	0.24084	0.07701	1.48611	1.80396	466

nm	X	Y	Z	X + Y + Z	nm
467	0.22985	0.08026	1.43952	1.74963	467
468	0.21840	0.08366	1.38988	1.69194	468
469	0.20681	0.08723	1.33873	1.63277	469
470	0.19536	0.09098	1.28764	1.57398	470
471	0.18421	0.09491	1.23742	1.51654	471
472	0.17332	0.09904	1.18782	1.46018	472
473	0.16268	0.10336	1.13876	1.40480	473
474	0.15228	0.10788	1.09014	1.35030	474
475	0.14210	0.11260	1.04190	1.29660	475
476	0.13217	0.11753	0.99419	1.24389	476
477	0.12257	0.12267	0.94734	1.19258	477
478	0.11327	0.12799	0.90145	1.14271	478
479	0.10429	0.13345	0.85661	1.09435	479
480	0.09564	0.13902	0.81295	1.04761	480
481	0.08730	0.14467	0.77051	1.00248	481
482	0.07930	0.15046	0.72944	0.95920	482
483	0.07171	0.15646	0.68991	0.91808	483
484	0.06458	0.16271	0.65210	0.87939	484
485	0.05795	0.16930	0.61620	0.84345	485
486	0.05186	0.17624	0.58232	0.81042	486
487	0.04628	0.18355	0.55041	0.78024	487
488	0.04115	0.19127	0.52033	0.75275	488
489	0.03641	0.19941	0.49196	0.72778	489
490	0.03201	0.20802	0.46518	0.70521	490
491	0.02791	0.21712	0.43992	0.68495	491
492	0.02414	0.22673	0.41618	0.66705	492
493	0.02068	0.23685	0.39388	0.65141	493
494	0.01754	0.24748	0.37294	0.63796	494
495	0.01470	0.25860	0.35330	0.62660	495
496	0.01216	0.27018	0.33485	0.61719	496
497	0.00992	0.28229	0.31755	0.60976	497
498	0.00796	0.29505	0.30133	0.60434	498
499	0.00629	0.30857	0.28616	0.60102	499
500	0.00490	0.32300	0.27200	0.59990	500
501	0.00377	0.33840	0.25881	0.60098	501
502	0.00294	0.35468	0.24648	0.60410	502
503	0.00242	0.37169	0.23477	0.60888	503
504	0.00223	0.38928	0.22345	0.61496	504
505	0.00240	0.40730	0.21230	0.62200	505
506	0.00292	0.42563	0.20116	0.62971	506
507	0.00383	0.44431	0.19012	0.63826	507
508	0.00517	0.46339	0.17922	0.64778	508
509	0.00698	0.48294	0.16856	0.65848	509
510	0.00930	0.50300	0.15820	0.67050	510
511	0.01214	0.52356	0.14813	0.68383	511
512	0.01553	0.54451	0.13837	0.69841	512
513	0.01947	0.56569	0.12899	0.71415	513
514	0.02399	0.58696	0.12007	0.73102	514
515	0.02910	0.60820	0.11170	0.74900	515
516	0.03481	0.62934	0.10390	0.76805	516
517	0.04112	0.65030	0.09666	0.78808	517

nm	X	Y	Z	X + Y + Z	nm
518	0.04798	0.67087	0.08998	0.80883	518
519	0.05537	0.69084	0.08384	0.83005	519
520	0.06327	0.71000	0.07825	0.85152	520
521	0.07163	0.72818	0.07320	0.87301	521
522	0.08046	0.74546	0.06867	0.89459	522
523	0.08974	0.76196	0.06456	0.91626	523
524	0.09945	0.77783	0.06078	0.93806	524
525	0.10960	0.79320	0.05725	0.96005	525
526	0.12016	0.80811	0.05390	0.98217	526
527	0.13111	0.82249	0.05074	1.00434	527
528	0.14236	0.83630	0.04775	1.02641	528
529	0.15385	0.84949	0.4489	1.04823	529
530	0.16550	0.86200	0.04216	1.06966	530
531	0.17725	0.87381	0.03950	1.09056	531
532	0.18914	0.88496	0.03693	1.11103	532
533	0.20116	0.89549	0.03445	1.13110	533
534	0.21336	0.90544	0.03208	1.15088	534
535	0.22575	0.91485	0.2984	1.17044	535
536	0.23832	0.92373	0.02771	1.18976	536
537	0.25106	0.93209	0.02569	1.20884	537
538	0.26399	0.93992	0.02378	1.22769	538
539	0.27710	0.94722	0.02198	1.24630	539
540	0.29040	0.95400	0.02030	1.26470	540
541	0.30389	0.96025	0.01871	1.28285	541
542	0.31757	0.96600	0.01724	1.30081	542
543	0.33143	0.97126	0.01586	1.31855	543
544	0.34548	0.97602	0.01458	1.33608	544
545	0.35970	0.98030	0.01340	1.35340	545
546	0.37408	0.98409	0.01230	1.37047	546
547	0.38864	0.98748	0.01130	1.38742	547
548	0.40337	0.99031	0.01037	1.40405	548
549	0.41831	0.99281	0.00952	1.42064	549
550	0.43345	0.99495	0.00875	1.43715	550
551	0.44879	0.99671	0.00803	1.45353	551
552	0.46433	0.99809	0.00738	1.46980	552
553	0.48006	0.99911	0.00678	1.48595	553
554	0.49597	0.99974	0.00624	1.50195	554
555	0.51205	1.00000	0.00575	1.51780	555
556	0.52829	0.99985	0.00530	1.53344	556
557	0.54469	0.99930	0.00490	1.54889	557
558	0.56120	0.99832	0.00453	1.56405	558
559	0.57782	0.99689	0.00420	1.57891	559
560	0.59450	0.99500	0.00390	1.59340	560
561	0.61122	0.99260	0.00362	1.60744	561
562	0.62797	0.98974	0.00337	1.62108	562
563	0.64476	0.98644	0.00314	1.63434	563
564	0.66157	0.98272	0.00293	1.64722	564
565	0.67840	0.97860	0.00275	1.65975	565
566	0.69523	0.97408	0.00258	1.67189	566
567	0.71205	0.96917	0.00243	1.68365	567
568	0.72882	0.96385	0.00230	1.69497	568

nm	X	Y	Z	X + Y + Z	nm
569	0.74551	0.95813	0.00219	1.70583	569
570	0.76210	0.95200	0.00210	1.71620	570
571	0.77854	0.94545	0.00201	1.72600	571
572	0.79482	0.93849	0.00194	1.73525	572
573	0.81092	0.93116	0.00189	1.74397	573
574	0.82682	0.92345	0.00184	1.75211	574
575	0.84250	0.91540	0.00180	1.75970	575
576	0.85793	0.90700	0.00176	1.76669	576
577	0.87308	0.89827	0.00173	1.77308	577
578	0.88789	0.88920	0.00171	1.77880	578
579	0.90231	0.87978	0.00168	1.78377	579
580	0.91630	0.87000	0.00165	1.78795	580
581	0.92979	0.85986	0.00161	1.79126	581
582	0.94279	0.84939	0.00156	1.79374	582
583	0.95527	0.83862	0.00151	1.79540	583
584	0.96721	0.82758	0.00145	1.79624	584
585	0.97860	0.81630	0.00140	1.79630	585
586	0.98938	0.80479	0.00133	1.79550	586
587	0.99954	0.79308	0.00127	1.79389	587
588	1.00908	0.78119	0.00120	1.79147	588
589	1.01800	0.76915	0.00114	1.78829	589
590	1.02630	0.75700	0.00110	1.78440	590
591	1.03398	0.74475	0.00106	1.77979	591
592	1.04098	0.73242	0.00104	1.77444	592
593	1.04718	0.72000	0.00103	1.76821	593
594	1.05246	0.70749	0.00102	1.76097	594
595	1.05670	0.69490	0.00100	1.75260	595
596	1.05979	0.68221	0.00096	1.74296	596
597	1.06179	0.66947	0.00093	1.73219	597
598	1.06280	0.65667	0.00088	1.72035	598
599	1.06291	0.64384	0.00084	1.70759	599
600	1.06220	0.63100	0.00080	1.69400	600
601	1.06073	0.61815	0.00076	1.67964	601
602	1.05844	0.60531	0.00072	1.66447	602
603	1.05522	0.59247	0.00068	1.64837	603
604	1.05097	0.57963	0.00064	1.63124	604
605	1.04560	0.56680	0.00060	1.61300	605
606	1.03903	0.55396	0.00054	1.59353	606
607	1.03136	0.54113	0.00049	1.57298	607
608	1.02266	0.52835	0.00043	1.55144	608
609	1.01304	0.51563	0.00038	1.52905	609
610	1.00260	0.50300	0.00034	1.50594	610
611	0.99136	0.49046	0.00030	1.48212	611
612	0.97933	0.47803	0.00028	1.45764	612
613	0.96649	0.46567	0.00026	1.43242	613
614	0.95284	0.45340	0.00025	1.40649	614
615	0.93840	0.44120	0.00024	1.37984	615
616	0.92319	0.42908	0.00023	1.35250	616
617	0.90724	0.41703	0.00022	1.32449	617
618	0.89050	0.40503	0.00021	1.29574	618
619	0.87292	0.39303	0.00020	1.26615	619

nm	X	Y	Z	X + Y + Z	nm
620	0.85445	0.38100	0.00019	1.23564	620
621	0.83508	0.36891	0.00017	1.20416	621
622	0.81494	0.35682	0.00015	1.17191	622
623	0.79418	0.34477	0.00013	1.13908	623
624	0.77295	0.33281	0.00011	1.10587	624
625	0.75140	0.32100	0.00010	1.07250	625
626	0.72958	0.30933	0.00008	1.03899	626
627	0.70758	0.29785	0.00007	1.00550	627
628	0.68560	0.28659	0.00006	0.97225	628
629	0.66381	0.27562	0.00005	0.93948	629
630	0.64240	0.26500	0.00005	0.90745	630
631	0.62151	0.25476	0.00004	0.87631	631
632	0.60111	0.24489	0.00003	0.84603	632
633	0.58110	0.23533	0.00003	0.81646	633
634	0.56139	0.22605	0.00003	0.78747	634
635	0.54190	0.21700	0.00003	0.75893	635
636	0.52260	0.20816	0.00002	0.73078	636
637	0.50354	0.19954	0.00002	0.70310	637
638	0.48474	0.19115	0.00002	0.67591	638
639	0.46619	0.18297	0.00002	0.64918	639
640	0.44790	0.17500	0.00002	0.62292	640
641	0.42986	0.16722	0.00001	0.59709	641
642	0.41209	0.15964	0.00001	0.57174	642
643	0.39464	0.15227	0.00001	0.54692	643
644	0.37753	0.14512	0.00001	0.52266	644
645	0.36080	0.13820	0.00001	0.49901	645
646	0.34445	0.13150	0.00000	0.47595	646
647	0.32851	0.12502	0.00000	0.45353	647
648	0.31301	0.11877	0.00000	0.43178	648
649	0.29800	0.11276	0.00000	0.41076	649
650	0.28350	0.10700	0.00000	0.39050	650
651	0.26954	0.10147	0.00000	0.37101	651
652	0.25611	0.09618	0.00000	0.35229	652
653	0.24319	0.09112	0.00000	0.33431	653
654	0.23072	0.08626	0.00000	0.31698	654
655	0.21870	0.08160	0.00000	0.30030	655
656	0.20709	0.07712	0.00000	0.28421	656
657	0.19592	0.07282	0.00000	0.26874	657
658	0.18517	0.06871	0.00000	0.25388	658
659	0.17483	0.06477	0.00000	0.23960	659
660	0.16490	0.06100	0.00000	0.22590	660
661	0.15536	0.05739	0.00000	0.21275	661
662	0.14623	0.05395	0.00000	0.20018	662
663	0.13749	0.05067	0.00000	0.18816	663
664	0.12914	0.04755	0.00000	0.17669	664
665	0.12120	0.04458	0.00000	0.16578	665
666	0.11364	0.04175	0.00000	0.15539	666
667	0.10646	0.03908	0.00000	0.14554	667
668	0.09969	0.03656	0.00000	0.13625	668
669	0.09333	0.03420	0.00000	0.12753	669
670	0.08740	0.03200	0.00000	0.11940	670

nm	X	Y	Z	X + Y + Z	nm
671	0.08190	0.02996	0.00000	0.11186	671
672	0.07680	0.02807	0.00000	0.10487	672
673	0.07207	0.02632	0.00000	0.09839	673
674	0.06768	0.02470	0.00000	0.09238	674
675	0.06360	0.02320	0.00000	0.08680	675
676	0.05980	0.02180	0.00000	0.08160	676
677	0.05628	0.02050	0.00000	0.07678	677
678	0.05297	0.01928	0.00000	0.07225	678
679	0.04981	0.01812	0.00000	0.06793	679
680	0.04677	0.01700	0.00000	0.06377	680
681	0.04378	0.01590	0.00000	0.05968	681
682	0.04087	0.01483	0.00000	0.05570	682
683	0.03807	0.01381	0.00000	0.05188	683
684	0.03540	0.01283	0.00000	0.04823	684
685	0.03290	0.01192	0.00000	0.04482	685
686	0.03056	0.01106	0.00000	0.04162	686
687	0.02838	0.01027	0.00000	0.03865	687
688	0.02634	0.00953	0.00000	0.03587	688
689	0.02445	0.00884	0.00000	0.03329	689
690	0.02270	0.00821	0.00000	0.03091	690
691	0.02108	0.00762	0.00000	0.02870	691
692	0.01960	0.00708	0.00000	0.02668	692
693	0.01823	0.00659	0.00000	0.02482	693
694	0.01698	0.00613	0.00000	0.02311	694
695	0.01584	0.00572	0.00000	0.02156	695
696	0.01479	0.00534	0.00000	0.02013	696
697	0.01383	0.00499	0.00000	0.01882	697
698	0.01294	0.00467	0.00000	0.01761	698
699	0.01212	0.00438	0.00000	0.01650	699
700	0.01135	0.00410	0.00000	0.01545	700
701	0.01062	0.00383	0.00000	0.01445	701
702	0.00993	0.00358	0.00000	0.01351	702
703	0.00928	0.00335	0.00000	0.01263	703
704	0.00867	0.00313	0.00000	0.01180	704
705	0.00811	0.00292	0.00000	0.01103	705
706	0.00758	0.00273	0.00000	0.01031	706
707	0.00708	0.00256	0.00000	0.00964	707
708	0.00662	0.00239	0.00000	0.00901	708
709	0.00619	0.00223	0.00000	0.00842	709
710	0.00579	0.00209	0.00000	0.00788	710
711	0.00541	0.00195	0.00000	0.00736	711
712	0.00505	0.00182	0.00000	0.00687	712
713	0.00471	0.00170	0.00000	0.00641	713
714	0.00440	0.00159	0.00000	0.00599	714
715	0.00410	0.00148	0.00000	0.00558	715
716	0.00383	0.00138	0.00000	0.00521	716
717	0.00357	0.00129	0.00000	0.00486	717
718	0.00333	0.00120	0.00000	0.00453	718
719	0.00310	0.00112	0.00000	0.00422	719
720	0.00289	0.00104	0.00000	0.00393	720
721	0.00270	0.00097	0.00000	0.00367	721

nm	X	Y	Z	X + Y + Z	nm
722	0.00252	0.00091	0.00000	0.00343	722
723	0.00235	0.00085	0.00000	0.00320	723
724	0.00219	0.00079	0.00000	0.00298	724
725	0.00204	0.00074	0.00000	0.00278	725
726	0.00191	0.00069	0.00000	0.00260	726
727	0.00178	0.00064	0.00000	0.00242	727
728	0.00166	0.00060	0.00000	0.00226	728
729	0.00154	0.00055	0.00000	0.00209	729
730	0.00144	0.00052	0.00000	0.00196	730
Sum	106.83514	106.84852	106.84504	320.52870	

APPENDIX F

Chromaticity Conversions

Computation of x, y from X, Y, Z values:

$$x = \frac{X}{X + Y + Z} \qquad y = \frac{Y}{X + Y + Z}$$

Computation of u', v' from X, Y, Z values:

$$u' = \frac{4X}{X + 15Y + 3Z} \qquad v' = \frac{9Y}{X + 15Y + 3Z}$$

Computation of x, y from u', v' values:

$$x = \frac{6.75u'}{4.5u' - 12v' + 9} \qquad y = \frac{3v'}{4.5u' - 12v' + 9}$$

Computation of u', v' from x, y values:

$$u' = \frac{4x}{-2x + 12y + 3} \qquad v' = \frac{9y}{-2x + 12y + 3}$$

APPENDIX G

Standard Illuminants (Normalized)

Wave-length (nm)	A Tungsten 2856 K	B Sunlight 4874 K	C Daylight 6774 K	D_{50} 5003 K	D_{55} 5503 K	D_{65} 6504 K	D_{75} 7504 K	E Equal Energy 5450 K
380	0.041	0.213	0.266	0.238	0.313	0.424	0.501	1.000
385	0.045	0.256	0.322	0.264	0.339	0.444	0.514	1.000
390	0.050	0.298	0.382	0.290	0.366	0.464	0.526	1.000
395	0.055	0.344	0.445	0.384	0.475	0.583	0.646	1.000
400	0.061	0.393	0.510	0.479	0.585	0.702	0.766	1.000
405	0.067	0.444	0.579	0.514	0.621	0.739	0.804	1.000
410	0.073	0.496	0.650	0.549	0.658	0.777	0.841	1.000
415	0.080	0.549	0.721	0.566	0.672	0.785	0.845	1.000
420	0.087	0.601	0.791	0.583	0.687	0.793	0.848	1.000
425	0.094	0.651	0.853	0.572	0.669	0.764	0.812	1.000
430	0.102	0.696	0.906	0.561	0.652	0.736	0.775	1.000
435	0.110	0.736	0.949	0.644	0.737	0.813	0.843	1.000
440	0.119	0.769	0.979	0.726	0.822	0.890	0.911	1.000
445	0.128	0.794	0.995	0.787	0.881	0.942	0.956	1.000
450	0.137	0.813	0.999	0.847	0.940	0.993	1.000	1.000
455	0.147	0.827	0.996	0.863	0.952	0.997	0.998	1.000
460	0.156	0.840	0.992	0.880	0.964	1.000	0.995	1.000
465	0.167	0.857	0.994	0.883	0.961	0.988	0.976	1.000
470	0.177	0.876	0.998	0.887	0.959	0.975	0.957	1.000
475	0.188	0.892	1.000	0.905	0.972	0.979	0.955	1.000
480	0.200	0.906	0.998	0.923	0.986	0.984	0.953	1.000
485	0.211	0.916	0.991	0.908	0.964	0.954	0.919	1.000
490	0.223	0.918	0.973	0.893	0.941	0.924	0.885	1.000
495	0.235	0.911	0.942	0.911	0.954	0.926	0.881	1.000
500	0.248	0.896	0.903	0.929	0.966	0.928	0.877	1.000
505	0.260	0.879	0.862	0.934	0.966	0.922	0.866	1.000
510	0.273	0.863	0.824	0.938	0.966	0.915	0.855	1.000
515	0.287	0.853	0.796	0.940	0.963	0.902	0.836	1.000
520	0.300	0.852	0.781	0.943	0.960	0.889	0.817	1.000
525	0.314	0.861	0.780	0.967	0.980	0.902	0.824	1.000
530	0.327	0.877	0.790	0.991	1.000	0.914	0.830	1.000
535	0.341	0.899	0.805	0.985	0.990	0.900	0.815	1.000
540	0.356	0.922	0.823	0.978	0.980	0.886	0.799	1.000
545	0.370	0.944	0.838	0.986	0.984	0.885	0.794	1.000

Wave-length (nm)	A Tungsten 2856 K	B Sunlight 4874 K	C Daylight 6774 K	D_{50} 5003 K	D_{55} 5503 K	D_{65} 6504 K	D_{75} 7504 K	E Equal Energy 5450 K
550	0.384	0.961	0.848	0.993	0.988	0.883	0.789	1.000
555	0.399	0.973	0.852	0.982	0.974	0.866	0.770	1.000
560	0.414	0.978	0.849	0.971	0.960	0.849	0.752	1.000
565	0.429	0.979	0.839	0.960	0.946	0.833	0.735	1.000
570	0.443	0.976	0.824	0.949	0.933	0.818	0.719	1.000
575	0.458	0.970	0.807	0.955	0.935	0.815	0.714	1.000
580	0.474	0.961	0.788	0.960	0.938	0.813	0.708	1.000
585	0.489	0.952	0.769	0.934	0.908	0.783	0.681	1.000
590	0.504	0.944	0.751	0.908	0.877	0.753	0.654	1.000
595	0.519	0.937	0.735	0.982	0.892	0.758	0.655	1.000
600	0.534	0.933	0.723	0.948	0.906	0.764	0.656	1.000
605	0.549	0.933	0.716	0.956	0.910	0.762	0.652	1.000
610	0.564	0.937	0.712	0.964	0.913	0.761	0.648	1.000
615	0.579	0.943	0.711	0.963	0.909	0.752	0.638	1.000
620	0.594	0.949	0.710	0.962	0.904	0.744	0.628	1.000
625	0.609	0.955	0.710	0.945	0.886	0.726	0.610	1.000
630	0.624	0.961	0.709	0.929	0.868	0.707	0.592	1.000
635	0.639	0.967	0.708	0.945	0.877	0.709	0.591	1.000
640	0.654	0.973	0.708	0.960	0.886	0.710	0.590	1.000
645	0.668	0.981	0.709	0.944	0.869	0.695	0.576	1.000
650	0.683	0.989	0.711	0.929	0.853	0.679	0.562	1.000
655	0.697	0.995	0.711	0.941	0.860	0.680	0.561	1.000
660	0.712	0.999	0.708	0.953	0.867	0.681	0.559	1.000
665	0.726	1.000	0.703	0.977	0.884	0.690	0.563	1.000
670	0.740	0.998	0.695	1.000	0.902	0.698	0.567	1.000
675	0.754	0.995	0.687	0.981	0.882	0.681	0.553	1.000
680	0.767	0.989	0.677	0.962	0.863	0.664	0.538	1.000
685	0.781	0.979	0.663	0.905	0.814	0.628	0.509	1.000
690	0.794	0.967	0.646	0.848	0.765	0.592	0.480	1.000
700	0.820	0.943	0.615	0.889	0.795	0.608	0.489	1.000
705	0.833	0.930	0.599	0.896	0.805	0.619	0.500	1.000
710	0.846	0.915	0.583	0.902	0.814	0.631	0.512	1.000
715	0.858	0.900	0.567	0.921	0.744	0.577	0.468	1.000
720	0.870	0.884	0.550	0.746	0.674	0.523	0.424	1.000
725	0.882	0.867	0.534	0.793	0.717	0.558	0.454	1.000
730	0.894	0.851	0.519	0.840	0.761	0.593	0.483	1.000
735	0.906	0.837	0.506	0.869	0.788	0.615	0.501	1.000
740	0.917	0.827	0.496	0.899	0.816	0.637	0.520	1.000
745	0.928	0.817	0.485	0.829	0.753	0.589	0.480	1.000
750	0.939	0.811	0.477	0.760	0.690	0.540	0.441	1.000
755	0.950	0.807	0.471	0.660	0.598	0.467	0.381	1.000
760	0.960	0.806	0.468	0.560	0.507	0.394	0.320	1.000
765	0.971	0.808	0.467	0.683	0.618	0.481	0.391	1.000
770	0.981	0.813	0.469	0.805	0.729	0.567	0.461	1.000
775	0.990	0.809	0.471	0.783	0.709	0.552	0.450	1.000
780	1.000	0.809	0.476	0.760	0.689	0.538	0.438	1.000

APPENDIX H

Standard Fluorescent Lamp References (Normalized)

Wave-length (nm)	F1 6430 K	F2 CW 4230 K	F3 WW 3450 K	F4 2940 K	F5 6350 K	F6 4150 K	F7 D 6500 K	F8 5000 K	F9 4150 K	F10 Tri-phosphor 5000 K	F11 Tri-phosphor 4000 K	F12 Tri-phosphor 3000 K
380	0.043	0.034	0.026	0.019	0.046	0.032	0.058	0.035	0.028	0.015	0.012	0.014
385	0.054	0.042	0.032	0.023	0.058	0.040	0.072	0.044	0.034	0.011	0.009	0.009
390	0.067	0.053	0.040	0.029	0.072	0.050	0.087	0.053	0.042	0.008	0.006	0.007
395	0.079	0.061	0.045	0.032	0.085	0.058	0.103	0.062	0.049	0.008	0.005	0.005
400	0.118	0.098	0.081	0.066	0.125	0.095	0.139	0.093	0.079	0.020	0.018	0.017
405	0.446	0.449	0.451	0.454	0.464	0.454	0.439	0.384	0.392	0.165	0.174	0.182
410	0.140	0.110	0.085	0.064	0.147	0.105	0.167	0.112	0.094	0.029	0.022	0.016
415	0.143	0.107	0.077	0.053	0.150	0.101	0.160	0.101	0.078	0.037	0.025	0.014
420	0.161	0.120	0.086	0.058	0.168	0.113	0.175	0.113	0.088	0.051	0.034	0.016
425	0.178	0.132	0.094	0.064	0.186	0.125	0.191	0.130	0.101	0.070	0.046	0.020
430	0.196	0.145	0.103	0.069	0.204	0.136	0.207	0.149	0.117	0.092	0.062	0.026
435	1.000	1.000	1.000	1.000	1.000	1.000	1.000	1.000	1.000	0.467	0.466	0.424
440	0.388	0.338	0.297	0.265	0.394	0.329	0.397	0.364	0.330	0.202	0.167	0.115
445	0.245	0.179	0.126	0.084	0.253	0.168	0.257	0.225	0.179	0.141	0.095	0.039
450	0.260	0.190	0.133	0.089	0.268	0.177	0.272	0.252	0.201	0.146	0.099	0.040
455	0.272	0.198	0.139	0.093	0.280	0.185	0.285	0.277	0.222	0.145	0.098	0.039
460	0.283	0.206	0.144	0.096	0.290	0.192	0.296	0.300	0.241	0.137	0.092	0.036
465	0.292	0.212	0.148	0.099	0.299	0.197	0.305	0.318	0.256	0.126	0.084	0.034
470	0.298	0.216	0.151	0.100	0.304	0.200	0.311	0.332	0.268	0.112	0.075	0.031
475	0.301	0.218	0.153	0.102	0.308	0.202	0.314	0.343	0.278	0.099	0.066	0.028
480	0.303	0.219	0.153	0.102	0.309	0.202	0.316	0.351	0.285	0.107	0.078	0.044
485	0.302	0.218	0.152	0.102	0.307	0.201	0.316	0.357	0.291	0.226	0.196	0.158
490	0.301	0.218	0.153	0.104	0.306	0.201	0.313	0.360	0.295	0.227	0.205	0.173
495	0.294	0.213	0.150	0.101	0.299	0.197	0.309	0.361	0.297	0.142	0.123	0.100
500	0.287	0.208	0.147	0.099	0.292	0.193	0.304	0.362	0.299	0.081	0.065	0.050
505	0.279	0.204	0.145	0.098	0.285	0.190	0.300	0.365	0.303	0.045	0.032	0.022
510	0.271	0.202	0.145	0.099	0.278	0.190	0.296	0.368	0.308	0.032	0.020	0.013
515	0.263	0.201	0.149	0.104	0.272	0.193	0.293	0.372	0.315	0.026	0.015	0.010

520	0.257	0.205	0.157	0.113	0.269	0.202	0.290	0.374	0.321	0.022	0.012	0.009
525	0.253	0.214	0.172	0.129	0.269	0.218	0.285	0.373	0.326	0.020	0.011	0.009
530	0.253	0.230	0.196	0.155	0.274	0.243	0.282	0.370	0.330	0.024	0.016	0.016
535	0.256	0.254	0.230	0.192	0.283	0.278	0.279	0.365	0.336	0.077	0.026	0.067
540	0.264	0.286	0.276	0.242	0.297	0.322	0.278	0.358	0.343	0.556	0.544	0.502
545	0.635	0.711	0.748	0.746	0.682	0.773	0.669	0.849	0.849	1.000	1.000	0.954
550	0.390	0.476	0.507	0.499	0.435	0.535	0.386	0.484	0.499	0.456	0.448	0.430
555	0.310	0.417	0.458	0.458	0.355	0.479	0.282	0.346	0.376	0.112	0.103	0.104
560	0.328	0.462	0.522	0.539	0.373	0.528	0.285	0.345	0.391	0.046	0.039	0.045
565	0.344	0.502	0.581	0.617	0.387	0.568	0.288	0.345	0.405	0.034	0.027	0.036
570	0.355	0.532	0.626	0.682	0.395	0.596	0.291	0.347	0.418	0.029	0.023	0.033
575	0.419	0.614	0.726	0.802	0.455	0.674	0.350	0.428	0.508	0.066	0.061	0.074
580	0.448	0.652	0.775	0.867	0.478	0.706	0.379	0.472	0.556	0.155	0.155	0.175
585	0.354	0.551	0.672	0.769	0.378	0.595	0.291	0.362	0.446	0.201	0.203	0.224
590	0.341	0.533	0.655	0.758	0.359	0.569	0.287	0.367	0.449	0.165	0.175	0.208
595	0.324	0.507	0.626	0.731	0.337	0.534	0.282	0.373	0.449	0.122	0.134	0.173
600	0.303	0.473	0.586	0.691	0.311	0.493	0.276	0.379	0.448	0.088	0.101	0.135
605	0.279	0.435	0.541	0.642	0.284	0.449	0.269	0.385	0.445	0.113	0.133	0.180
610	0.255	0.395	0.491	0.586	0.256	0.403	0.263	0.391	0.441	0.599	0.759	1.000
615	0.230	0.353	0.441	0.528	0.229	0.358	0.257	0.399	0.441	0.469	0.585	0.774
620	0.205	0.313	0.391	0.476	0.203	0.314	0.252	0.407	0.444	0.164	0.181	0.214
625	0.182	0.276	0.344	0.415	0.180	0.274	0.248	0.413	0.448	0.165	0.181	0.210
630	0.161	0.240	0.299	0.361	0.157	0.237	0.244	0.416	0.451	0.143	0.168	0.215
635	0.142	0.209	0.260	0.314	0.138	0.205	0.236	0.415	0.446	0.060	0.070	0.094
640	0.124	0.180	0.223	0.270	0.120	0.175	0.229	0.414	0.441	0.026	0.028	0.038
645	0.108	0.155	0.191	0.232	0.105	0.149	0.227	0.421	0.447	0.030	0.032	0.040
650	0.095	0.134	0.164	0.198	0.091	0.128	0.227	0.425	0.456	0.043	0.049	0.061
655	0.083	0.115	0.140	0.169	0.080	0.109	0.229	0.424	0.475	0.038	0.041	0.050
660	0.073	0.099	0.119	0.144	0.069	0.093	0.224	0.411	0.474	0.031	0.034	0.041
665	0.064	0.085	0.101	0.122	0.061	0.079	0.196	0.369	0.405	0.027	0.029	0.035
670	0.057	0.073	0.086	0.103	0.054	0.067	0.165	0.322	0.324	0.021	0.021	0.024
675	0.050	0.063	0.073	0.087	0.047	0.057	0.146	0.293	0.281	0.018	0.018	0.020
680	0.044	0.054	0.062	0.074	0.042	0.049	0.132	0.270	0.253	0.020	0.020	0.022

Wave-length (nm)	F1 6430 K	F2 CW 4230 K	F3 WW 3450 K	F4 2940 K	F5 6350 K	F6 4150 K	F7 D 6500 K	F8 5000 K	F9 4150 K	F10 Tri-phosphor 5000 K	F11 Tri-phosphor 4000 K	F12 Tri-phosphor 3000 K
685	0.039	0.047	0.053	0.063	0.037	0.042	0.123	0.253	0.232	0.024	0.027	0.031
690	0.038	0.044	0.049	0.056	0.036	0.040	0.114	0.237	0.216	0.024	0.027	0.034
695	0.033	0.036	0.040	0.046	0.031	0.032	0.104	0.217	0.195	0.014	0.016	0.021
700	0.030	0.031	0.034	0.039	0.028	0.028	0.093	0.197	0.175	0.015	0.019	0.023
705	0.027	0.028	0.030	0.034	0.026	0.025	0.085	0.181	0.161	0.045	0.056	0.074
710	0.025	0.025	0.026	0.029	0.024	0.022	0.078	0.165	0.147	0.061	0.077	0.102
715	0.022	0.022	0.022	0.024	0.021	0.019	0.070	0.148	0.132	0.028	0.034	0.047
720	0.020	0.019	0.019	0.021	0.019	0.017	0.062	0.131	0.116	0.007	0.008	0.010
725	0.019	0.017	0.017	0.018	0.018	0.015	0.056	0.118	0.105	0.003	0.004	0.004
730	0.018	0.016	0.015	0.016	0.017	0.013	0.051	0.107	0.096	0.003	0.003	0.004
735	0.017	0.015	0.014	0.015	0.016	0.013	0.047	0.099	0.088	0.003	0.003	0.003
740	0.017	0.015	0.014	0.014	0.016	0.012	0.043	0.091	0.081	0.003	0.003	0.004
745	0.016	0.013	0.012	0.012	0.015	0.011	0.040	0.084	0.074	0.003	0.003	0.004
750	0.016	0.013	0.012	0.012	0.015	0.012	0.037	0.078	0.069	0.003	0.003	0.003
755	0.015	0.012	0.011	0.011	0.014	0.011	0.035	0.074	0.066	0.002	0.003	0.002
760	0.016	0.013	0.012	0.012	0.015	0.012	0.033	0.070	0.062	0.003	0.004	0.003
765	0.016	0.013	0.012	0.012	0.016	0.013	0.030	0.063	0.056	0.003	0.004	0.003
770	0.014	0.011	0.010	0.010	0.013	0.010	0.027	0.055	0.049	0.002	0.002	0.002
775	0.012	0.009	0.009	0.009	0.012	0.008	0.022	0.047	0.042	0.002	0.002	0.001
780	0.010	0.008	0.007	0.006	0.010	0.006	0.018	0.039	0.034	0.001	0.001	0.001

APPENDIX I

WTDS Phosphor Designations

First Letter

A	Reddish-purple or purple
B	Blue, purplish-blue, or greenish blue
D	Tricolor screens where one or more phosphor is significantly different than phosphor X or XX used for home entertainment; used only for desaturated blue
G, H	Bluish-green, green, or yellowish-green
K	Yellow-green
L	Orange or yellowish-pink
M	Custom phosphor controlled by individual manufacturer as identified by second letter
R	Reddish-orange, red, purplish-red, pink, or purplish pink
S	Screens intended for two-color displays
V	Multicolor voltage-dependent screens
W	White
X	(Two letter) tricolor screens for color entertainment
X	(Three letter) tricolor data display screen differing slightly from color entertainment screens
Y	Greenish-yellow, yellow, or orange-yellow
Z	CRT screens that do not fit the above categories

Second letter for M phosphors where it is an identification of the manufacturer of custom phosphors.

Second Letter for Type "M" Phosphors Only

A	AEG		M	Matsushita
B	Mitsubishi		N	NEG
C	Clinton		P	Philips (NV or NAP)
D	Thomson-CSF		R	Raytheon
E	Toshiba		S	Sony
G	GE/RCA		T	Thomas Electronics
H	Hitachi		W	Westinghouse
I	ISTC		Z	Zenith/Rauland

Derived from *Optical Characteristics of Cathode-Ray Tube Screens*, EIA publication TEP-116C, February 1993.

APPENDIX J

JEDEC-To-WTDS Phosphor Equivalents

JEDEC	WTDS	JEDEC	WTDS	JEDEC	WTDS
P1	GJA	P21	RDA	P41	YDA
P2	GLA	P22	X or XX*	P42	GWA
P3	YBA	P23	WGA	P43	GYA
P4	WWA	P24	GESA	P44	GXA
P5	BJA	P25	LJA	P45	WBA
P6	Obsolete	P26	LCA	P46	KGA
P7	GMA	P27	REA	P47	BHA
P8	Obsolete	P28	KEA	P48	KHA
P9	Obsolete	P29	SAA	P49	VAA
P10	ZAA	P30	Cancelled	P50	VBA
P11	BEA	P31	GHA	P51	VCA
P12	LBA	P32	GBA	P52	BLA
P13	RCA	P33	LDA	P53	KJA
P14	YCA	P34	ZBA	P54	None
P15	GGA	P35	BGA	P55	BMA
P16	AAA	P36	KFA	P56	RFA
P17	WFA	P37	BKA	P57	LLA
P18	None	P38	LKA		
P19	LFA	P39	GRA		
P20	KAA	P40	CCA		

*Several variants from XXA to XXG are equivalent to the earlier P22 variants that were differentiated by chemical composition.

Derived from *Optical Characteristics of Cathode-Ray Tube Screens*, EIA publication TEP-116C, February 1993. Note that one- and two-letter designations existed from inception of the Worldwide Type Designation System for phosphor registrations, which replaced P numbers until 1993 when an additional letter was added to the two-letter designations. For identifying two-letter phosphor designations during that period, just omit the third letter. Single-letter designations and some other two-letter designations were converted to somewhat different three-letter designations. See TEP-116C for equivalent three-letter designations.

APPENDIX K

Instrument Manufacturers

Photometers and Radiometers

Gamma Scientific
8581 Aero Drive
San Diego, CA 92123
Phone: (619 279-8034
Fax: (619) 576-9286

Graseby Optronics
12151 Research Parkway
Orlando, FL 32826-3207
Phone: (407) 282-1408
Fax: (407) 273-9046

Hoffman Engineering
P.O. Box 4430
Stamford, CT 06907-0430
Phone: (203) 425-8900
Fax: (203) 425-8910

International Light
17 Graf Road
Newburyport, MA 01950-4092
Phone: (508) 465-5923
Fax: (508) 462-0759

LMT Lichtmesstechnik GMBH Berlin
Helmholtzstrasse 9
D-10587 Berlin, Germany
Phone: 49-30-3934028
Fax: 49-30-3918001

Minolta Camera Co., Ltd.
3-13-Chome
Azuchi-Machi
Chuo-Ku
Osaka 541, Japan

Minolta Corporation
101 Williams Drive
Ramsey, NJ 07446-1293
Phone: (201) 825-4000
Fax: (201) 423-0590

Photo Research
9330 DeSoto Avenue
Chatsworth, CA 91311-4926
Phone: (818) 341-5151
Fax: (818) 341-7070

Tektronix, Inc.
P.O. Box 500
Beaverton, OR 97077-0001
Phone: (800) 426-2200 (U.S.)
       (503) 627-7111 (Intl.)

Topcon Instrument Corporation of America
65 West Century Road
Paramus, NJ 07652
Phone: (201) 261-9450
Fax: (201) 387-2710

Photometers, Video

Microvision
180 Knowles Drive, Suite 209
Los Gatos, CA 95030
Phone: (800) 931-3188
Fax: (408) 374-9394

Photo Research
9330 DeSoto Avenue
Chatsworth, CA 91311-4926
Phone: (818) 341-5151
Fax: (818) 341-7070

While this list is believed to be accurate as of 1997, the author and publisher assume no responsibility for its accuracy and the quality of products and services listed.

Goniophotometers

ELDIM
4, Rue Alfred Kastler
14000 Caen, France
Phone: 33-31-94-7600
Fax: 33-31-47-3777

LMT Lichtmesstechnik GMBH Berlin
Helmholtzstrasse 9
D-10587 Berlin, Germany
Phone: 49-30-3934028
Fax: 49-30-3918001

Colorimeters

Graseby Optronics
12151 Research Parkway
Orlando, FL 32826-3207
Phone: (407) 282-1408
Fax: (407) 273-9046

Hoffman Engineering
P.O. Box 4430
Stamford, CT 06907-0430
Phone: (203) 425-8900
Fax: (203) 425-8910

LMT Lichtmesstechnik GMBH Berlin
Helmholtzstrasse 9
D-10587 Berlin, Germany
Phone: 49-30-3934028
Fax: 49-30-3918001

Minolta Camera Co., Ltd.
3-13-Chome
Azuchi-Machi
Chuo-Ku
Osaka 541, Japan

Minolta Corporation
101 Williams Drive
Ramsey, NJ 0744-1293
Phone: (201) 825-4000
Fax: (201) 423-0590

Philips
85 McKee Street
Mahwah, NJ 07430
Phone: (201) 529-2188
Fax: (201) 529-2109

Tektronix, Inc.
P.O. Box 500
Beaverton, OR 97077-0001
Phone: (800) 426-2200 (U.S.)
       (503) 627-7111 (Intl.)

Topcon Instrument Corporation of America
65 West Century Road
Paramus, NJ 07652
Phone: (201) 261-9450
Fax: (201) 387-2710

Spectroradiometers

Gamma Scientific
8581 Aero Drive
San Diego, CA 92123
Phone: (619) 279-8034
Fax: (619) 576-9286

Instrument Systems
Tassiloplatz 7
81541 Munich, Germany
Phone: 49-89-4129-4601
Fax: 49-89-4129-4602

Minolta Camera Co., Ltd.
3-13-Chome
Azuchi-Machi
Chuo-Ku
Osaka 541, Japan

Minolta Corporation
101 Williams Drive
Ramsey, NJ 07446-1293
Phone: (201) 825-4000
Fax: (201) 423-0590

Optronic Laboratories, Inc.
4470 35th Street
Orlando, FL 32811
Phone: (800) 899-3171, (407) 422-3171
Fax: (407) 648-5412

Photo Research (Division of Kollmorgan
   Instruments)
9330 DeSoto Avenue
Chatsworth, CA 91311-4926
Phone: (818) 341-5151
Fax: (818) 341-7070

Convergence Measurement Instruments

Klein Optical Instruments
8948 SW Barbur Boulevard
Portland, OR 97219
Phone: (503) 245-1012
Fax: (503) 245-8166

Minolta Camera Co., Ltd.
3-13-Chome
Azuchi-Machi
Chuo-Ku
Osaka 541, Japan

Minolta Corporation
101 Williams Drive
Ramsey, NJ 07446-1293
Phone: (201) 825-4000
Fax: (201) 423-0590

Quantum Data
2111 Big Timber Road
Elgin, IL 60123
Phone: (847) 888-0450
Fax: (847) 888-2802

Integrating Spheres and Systems

Labsphere
P.O. Box 70
North Sutton, NH 03260
Phone: (603) 927-4266
Fax: (603) 927-4694

Test Pattern Generators

Quantum Data
2111 Big Timber Road
Elgin, IL 60123
Phone: (847) 888-0450
Fax: (847) 888-2802

Team Systems
2934 Corvin Drive
Santa Clara, CA 95051
Phone: (408) 720-8877
Fax: (408) 720-9643

Tektronix, Inc.
P.O. Box 500
Beaverton, OR 97077-0001
Phone: (800) 426-2200 (U.S.),
(503) 627-7111 (Intl.)

Video Instruments
P.O. Box 33
2155 Bellbrook Ave.
Xenia, OH 45385-0033
Phone: (800) 962-8905
Fax: (513) 376-2802

Calibration Light Sources

Hoffman Engineering
P.O. Box 4430
Stamford, CT 06907-0430
Phone: (203) 425-8900
Fax: (203) 425-8910

Optonic Laboratories, Inc.
4470 35th Street
Orlando, FL 32811
Phone: (800) 899-3171 (U.S.)
(407) 422-3171 (Intl.)
Fax: (407) 648-5412

Silicon Photodiodes

Centronic Inc.
2088 Anchor Court
Newbury Park, CA 91320
Phone: (805) 499-5902
Fax: (805) 499-7770

EG&G Optoelectronics
22001 Dumberry
Vaudreuil, Quebec J7V 8P7, Canada
Phone: (514) 424-3300
Fax: (514) 424-3411

EG&G Vactec Optoelectronics
10900 Page Boulevard
St. Louis, MO 63132
Phone: (314) 423-4900
Fax: (314) 423-3956

Hamamatsu Corporation
360 Foothill Road
P.O. Box 6910
Bridgewater, NJ 08807-0910
Phone: (908) 231-0960
Fax: (908) 231-1539

UDT Sensors, Inc.
12525 Chadron Ave.
Hawthorne, CA 90250
Phone: (310) 978-0516
Fax: (310) 644-1727

Silicon Photodiodes (Photopic)
Inphora, Inc.
1926 Contra Costa Boulevard No. 163
Pleasant Hill, CA 94523
Phone: (510) 689-2039
Fax: (510) 689-2788

UDT Sensors, Inc.
12525 Chadron Avenue
Hawthorne, CA 90250
Phone: (310) 978-0516
Fax: (310) 644-1727

Photomultiplier Tubes and Accessories
Burle Electron Tubes
1000 New Holland Ave.
Lancaster, PA 17601-5688
Phone: (800) 366-2875 (U.S.)
       (717) 295-6000 (Intl.)
Fax: (717) 295-6096

Thorn EMI/Electron Tubes Limited
Bury Street
Ruislip, Middlesex HA4 7TA
Phone: 44-895-630771

Thorn EMI/Electron Tubes Inc.
100 Forge Way, Unit F
Rockaway, NJ 07866
Phone: (800) 521-8382 (U.S. only)
Fax: (201) 586-9771

Hamamatsu Corporation
360 Foothill Road
P.O. Box 6910
Bridgewater, NJ 08807-0910
Phone: (908) 231-0960
Fax: (908) 231-1539

Filters (Glass)
Hoya Corporation
Colored Glass Dept., Optical Division
572 Miyazawa-cho
Akishima-shi
Tokyo 196, Japan
Phone: (0425) 41-3131

Hoya Optics, Inc.
3400 Edison Way
Fremont, CA 94538
Phone: (415) 490-1880
Fax: (415) 490-1988

Kopp Glass, Inc.
2108 Palmer Street
Pittsburgh, PA 15218
Phone: (412) 271-0190
Fax: (412) 271-4103

Schott Glaswerke
Abt. ZFP
Postfach 2480
D-6500 Mainz
Germany

Schott Glass Technologies, Inc.
400 York Avenue
Duryea, PA 18642
Phone: (717) 457-7485
Fax: (717) 457-7330

Filters (Thin Film)
Balzers AG
Dünnschicht-Komponenten
FL-9496 Balzers
Furstentum
Liechtenstein
Phone: +41 (0) 75-388 4111
Fax: +41 (0) 75-388 5405

Balzers
Thin Film Components
46249 Warm Springs Blvd.
Fremont, CA 94539
Phone: (510) 651-1232
Fax: (510) 651-1669

Corion Corporation
73 Jeffrey Avenue
Holliston, MA 01746
Phone: (508) 429-5065
Fax: (508) 429-8983

OCLI
2789 Northpoint Parkway
Santa Rosa, CA 95407-7397
Phone: (707) 545-6440
Fax: (707) 525-7410

Oriel Corporation
250 Long Beach Boulevard
Stratford, CT 06497-0872
Phone: (203) 377-8282
Fax: (203) 378-2457

Reynard Corporation
1020 Calle Sombra
San Clemente, CA 92673-6227
Phone: (714) 366-8866
Fax: (714) 498-9528

Positioners, X–Y
CELCO
70 Constantine Drive
Mahwah, NJ 07430
Phone: (201) 327-1123
Fax: (201) 327-7047

Gamma Scientific
8581 Aero Drive
San Diego, CA 92123
Phone: (619) 279-8034
Fax: (619) 576-9286

Hoffman Engineering
P.O. Box 4430
Stamford, CT 06907-0430
Phone: (203) 425-8900
Fax: (203) 425-8910

Microvision
180 Knowles Drive, Suite 209
Los Gatos, CA 95030
Phone: (800) 931-3188
Fax: (408) 374-9394

APPENDIX L

Calibration Services

Hoffman Engineering
P.O. Box 4430
Stamford, CT 06907-0430
Phone: (203) 425-8900
Fax: (203) 425-8910

National Institute of Standards and
   Technology (NIST)
Physical Measurement Services Program
Gaithersburg, MD 20899
Phone: (301) 975-2002

Opto-Cal, Inc.
13891 Deanly Court
Lakeside, CA 92040
Phone: (619) 561-9983
Fax: (619) 561-8810

Optronic Laboratories
4470 35th Street
Orlando, FL 32811-6590
Phone: (407) 422-3171, (800) 899-3171
Fax: (407) 648-5412

APPENDIX M

Standards Organizations

American National Standards Institute (ANSI)
11 West 42nd Street
New York, NY 10036
Phone: (212) 642-4900
Fax: (212) 398-0023

ASTM
100 Barr Harbor Drive
West Conshohocken, PA 19428-2959
Phone: (610) 832-9500
Fax: (610) 832-9555

Commission Internationale De L'Eclairage (CIE)
CIE Central Bureau
Kegelgasse 27
A-1030 Vienna, Austria
Phone: 43-1-714-31-870
Fax: 43-1-713-08-38-18
e-mail: x0401daa@vm.unvie.ac.at

Council for Optical Radiation Measurements (CORM)
c/o Labsphere, Inc.
P.O. Box 70
North Sutton, NH 03260-0070
Phone: (603) 927-4266
Fax: (603) 927-4694

Deutsches Institut fur Normung (DIN)
Burggrafenstrasse 6
D-1000 Berlin 30, Germany
Phone: 49-30-2601-1
Fax: 49-30-2601-1231

European Broadcasting Union (EBU)
CP67
CH-1218 Grand-Saconnexge, Switzerland
Phone: 41-22-717-2111
Fax: 41-22-717-2481
e-mail: ebu@ebu.ch

Electronic Industries Association (EIA)
2500 Wilson Boulevard
Arlington, VA 22201-3834
Phone: (703) 907-7500
Fax: (703) 907-7501
Note: Documents may be ordered from Global Engineering Documents. See separate listing.

Global Engineering Documents
15 Inverness Way
Englewood, CO 80112-5704
Phone: (800) 854-7179 (U.S., Canada)
(303) 397-7956 (Intl.)

Human Factors Engineering Society (HFES)
P.O. Box 1369
Santa Monica, CA 90406-1369
Phone: (310) 394-1811
Fax: (310) 394-2410
e-mail: 72133.1474@compuserve.com

International Electrotechnical Commission (IEC)
3 Rue de Varembe
P.O. Box 131
1211 Geneva 20, Switzerland
Phone: 41-22-7340150
Fax: 41-22-7333843

Illuminating Engineering Society of North America
120 Wall Street, Floor 17
New York, NY 10005-4001

Phone: (212) 248-5000
Fax: (212) 248-5017

Institute of Electrical and Electronics
  Engineers (IEEE)
445 Hoes Lane
P. O. Box 1331
Piscataway, NJ 08855-1331
Phone: (800) 678-4333 (U.S.)
       (908) 981-0060 (Intl.)
Fax: (908) 981-9667

International Organization for
  Standardization (ISO)
Burggrafenstrasse 6
D-1000 Berlin 30, Germany
Phone: 49-30-2601-1
Fax: 49-30-2601-1231

Intersociety Color Council (ISCC)
11491 Sunset Hills Road, Suite 301
Reston, VA 22090
Phone: (703) 318-0263

National Information Display Laboratory
  (NIDL)
P.O. Box 8619
Princeton, NJ 08543-8619
Phone: (609) 951-0150
Fax: (609) 734-2313
e-mail: nidl@maca.sarnoff.com

National Institute of Standards and
  Technology (NIST)
Flat Panel Display Laboratory
B344 Metrology Building (220)
Gaithersburg, MD 20899
Phone: (301) 975-3842

Fax: (301) 975-4091
e-mail: kelley@eeel.nist.gov

Society of Automotive Engineers
400 Commonwealth Drive
Warrendale, PA 15096-0001
Phone: (412) 776-4841
Fax: (412) 776-0243

Society for Imaging Science and
  Technology (IS&T)
7003 Kilworth Lane
Springfield, VA 22151
Phone: (703) 642-9090
Fax: (703) 642-9094

Society for Information Display
1526 Brookhollow Drive, Suite 82
Santa Ana, CA 92705-5421
Phone: (714) 545-1526
Fax: (714) 545-1547
e-mail: socforinfodisplay@mcimail.com

Society of Motion Picture and Television
  Engineers (SMPTE)
595 West Hartsdale Avenue
White Plains, NY 10607
Phone: (914) 761-1100
Fax: (914) 761-3115

Video Electronics Standards Association
2150 North First Street, Suite 440
San Jose, CA 95131-2029
Phone: (408) 435-0333
Fax: (408) 435-8225
Phone: (203) 425-8900
Fax: (203) 425-8910

INDEX

Absorptance, filter, 58, 59
Addressability vs. resolution, 89–90, 192
Ambient light, 158
American National Standards Institute (ANSI), 269
American Society for Testing and Materials, 269–270
Angstroms vs. nanometers, 3
Anisotropy, LCD, 189
ANSI, *see* American National Standards Institute
Antireflection coatings, 59, 63, 66, 97
Anti-Stokes fluorescence, 47
Aspect ratio, 88, 225–226
Astigmatism, 98
ASTM, *see* American Society for Testing and Materials
Avalanche photodiodes, 79–81, 246

Barrel distortion, 101, 231–232
Blackbody radiation, 22, 29–30
Blind spot, 5
Boltzmann's constant, 288
Braun, Karl Ferdinand, 90
Brightness, *see* Luminance

Calibration:
    chromaticity measurements, 266–267
    illuminance measurements, 260–263
    list of light source manufacturers, 313
    list of service companies, 316
    luminance measurements, 264–266
    photometric instruments, 260–265
Candela, 15
Cathode ray tubes, *see also* CRT displays
    display screen, 96–97
    elements of, 90–97
    history of, 90
    overview, 90–92
CEI, *see* International Electrotechnical Commission
Charge-coupled devices:
    for measuring convergence, 234
    overview, 81–83
    in resolution measurement, 209, 212–213
Chromaticity, *see also* Colorimetry
    calibrating measurements, 266–267
    CIE diagrams, 17–21

    color measurement, 182–183
    conversions, 302
    filter colorimeters, 147–149
    overview, 184–185
CIE, *see* Commission Internationale de l'Eclairage
Cold mirror, 65
Color, *see also* Chromaticity
    anisotropy in LCDs, 189
    vs. CRT display resolution, 98
    and human eye, 4–5
    human eye sensitivity, 6–8
    measurement, 182–189
Color blindness, 9–10
Color gamut, 18, 19, 187, 188
Colorimeters:
    calibrating, 266–267
    filter colorimeters, 147–149
    list of manufacturers, 312
    overview, 147
    scanning, *see* Spectroradiometers
Colorimetry:
    chromaticity, 184–185
    CIE chromaticity diagrams, 17–21
    color measurement basics, 184–185
    color purity, 183–184
    color tracking, 185–186
    color uniformity, 186–187
    just-noticeable difference (JND), 19–21
    overview, 17
    standards, 184, 186–187
    tristimulus curves, 19, 294–301
Color purity, 183–184
Color shutter, liquid crystal, 114–115
Color space, 18, 19–20, 182
Color temperature:
    correlated, 23, 24
    overview, 21–23
    and Planckian locus, 22–24
Color tracking, 185–186
Color uniformity, 186–187
Commission Internationale de l'Eclairage (CIE):
    chromaticity diagrams, 17–21
    defined, 270
    tristimulus values, 19, 294–301
Comparators, optical, 153

Contrast, display:
  enhancement filters, 65–68
  faceplate glass, 59–60, 97
  measuring, 167–169
Convergence measurement, 153–155, 232–236, 313
Conversions, *see also* Units of measurement
  chromaticity, 302
  photometry, 293
  radiometry, 293
CORM, *see* Council for Optical Radiation Measurements (CORM)
Cosine law, 25
Council for Optical Radiation Measurements (CORM), 270–271
CRT displays:
  aspect ratio, 88, 225–226
  barrel distortion, 101, 231–232
  cathode-ray tube elements, 90–97
  centering, 226
  chromaticity, 184–185
  color measurement, 182–188
  color purity, 183–184
  color tracking, 185–186
  color uniformity, 186–187
  color vs. resolution, 98
  convergence measurement, 153–155, 232–236
  flat, 104–105
  format, 87–88
  and gamma, 88, 166
  glass faceplates, 59–60, 97
  glass filter materials, 59–60
  and gray scale, 88
  history of CRT, 90
  hook or flagging distortion, 102
  interlacing, 10, 11, 99
  jaggies, 236–237
  jitter, 237
  keystone distortion, 101, 230–231
  line pairing distortion, 103
  magnetic aberrations, 237–238
  measuring color, 182–189
  measuring size, 225
  moire patterns, 236
  nonlinearity, 103, 226–228
  orthogonal distortion, 102, 229–230
  overscanned vs. underscanned, 225
  phosphors, 100–101
  pincushion distortion, 101, 231–232
  positioning, 226
  projection displays, 103–104
  raster scanning, 10, 11, 95, 98–99, 225
  resolution measurement, 192–220
  resolution vs. addressability, 89–90, 192
  ringing distortion, 103
  S distortion, 227
  shadow mask, 96–97, 232–236
  signal inputs, 100
  sizes of display, 88–89, 225

  spectral output curve, 187–188
  tilt, 102, 228–229
  trapazoidal distortion, 101, 230–231
  tube overview, 90–92
  types of distortion, 101–103, 224, 226–238
  video bandwidth, 100, 292

Dark adaption, 5, 6
Deflection, electron beam, 95–96, 98
Deutsche Industrie Norms (DIN), 271
Dichroic filters, 57, 65
Diffuse reflectance, 65, 169–173
Digital clock frequency, 292
Digital micromirror displays, 124–125
DIN, *see* Deutsche Industrie Norms
Diodes, *see* Charge-coupled devices; Light-emitting diodes; Photodiodes
Displays, *see also* CRT displays
  aspect ratio, 88, 225–226
  characteristics, 88–90
  color vs. monochrome, 10, 11
  contrast enhancement, 59–60, 65–68, 97
  and correlated color temperature, 23, 24
  digital micromirror displays, 124–125
  electroluminescent displays, 119–120
  field emission displays, 105, 106
  flat CRT displays, 104–105
  and flicker, 10–11, 95, 254
  formats, 87–88
  and gamma, 88, 166
  glass faceplates, 59–60, 97
  and gray scale, 88
  and human eye limitations, 8–10
  instruments for measuring, 129–155
  interlaced scanning, 10, 11, 99
  jumbo, 125
  and just-noticeable difference, 20–21
  light-emitting diode displays, 120–121
  light valve displays, 122–123
  liquid crystal displays, 107–115
  measuring color, 182–189
  measuring contrast, 167–169
  modulation transfer function measurement, 213–220
  plasma displays, 115–118
  projection CRT displays, 103–104
  refresh rate, 10, 11, 135
  resolution measurement, 192–220
  resolution vs. addressability, 89–90, 192
  sizes of display, 88–89, 225
  spatial measurement, overview, 224–225
  vacuum fluorescent displays, 105–107
Distortion, CRT displays, 101–103, 224, 226–238

EBU, *see* European Broadcast Union
EIA, *see* Electronic Industries Association
EIAJ, *see* Electronic Industries Association of Japan
Eidophor system, 122

Electroluminescent displays:
  alternating current, 119–120
  direct current, 119
  full-color, 120
  as light source, 39–42
  multicolor, 120
Electromagnetic radiation, 2–3
Electromagnetic spectrum:
  human eye sensitivity, 6–8
  light vs. radio waves, 1
  overview, 2–3
Electron beam:
  cathode-ray tube basics, 90–97
  Coulomb aging of screen phosphor, 52
  spot characteristics, 194–196
Electron gun, 91–94, 96, 97
Electronic Industries Association (EIA), 271
Electronic Industries Association of Japan (EIAJ), 272
EN, *see* European Norms
European Broadcast Union (EBU), 271
European Norms (EN), 139, 272
Eye, *see* Human eye

Faceplates, glass, 59–60, 97
Field emission displays, 105, 106
Filter colorimeters, 147–149
Filters, *see* Glass filters; Light filters; Thin-film filters
Flagging distortion, 102
Flat CRT displays, 104–105
Flat-panel displays:
  geometric measurement, 224
  jaggies, 236–237
  moire patterns, 236
  in projection systems, 224
Flicker, 10–11, 95, 254
Fluorescent lamps:
  as light source, 36–39
  measuring flicker, 254
  standard references, 305–308
  time-related measurements, 241–242
Footcandles, 15
Footlamberts, 16
Full-field luminance, 159–161

Gamma:
  as CRT characteristic, 88
  formula for, 288
  measuring, 166
Gas discharge lamps, as light source, 35–36. *See also* Fluorescent lamps; Plasma displays
Gas photodiodes, 71
Gelatin filters:
  vs. glass filters, 60–61
  overview, 62–63
Geometric distortion, 101–103, 224, 226–238
Glare reduction, 65–66

Glass filters:
  construction methods, 61–62
  vs. gelatin filters, 60–61
  list of manufacturers, 314
  neutral-density, 54
  overview, 59–60
  spectral transmission curves, 54, 57
  stacked vs. mosaic, 61–62
Goniophotometers, 151–152, 312
Grassman's laws, 288
Gray scale, 88

Halogen lamps, as light source, 34–35
HFES, *see* Human Factors and Ergonomics Society
Hook distortion, 102
Human eye:
  20/20 vision, 8
  accommodation by, 9
  color blindness, 9–10
  correcting focusing defects, 8–9
  discrimination abilities, 10
  and flicker, 10–11
  how it works, 3–5
  as light detector, 3
  limitations, 8–10
  physiology, 3–6
  Schade model, 3
  spectral sensitivity, 6–8
Human Factors and Ergonomics Society (HFES), 272

ICI, *see* International Commission on Illumination
IEC, *see* International Electrotechnical Commission
IEEE, *see* Institute of Electrical and Electronics Engineers
IESNA, *see* Illuminating Engineering Society of North America
Illuminance:
  calibrating measurements, 260–263
  measuring CRT output linearity, 165
  overview, 15–16
  photometer characteristics, 129, 140–143
  table of common levels, 16
Illuminating Engineering Society of North America (IESNA), 273
Image dissector tubes, 206, 207
Incandescent lamps:
  chromaticity measurements, 266–267
  as light source, 33–34
  list of formulas, 290
  typical visible spectrum, 33
Incident light, illustrated, 58
Institute of Electrical and Electronics Engineers (IEEE), 273
Instruments:
  calibrating, 258–268
  for convergence measurement, 153–155
  filter colorimeters, 147–149
  goniophotometers, 151–152

Instruments (*Continued*)
 integrating spheres, 152–153
 list of manufacturers, 311–315
 for luminance measurement, 156–158
 optical comparators, 153
 photometers, 129–146
 radiometers, 146–147
 spectrophotometers, 151
 spectroradiometers, 149–151
Integrating spheres, 152–153, 313
Interference filters, 64–65
Interlacing, 10, 11, 99
International Commission on Illumination (ICI), 6, 7, 8
International Electrotechnical Commission (IEC), 272
International Organization for Standardization (ISO), 274
International Society for Optical Engineering (SPIE), 277
International System of Units (SI), 258–259
Intersociety Color Council (ISCC), 273–274
Inverse square law:
 and illuminance, 15–16
 and irradiance, 13
 overview, 25–26, 288
Irradiance, 13
ISCC, *see* Intersociety Color Council
ISO, *see* International Organization for Standardization
Isotropic radiators, 12, 15
IS&T, *see* Society for Imaging Science and Technology

Jaggies, 236–237
Jitter, 237
JND, *see* Just-noticeable difference
Jumbo displays, 125
Just-noticeable difference (JND), 19–21

Kell factor, 196
Keystone distortion, 101, 230–231
Kirchoff's law, 289
Klein gauge, 153, 154, 234

Lambertian surface, 25, 289
Lambert's law, 25, 289
Lamps, *see also* Fluorescent lamps; Incandescent lamps
 gas discharge, 35–36
 halogen, 34–35
 as light sources, 33–42
 luminous efficacy, 14–15
 and Planckian locus, 22–23
 radiant flux, 15
 radiant power, 12
Landscape mode, 87
LCDs, *see* Liquid crystal displays
LEDs, *see* Light-emitting diodes
Light:
 absorbed, 58
 and human eye, 3–5
 incident, 58
 vs. radio waves, 1
 reflected, 58, 170, 171
 table of common illuminance levels, 16
 table of common luminance levels, 16
 transmitted, 58
 velocity in vacuum, formula, 289
Light bulbs, *see also* Incandescent lamps
 for calibrating instruments, 261–263
 luminous efficacy, 14–15
 radiant flux, 15
 radiant power, 12
Light detectors:
 human eye as, 3
 photoconductive sensors, 69–71
 photodiodes, 71–73
 photomultiplier tubes, 73–76
 selenium cells, 76
 silicon detectors, *see* Silicon photodiodes
 thermal detectors, 69, 146
Light-emitting diodes (LEDs):
 alphanumeric LEDs, 179
 calibration issues, 267–268
 discrete, 177–179
 in displays, 120–121
 dot-matrix, 179
 infrared, 12
 as light source, 42–46
 measuring luminous intensity, 46, 177–179
 overview, 42–46
 typical colors, 43
 visible, 15, 177–179
Light filters, *see also* Filter colorimeters
 absorptance, 58, 59
 for contrast enhancement, 65–68
 dichroic, 57, 65
 gelatin, 62–63
 glass, 54, 57, 59–62
 interference, 64–65
 light attenuation, 53–54
 long-wave pass, 56, 65
 louvered, 68
 mosaic, 61, 62
 narrow-bandpass, 55, 64–65
 neutral-density, 53, 54, 59, 63–64
 optical density, 59
 overview, 52–53
 percent transmission, 59
 plastic, 53, 62–63
 polarized, 68
 short-wave pass, 56, 65
 specifying, 58–59
 spectral shaping, 57, 60–62
 spectral transmission curves, 53–57
 surface reflection, 59, 65–66, 169, 171, 173, 291
 terminology, 58–59

INDEX 323

thin-film, 53, 54, 63–65
transmissivity, 59
transmittance, 58, 59
types of filters, 53–65
wavelength separation, 57, 65
wide bandpass, 55
Wratten, 62, 63
Light sources:
  characteristics, 29
  electroluminescent lamps, 39–42
  fluorescent lamps, 36–39
  gas discharge lamps, 35–36
  halogen lamps, 34–35
  incandescent lamps, 33–34
  light-emitting diodes, 42–46
  moon, 32
  planets, 32
  role of phosphors, 46–52
  sky, 32–33
  stars, 30–32
  sun, 30–32
  thermal sources, 29–30
Light valve displays, 122–123
Line pairing distortion, 103
Liquid crystal displays (LCDs):
  active matrix-addressed, 112–113
  addressing, 108–113
  color anisotropy, 189
  color shutter, 114–115
  overview, 107–108
  passive matrix-addressed, 112
  plasma addressed, 118
  projection displays, 113, 122–123
  supertwisted nematic, 110–112
  twisted nematic, 109–110
Long-wave pass filters, 56, 65
Louvered filters, 68
Luinous flux, 14–15
Lumen, 14
Luminance:
  aging, 162
  and ambient light, 158
  calibrating measurements, 264–266
  full-field, 159–161
  linearity, 164–166
  measurement overview, 158–161
  measuring projection equipment, 173–176
  measuring uniformity, 162, 163
  overview, 16–17
  photometer characteristics, 144–145
  photometer selection, 156
  stability, 163–164
  table of common levels, 16
  warmup, 162
Luminance microphotometer, measuring CRT phosphor linearity, 165–166
Luminous exitance, 17
Luminous intensity, 15, 46, 129, 177

MacAdam ellipses, 20
Magnetic deflection, 95–96, 98
Magnetic fields, CRT display problems, 237–238
Measurement units, *see also specific units by name*
  list of abbreviations, 286
  list of conversions, 293
  list of SI prefixes, 287
  overview, 26, 258–259
Mercury vapor lamps, as light source, 35–36
Metrology:
  overview, 258–260
  standards, 259
  terminology, 259–260
  units of measurement, 26, 258–259, 286, 287, 293
Micrometers, 3
Microns, 3
Microphotometers:
  measuring convergence, 235–236
  measuring CRT phosphor linearity, 165–166
Military Standards (MIL), 275
Misconvergence, 153–155, 232–236
Modulation transfer function (MTF), 213–220
Moire patterns, 236
Moon, as light source, 32
Mosaic filters, 61, 62
MTF, *see* Modulation transfer function
Multilayer filters, 64–65

Nanometer, 3
NAPM, *see* National Association of Photographic Manufacturers
Narrow-bandpass filters, 55, 64–65
National Association of Photographic Manufacturers (NAPM), 275
National Bureau of Standards, *see* National Institute of Standards and Technology (NIST)
National Information Display Laboratory (NIDL), 275–276
National Institute of Standards and Technology (NIST), 139, 276
Neon lamps, 116–117
Neutral-density filters, 53–54, 59, 63–64
NIDL, *see* National Information Display Laboratory
Night vision, 5
NIST, *see* National Institute of Standards and Technology
Nit, 16, 159
Nixie tubes, 116–117
Nonlinearity, CRT displays, 103, 226–228

Optical comparators, 153
Optical density, 59, 291
Orthogonal distortion, 102, 229–230

PDPs, *see* Plasma displays
Peltier effect, 69

INDEX

Phosphor:
  burning, 51–52
  for CRTs, 46–47
  decay times, 48, 51, 251–254
  for display screens, 100–101
  in electroluminescent lamps, 39–40
  in fluorescent lamps, 36–37, 39
  in gas discharge lamps, 36
  JEDEC-to-WTDS equivalents, 310
  light source chromaticity measurements, 266–267
  measuring persistence, 251–254
  WTDS designations, 309
Photoconductive sensors, 69–71
Photodiodes:
  gas, 71
  silicon, *see* Silicon photodiodes
  vacuum, 71, 72
Photographic measurement method, 199–200
Photometers:
  aging, 136–137
  analog output, 139–140
  aperture/slit-scan resolution measurement methods, 202–206
  calibrating for illuminance measurement, 158, 260–263
  calibrating for luminance measurement, 264–266
  common characteristics, 130–140
  convenience features, 140
  environmental effects, 137–139
  European Community certification, 139
  fatigue, 134–135
  humidity effects, 138–139
  illuminance measurement overview, 140–143
  linearity, 133–134
  list of manufacturers, 311
  luminance measurement overview, 144–145
  measuring LED output, 267–268
  overview, 129–146
  portability, 140
  power supply variations, 136
  range, 133–134
  resolution measurement methods, 202–213
  response time, 135
  safety certification, 139
  selecting for luminance measurement, 156–158
  spectral characteristics, 130–133
  standards, 139
  temperature effects, 137–138
  using CCD displays, 212–213
  video, 146, 212–213
  zeroing, 137
Photometry, *see also* Photometers
  defined, 14
  fundamental unit, 15
  illuminance, 15–16
  list of conversions, 293
  luminance, 16–17
  luminous exitance, 17
  luminous flux, 14–15
  luminous intensity, 15, 46, 129, 177
Photomultiplier tubes:
  intensity-versus-time measurements, 246–250
  as light detectors, 73–76
  list of manufacturers, 314
  photopic correction, 249
  in resolution measurement, 203, 206–207, 209
  when to use, 135
Photopic response curve, 6, 7, 130
Phototransistors, 77
Photovoltaic cells, 76, 77
Pixel, as unit of display measurement, 224
Planckian locus, 22–23, 24
Planck's constant, 289
Planck's radiation law, 29–30, 289
Planets, as light source, 32
Plasma-addressed liquid crystal displays, 118
Plasma displays, 115–118
Plastic filters:
  for contrast enhancement, 66–68
  gelatin, 62–63
  vs. glass filters, 60–61
  neutral density, 54
  sheet plastic, 63
  spectral transmission curves, 53, 67
Polarized filters, 68
Positioners, x-y, list of manufacturers, 315
Power supplies, 136
Projection displays:
  cathode-ray tube, 103–104
  flat-panel distortion, 224
  liquid crystal, 113, 122–123
  test and measurement methods, 173–176

Quartz-iodine lamps, as light source, 34–35

Radiance, 14
Radiant energy, 14
Radiant exitance, 13
Radiant flux, 12
Radiant intensity, 12–13
Radiometers, *see also* Spectroradiometers
  list of manufacturers, 311
  overview, 14, 146
  semiconductor-photodiode-based, 146–147
  thermal-detector-based, 146
Radiometry, *see also* Photometry
  defined, 12
  fundamental unit, 12
  list of conversions, 293
  overview, 12–14
  radiometric detectors, 14, 146–147
Raster:
  changes in size, 225
  and interlacing, 10, 11, 99
  scanning, 10, 11, 95, 98–99

shrinking-raster resolution measurement method, 198–199
Reflectance, defined, 59
Reflected light, illustrated, 58, 170, 171
Reflection:
    antireflection coatings, 59, 63; 66, 97
    diffuse, 65, 169–172
    measuring, 169–173
    reducing, 65–66
    surface, 59, 65–66, 169, 171, 173, 291
Resolution:
    vs. addressability, 89–90, 192
    aperture/slit-scan measurement methods, 202–206
    diode array scanning measurement methods, 209, 212–213
    limitations of color CRTs, 98
    list of measurement standards, 220
    measurement overview, 192–194
    measuring, 192–220
    methods of expressing, 196–197
    microscope trace width measurement, 197–198
    modulation transfer function measurement, 213–220
    photographic measurement method, 199–200
    photometric measurement methods, 202–213
    shrinking-raster measurement method, 198–199
    spot characteristics, 194–196
    spot contour plotting measurement methods, 206–212
    television camera measurement method, 200
    test patterns for measurement, 194, 200–202
    vernier line measurement method, 199
    visual measurement methods, 197–202
Response time, 135, 254–256
Rhodopsin, 5
Ringing distortion, 103

SAE, *see* Society of Automotive Engineers
Safety standards, 139, 278
Scan frequencies formula, 292
Scanning colorimeters, *see* Spectroradiometers
Schade, Otto H., 3
Scotopic vision, 8
Screens, *see* CRT displays
S distortion, 227
Secondary emission, 73
Seebeck effect, 69
Selenium cells, 76
Shadow masks, 96–97, 232–236
Short-wave pass filters, 56, 65
Shrinking-raster resolution measurement, 198–199
SID, *see* Society for Information Display
Silicon photodiodes:
    avalanche mode, 79–81, 246
    charge-coupled devices, 81–83
    list of manufacturers, 313–314
    modes of operation, 77–81
    overview, 77, 78

photoamperic mode, 77, 79, 241–244, 254
photoconductive mode, 79, 80, 244–246
photopic correction, 131–132, 244
photovoltaic mode, 77
response time, 254–256
spectral sensitivity, 81, 130–133
time-related measurements, 241–244
Skiatron system, 122
Sky, as light source, 32–33
SMPTE, *see* Society of Motion Picture and Television Engineers
Society for Imaging Science and Technology (IS&T), 274–275
Society for Information Display (SID), 277
Society of Automotive Engineers (SAE), 276–277
Society of Motion Picture and Television Engineers (SMPTE), 277
Solar constant, 13
Sony Jumbotron, 125
Spectral sensitivity, 6–8, 19, 130–133
Spectrophotometers, 151
Spectroradiometers:
    calibrating for chromaticity measurements, 266–267
    list of manufacturers, 312
    overview, when to use, 149–151
    spectral output curves, 187–189
    using diode arrays, 82, 150–151
Spectrum, *see* Electromagnetic spectrum
Specular gloss, 69, 169, 171, 173
SPIE, *see* International Society for Optical Engineering
Standard illuminants, 303–304
Standards, *see also* Standards organizations
    chromaticity, 184, 186–187
    colorimetry, 17–21, 270
    convergence measurement, 236
    metrology and calibration, 259
    overview, 269
    resolution measurement, 220
    safety and regulation, 139, 278
    spectral sensitivity, 6–8, 19, 130–133
Standards organizations:
    American National Standards Institute (ANSI), 269
    American Society for Testing and Materials (ASTM), 269–270
    Commission Internationale de l'Eclairage (CIE), 270
    Council for Optical Radiation Measurements (CORM), 270–271
    Deutsche Industrie Norms (DIN), 271
    Electronic Industries Association (EIA), 271
    Electronic Industries Association of Japan (EIAJ), 272
    European Broadcast Union (EBU), 271
    European Norms (EN), 272
    Human Factors and Ergonomics Society (HFES), 272

326 INDEX

Standards organizations (*Continued*)
  Illuminating Engineering Society of North America (IESNA), 273
  Institute of Electrical and Electronics Engineers (IEEE), 273
  International Electrotechnical Commission (IEC), 272
  International Organization for Standardization (ISO), 274
  International Society for Optical Engineering (SPIE), 277
  Intersociety Color Council (ISCC), 273–274
  list of organization addresses, 317–318
  Military Standards (MIL), 275
  National Association of Photographic Manufacturers (NAPM), 275
  National Information Display Laboratory (NIDL), 275–276
  National Institute of Standards and Technology (NIST), 276
  Society for Imaging Science and Technology (IS&T), 274–275
  Society for Information Display (SID), 277
  Society of Automotive Engineers (SAE), 276–277
  Society of Motion Picture and Television Engineers (SMPTE), 277
  Video Electronics Standards Association (VESA), 278
Stars, as light source, 30–32
Stefan-Boltzmann constant, 289
Stefan-Boltzmann law, 29–30, 289
Stiles ellipses, 20
Stokes's law, 47
Sun, as light source, 30–32
Supertwisted nematic LCD, 110–112
Surface reflection, 59, 65–66, 173, 291

Talbott's law, 289
Television:
  camera resolution measurement method, 200
  test pattern resolution measurement method, 200–202
Temperature, color, *see* Color temperature
Test patterns:
  for convergence measurement, 233
  generators, list of manufacturers, 313
  for luminance measurement, 157
  for resolution measurement, 194, 200–202

Thermal detectors, 69, 146
Thermal light sources, 29–30
Thermocouples, 69
Thermopiles, 69, 70
Thin-film filters:
  list of manufacturers, 314–315
  multilayer, 64–65
  neutral-density, 54, 59, 63–64
  overview, 63
  spectral transmission curves, 54, 55
Tilt, 102, 228–229
Transmittance:
  defined, 58, 59
  filters in series, 291
  vs. filter thickness, 291
  formulas, 291
Trapazoidal distortion, 101, 230–231
Tristimulus values, CIE, 19, 294–301
Tungsten-halogen lamps, as light source, 34–35
Twisted nematic LCDs, 109–110

Units of measurement, *see also specific units by name*
  list of abbreviations, 286
  list of conversions, 293
  list of SI prefixes, 287
  overview, 26, 258–259

Vacuum fluorescent displays, 105–107
Vacuum photodiodes, 71, 72
Vernier line resolution measurement, 199
VESA, *see* Video Electronics Standards Association
Video bandwidth, 100, 292
Video Electronics Standards Association (VESA), 278
Video photometers, 146, 212–213
Viewing angle:
  electroluminescent lamps, 41
  light-emitting diodes, 44–45
  liquid-crystal displays, 185, 189
Visual purple, 5

Watt, 12
Wien displacement law, 290
Wratten filters, 62, 63

Zeroing photometers, 137
Zworykin, Vladimir K., 90